SYMBOL	DEFINITION	PAGE
	Functions and Graphs	
$\mathcal{R}$	relation	234
$\mathcal{R}^{-1}$	inverse relation	234
f	function	170
$f(x)$	f of x, the image of x under f	170
$P(x)$	P of x, a polynomial in x	93
$x \to 3$	x increasing and approaching 3	240
$3 \leftarrow x$	x decreasing and approaching 3	240
(a, b)	ordered pair a, b	7
P	point	187
$\|P_1 P_2\|$	length of the line segment $\overline{P_1 P_2}$	187
	General	
$= (\neq)$	equals (does not equal)	3 (3)
$<$	is less than	19
$>$	is greater than	19
$\leq (\geq)$	is less (is greater) than or equal to	50 (50)
$\|a\|$	absolute value of a	22
$\sqrt{a}$	square root of a	32
$0.123\ldots$	nonrepeating or nonperiodic decimal	30
$0.12\overline{12}$	repeating or periodic decimal	30
LCD	least common denominator	147
LCM	least common multiple	147
$n!$	n factorial	357
$\mathbf{v}$	vector	296
$[a, b]$	vector from $(0, 0)$ to (a, b)	296
$\|\mathbf{v}\|$	norm of $\mathbf{v}$	297
$A_{m \times n}$, $[a_{ij}]_{m \times n}$	m by n matrix	315
$\delta(A)$	determinant of the matrix A	323

INTERMEDIATE ALGEBRA

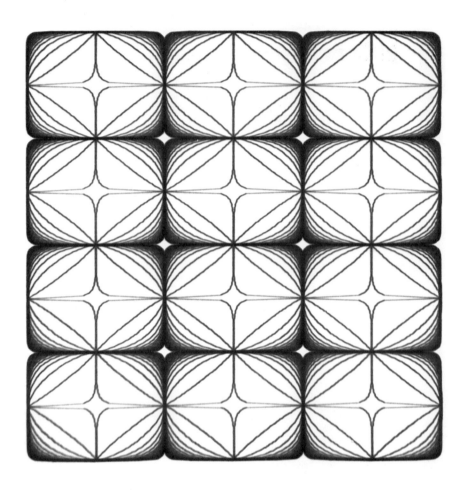

INTERMEDIATE ALGEBRA

FRANK J. FLEMING

Los Angeles Pierce College

HARCOURT BRACE JOVANOVICH, INC.
New York / Chicago / San Francisco / Atlanta

COVER AND TITLE PAGE: Adaptations of sub- and superellipses, plotted on a Stromberg Datagraphics S-C 4020 plotter. Courtesy of (Mrs.) Leigh Hendricks, Albuquerque, New Mexico.

DRAWINGS: Bertrick Associate Artists, Inc.

© 1972 by Harcourt Brace Jovanovich, Inc.

All rights reserved. No part of this publication may be reproduced or transmitted in any form or by any means, electronic or mechanical, including photocopy, recording, or any information storage and retrieval system, without permission in writing from the publisher.

ISBN: 0-15-541541-7

Library of Congress Catalog Card Number: 76-172876

Printed in the United States of America

Preface

A course in intermediate algebra should help the student improve his understanding of the fundamental principles of algebra and upgrade his manipulation of mathematical symbols. To accomplish these aims, *Intermediate Algebra* first reviews the topics of elementary algebra, placing greater emphasis on the properties of real numbers, and extends them beyond the elementary level so the student is prepared for the new concepts and techniques that follow.

Elementary topics are reviewed primarily in Chapters 1 through 7 and part of Chapter 10. Chapter 8 is a separate unit on conic sections, with each curve defined as a set of points in terms of distances from fixed points or lines. Most of the properties of these curves are discussed, but the principal emphasis is on the development of quick sketching techniques rather than tedious point plotting.

The discussion of the techniques of graphing, introduced in Chapter 7, is continued in Chapter 9 in the treatment of graphs of polynomial and rational functions. Complex numbers, logarithms, and sequences are the subjects of Chapters 11, 14, and 15, respectively. Chapters 12 and 13 introduce vectors and matrices and give the student some idea of the nature and use of these quantities.

Above all, this book is written for the student. Each idea is presented in a way the student can understand and is then illustrated by an example worked out in detail, with additional explanation where it seems desirable. However, there is no loss of accuracy or preciseness.

With the exception of a few elementary ideas, each concept is presented in a separate brief section followed by an exercise set of graded problems. Many of the exercise sets contain questions or problems that can be answered orally, in order that the student can be involved in the discussion and development of ideas. In situations where manipulative techniques are important, the exercise sets contain large numbers of problems so that the student can have ample practice. Problems requiring more than the usual amount of thought and analysis are marked with asterisks.

Intermediate Algebra emphasizes completing the square, using it at every opportunity. The factorization of quadratic trinomials by completing the square eliminates the guesswork entailed in other methods. For those who prefer trial-and-error methods, however, one is given in an appendix.

Emphasis is also placed on the function concept, which is introduced in Chapter 7 with a discussion of binary relations and is used throughout the remainder of the book wherever applicable.

Intermediate algebra courses vary greatly in both length and content, and this volume offers the instructor wide latitude in his choice of course material. For example, Chapter 8, on conic sections, can be omitted without loss of continuity. Vectors and matrices are optional topics at this level, so Chapters 12 and 13 may also be omitted if class time is limited.

The instructor may also choose the rigor of the course he wishes to present. Formal proofs of a number of theorems are given for those who wish to concentrate on this aspect. On the other hand, the many illustrative examples make it possible to ignore formal proof completely and concentrate on the application of the ideas and the development of manipulative skills.

I am grateful for the helpful criticism offered by the reviewers of the manuscript: Joe K. Bryant, Monterey Peninsula College, Monterey, California; Douglas B. Crawford, College of San Mateo, San Mateo, California; Warren Davis, Bucks County Community College, Newtown, Pennsylvania; John N. Fujii, Merritt College, Oakland, California; and W. E. Fulwood and H. Eugene Hall, DeKalb College, Clarkston, Georgia. Finally, I wish to express my appreciation to my wife, Marilyn, who typed the manuscript, and to Mrs. Pat Sigman, who supplied the answers to the problem sets.

Frank J. Fleming

Contents

1 Real Numbers — 1

1.1 Sets — 1
Set Membership — 2
Number of a Set — 2
Equality of Sets — 3
Subsets — 3
The Universe of Discourse or Universal Set — 3

1.2 Operations on sets — 4
Union — 5
Venn Diagrams — 5
Set-Builder Notation — 5
Intersection — 6
Complementation — 6
Ordered Pairs — 7
Cartesian Product — 7
Multiple Operations — 8

1.3 Properties of Real Numbers — 9
Substitution Principle — 9
Binary Operations on the Real Numbers — 10
More About Axioms — 11

1.4 Theorems and Deductive Proof — 13
Deductive Proof — 15

1.5 Order and Absolute Value — 19
The Number Line — 19
Order — 19
Positive and Negative Numbers — 19
Axioms of Order — 20

	The Less Than Relation	20
	Absolute Value	22
1.6	**Review of Addition**	24
	Sums	25
	Differences	25
	Least Common Multiple	26
1.7	**Review of Multiplication**	28
	Products	28
	Quotients	28
1.8	**Decimal and Radical Notation**	29
	Infinite Decimals	30
	Terminating Decimals	30
	Periodic Decimals	30
	Changing a Decimal to the Quotient of Two Integers	31
	Radical Notation	32
	Summary	35
	Review Exercises	36

2 *First-Degree Open Sentences with One Variable* 39

2.1	**Equivalent Equations and Inequalities**	40
	Use of Axioms in Generating Equivalent Equations	41
	Use of Theorems in Generating Equivalent Open Sentences	41
	Eliminating Fractions from an Open Sentence	42
2.2	**Solutions of First-Degree Open Sentences**	45
	Permissible Replacements	45
	Types of Open Sentences	45
	Solution Sets of Equations	46
	Solution Sets of Inequalities	47
2.3	**Open Sentences with Absolute Value**	51
	Equations	52
	Inequalities	52
2.4	**Word Problems**	55
	Summary	60
	Review Exercises	61

CONTENTS ix

Exponents, Roots, and Radicals   63

       Powers   63

3.1 Integers as Exponents   64
       Zero as an Exponent   64
       Negative Integers as Exponents   64

3.2 Other Rational Numbers as Exponents   67
       Roots of Real Numbers   67
       Exponents of the Form m/n 68

3.3 Sums and Differences Involving Powers   70
       Algebraic Expressions   70
       Sums of Expressions   71
       Differences of Expressions   71

3.4 Further Applications of the Distributive Law   73

3.5 Changing the Form of Radical Expressions   77
       Simplest Radical Form   77
       Rationalizing Denominators   79
       Rationalizing Numerators   79

3.6 Products and Quotients Involving Radicals   82
       Quotients   83
       Binomial Expressions   83
       Radicals with Different Indices   85

3.7 Sums and Differences Involving Radicals   88

       Summary   89
       Review Exercises   90

Polynomials   92

4.1 Some Characteristics of Polynomials   92
       Zero of a Polynomial   93
       Degree of a Polynomial   94

4.2 Binary Operations with Real Polynomials   95
       Sums and Differences   95
       Products   95

4.3 Quotients—Synthetic Division   97
       Division Algorithm   98
       Synthetic Division   98

4.4	**Completing the Square**	102
	Recognizing Perfect Square Trinomials	102
	Forming Perfect Square Trinomials	102
	Completing the Square	103
4.5	**Factoring Quadratic Trinomials**	106
	Factoring the Difference of Two Squares	106
	Factoring Trinomials as a Difference of Squares	106
	Factoring in the Set of Irrational Numbers	107
4.6	**Other Methods of Factoring**	109
	Sum or Difference of Two Cubes	109
	Fourth-Degree Trinomials as a Difference of Squares	110
	Factoring by Grouping	110
	Summary	112
	Review Exercises	113

5 Quadratic Equations and Inequalities in One Variable — 115

	Standard Form	115
5.1	**Solution of Quadratic Equations by Factoring**	116
5.2	**The Quadratic Formula**	118
	Using the Quadratic Formula	119
	The Discriminant	120
5.3	**Equations Involving Radicals**	122
	Clearing Radicals from Equations	123
	Extraneous Solutions	125
5.4	**Equations Quadratic in Form**	126
5.5	**Quadratic Inequalities**	128
5.6	**Word Problems**	132
	Summary	136
	Review Exercises	136

6 Fractions — 138

6.1	**Equivalent Fractions**	138
	Simplest Form of a Fraction	140
	Building Fractions to Higher Terms	141
	Changing Signs	141

6.2	**Sums and Differences**	145
	Least Common Multiple	146
	Sums of Fractions with Different Denominators	148
	Differences of Fractions	149
6.3	**Products and Quotients**	153
	Simplifying Products of Fractions	153
	Simplifying Quotients	154
6.4	**Equations Involving Fractions**	157
	Clearing an Equation of Fractions	158
6.5	**Complex Fractions**	161
	Simplifying Complex Fractions	162
	Summary	166
	Review Exercises	166

7 *Relations, Functions, and Graphs—I* 168

7.1	**Relations and Functions**	168
	Domain and Range	169
	Notations	170
	Restrictions	170
	Specifying the Domain and the Range of a Relation	170
	Functions	171
7.2	**Graphs of Relations in $R \times R$**	173
	Coordinates	173
	Graphs in $U \times U$	174
	Graphs in $R \times R$	174
7.3	**Linear Functions**	176
	Graphs of Linear Functions	176
	Intercepts	177
	Lines Parallel to an Axis	178
7.4	**Equations for Lines**	179
	Distance between Two Points	180
	Slope of a Line	181
	Line Containing Two Given Points	182
	Line with Given Slope Containing a Given Point	183
7.5	**Quadratic Functions**	185
	Intercepts	185
	Vertex of a Parabola	186

Symmetry	187
Axis of Symmetry	188
Graphs of Quadratic Functions	188
Summary	192
Review Exercises	193

8 Conic Sections   195

8.1 The Circle	196
Equations for Circles	196
Standard Form of the Equation for a Circle	196
Center and Radius	197
8.2 The Parabola	198
Equation for a Parabola	199
Parabola with Vertex Not at the Origin	201
Properties	202
8.3 The Ellipse	205
Equation for an Ellipse	205
Eccentricity	209
8.4 The Hyperbola	210
Equation for a Hyperbola	211
Eccentricity	213
8.5 Sketching the Conic Sections	215
Circles	215
Parabolas	215
Ellipses	216
Hyperbolas	217
Summary	219
Review Exercises	220

9 Relations, Functions, and Graphs—II   222

9.1 Inequalities	222
Linear Inequalities	223
Graphs of Linear Inequalities	224
Quadratic Inequalities	225

9.2	Variation	229
	Direct Variation	229
	Inverse Variation	229
	Joint Variation	231
	Combined Variation	231
9.3	Inverse Relations	234
	Graphs of Inverse Relations	234
	Inverse of a Function	234
	Inverse Functions	235
9.4	Graphs of Polynomial Functions	237
9.5	Graphs of Rational Functions	239
	Asymptotes	240
	Summary	245
	Review Exercises	246

10 Systems of Equations and Inequalities — 247

10.1	Systems of Linear Equations—Substitution	247
	Testing for Inconsistent and Dependent Equations	248
	Equivalent Systems of Equations	248
	Solution by Substitution	249
	Solution Sets of Systems Involving Three Variables	249
10.2	Systems of Linear Equations—Linear Combinations	253
	Linear Combinations	253
	Solution of a System of Two Equations	253
	Solution of a System of Three Equations	254
10.3	Systems Involving Quadratic Equations—I	257
	Solution by Substitution	258
10.4	Systems Involving Quadratic Equations—II	261
	Method 1	261
	Method 2	262
10.5	Systems of Inequalities	264
10.6	Word Problems	268
	Summary	276
	Review Exercises	277

11 Complex Numbers — 279

11.1 An Extension of the Real Number System — 279
- Powers of i — 280
- Square Roots of Negative Numbers — 280
- Equality of Complex Numbers — 281

11.2 Sums and Differences — 283
- Negative of a Complex Number — 283
- Difference of Two Complex Numbers — 283

11.3 Products — 284
- Multiplicative Inverses — 285

11.4 Quotients — 286
- Multiplicative Inverse — 287
- Conjugates — 287
- Quotients as Fractions — 288

11.5 Solutions of Quadratic Equations — 290
- The Quadratic Formula — 290
- Conjugate Solutions — 290
- Sum and Product of the Solutions — 291

Summary — 293

Review Exercises — 294

12 Vectors — 295

12.1 Vectors in the Plane — 295
- Symbolic Representation — 296
- Equality of Vectors — 297
- Magnitude and Direction — 297
- Scalar Multiplication — 297
- Unit Vectors — 298
- Vector as a Sum — 299

12.2 Operations with Vectors — 302
- Sums — 302
- Differences — 303

12.3 The Parallelogram Law—Applications — 305
- The Parallelogram Law — 306
- Sums — 307
- Resultant of Forces — 307
- Differences — 308

Summary		311
Review Exercises		312

13 Introduction to Matrices and Determinants — 313

13.1	Basic Definitions	313
	Dimension or Order	314
	Entries and Addresses	314
	Matrix Notation	314
	Equal Matrices	315
13.2	Matrix Addition	317
	Sums	317
	Negative of a Matrix	318
	Differences	318
13.3	Matrix Multiplication	320
13.4	Determinants	323
	The Determinant Function	323
	Second-Order Determinants	324
	Third-Order Determinants	324
13.5	Cramer's Rule	326
	Summary	328
	Review Exercises	329

14 Logarithms — 331

14.1	Properties of Logarithms	332
	Properties of Logarithms to the Same Base	332
14.2	Logarithms to the Base 10	335
	Scientific Notation	335
	Logarithms to the Base 10	336
	Using a Table of Logarithms	336
	Antilogarithms	338
	Linear Interpolation	338
14.3	Computations	340
	Summary	342
	Review Exercises	343

15 Sequences 345

15.1 Sequence Functions — 345
    Finding the Terms of a Sequence — 346
    Equation for a Sequence — 346

15.2 Arithmetic Sequences — 347
    The nth Term of an Arithmetic Sequence — 347
    First Term and Common Difference — 347
    Arithmetic Means — 348
    Sum of a Number of Terms — 348

15.3 Geometric Sequences — 351
    Finding the Common Ratio and First Term — 351
    Geometric Means — 352
    Sum of n Terms — 352
    Infinite Geometric Sequences — 353

15.4 Powers of Binomials — 356
    Some Specific Powers — 356
    Some Observable Features — 356
    Factorial Notation — 357
    General Term of a Binomial Expansion — 357

    Summary — 360
    Review Exercises — 361

Appendix 1
Factoring Quadratic Trinomials over the Set of Integers — 363

Appendix 2
Table of Squares and Square Roots — 367

Answers to Odd-Numbered Exercises — 369

Index — 423

INTERMEDIATE ALGEBRA

Real Numbers

This course is primarily concerned with operations involving real numbers and with the properties of real numbers with respect to these operations. Although you have studied the development of the real number system in elementary algebra, a restatement of the basic principles will be helpful.

1.1 Sets

The real numbers form a *set*. Before we begin a discussion of the real numbers themselves, let us consider sets in general.

We use the word **set** to refer to a collection of items that are called **elements** or **members** of the set. A set is said to be **well-defined** if it is possible to determine whether a particular item does or does not belong to the set. For example, the members of the Green Bay Packers football team form a well-defined set because it can be easily determined whether a particular person is or is not a member of the team. On the other hand, a list of "good books" does not form a well-defined set because such a list depends on the personal preference of the person who made the list.

In a general discussion of sets, capital letters are frequently used to represent, or identify, sets. In other cases we need to be more specific, so the set is indicated by a pair of braces, { }, enclosing either a list of the elements or a brief description of the members of the set.

Example   The set composed of the first five letters of the English alphabet can be written in **list** or **roster form** as

$\{a, b, c, d, e\}$,

which is read, "the set whose elements are *a, b, c, d,* and *e.*"

1

Another method of specifying a set, called *set-builder notation,* in which the elements are described symbolically, will be discussed later.

Set Membership

The symbol ∈ indicates that a particular element belongs to a specified set, and the symbol ∉ indicates that an element is not a member of the set.

Examples (a) $x \in A$ is read:
   x is a member of A,
   x is an element in A, or
   x is contained in A.
 (b) $x \notin B$ is read:
   x is not a member of B, etc.

When a letter, such as x in the examples above, is used to represent an unspecified member of a set, it is called a **variable**, and the set is called the **replacement set** for the variable. For example, if $A = \{a, b, c, d\}$ and $x \in A$, then x is a variable which may represent a, b, c, or d.

Number of a Set

It seems likely that the concept of number arose from a primitive use of sets. One possibility is that some early man used a pile of pebbles to correspond to the members of his flock of sheep, using a pebble for each sheep. Such a pairing of the elements of one set with those of another is called a **one-to-one correspondence**. Two sets that can be placed in one-to-one correspondence with each other have the same number of elements and are also said to be **equivalent** sets.

This use of number relates to the concept of "how many." Numbers that are used to indicate how many elements there are in a set are called **counting numbers** or **natural numbers**. The set of natural numbers is represented by N. A set which contains no members is called the **empty set** or the **null set** and is represented by ∅. The number of elements in the empty set is *zero.*

Generally, the order in which the elements of a set are listed is not important. However, if you list the counting numbers in increasing magnitude $\{1, 2, 3, \ldots\}$, you have an example of an **ordered set**.

The preceding discussion suggests that it is sometimes possible to count the elements in a set. For example, if the elements of a given set are placed in one-to-one correspondence with the ordered set of counting numbers, starting with 1 as shown in Figure 1.1-1, then if there is a last element in the given set, the set is **finite**. If there is no last element, the set is **infinite**. To specify an ordered set

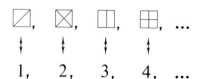

Figure 1.1-1

that is either infinite or contains a very large number of elements, three dots (. . .) are used to represent the unlisted elements and are usually read as "and so on."

Examples (a) $\{1, 2, 3, \ldots\}$ is an infinite set.
(b) $\{1, 2, 3, \ldots, 100\}$ is the set of all counting numbers from 1 to 100 inclusive.

Equality of Sets

Two sets are said to be **equal** if they contain the same elements. Thus, if A represents $\{a, b, c\}$ and B represents $\{a, b, c\}$, then $A = B$. In other words, the equality of two sets implies that two different representations, or names, are used for the same set. $A \neq B$ (read "A is not equal to B") implies that A and B represent different sets.

Subsets

If A and B are two sets such that every element of A is also an element of B, then A is a **subset** of B. This relationship is designated by $A \subseteq B$, which can be read as "A is contained in B," or "A is a subset of B." $A \not\subseteq B$ is read "A is not a subset of B." If A is a subset of B, and B contains at least one element that is not a member of A, then A is said to be a **proper subset** of B. The symbol $\subset$ is used to indicate that one set is a proper subset of another.

It follows from the definition of subset that every set is a subset of itself. Also, if $A \subseteq B$, then A cannot contain an element that is not a member of B. Note that the empty set contains no elements that are not contained in every other set, and therefore *the empty set is a subset of every set,* including itself.

Example List all possible subsets of $\{a, b, c\}$.

Solution $\{a, b, c\}, \{a, b\}, \{a, c\}, \{b, c\}, \{a\}, \{b\}, \{c\}, \emptyset$.

The Universe of Discourse or Universal Set

When we work with sets it is often convenient to define the set of all elements that may be considered in the discussion. Such a set is called the **universe of discourse** or, more commonly, the **universal set**. For example, if $A = \{a, b, c\}$ and $B = \{a, c, d, e\}$, we can consider the universal set, designated by U, to be the set of letters of the English alphabet. The universal set may be different for different problems, but for a discussion of sets in general, we consider every set to be a subset of U.

EXERCISE 1.1

ORAL

1. Explain one-to-one correspondence.

4 REAL NUMBERS

Give definitions for each of the following:

2. Equivalent sets
3. Finite set
4. Infinite set
5. Equal sets
6. Subset
7. Proper subset

Read each of the following symbolic statements:

8. $a \in \{a, b, c\}$
9. $t \notin \{a, b, c\}$
10. $5 = 2 + 3$
11. $7 \neq 2 + 3$
12. $\{x, y\} \subset \{x, y, z\}$
13. $\{a, b\} \not\subset \{x, y, z\}$

WRITTEN

In Problems 1–8, replace the comma between each pair of symbols with the proper symbol $\in$ or $\notin$.

Examples   (a) x, {letters of the alphabet}   (b) 8, $\{1, 3, 5, \ldots\}$

Solutions   (a) $x \in$ {letters of the alphabet}   (b) $8 \notin \{1, 3, 5, \ldots\}$

1. a, $\{a, b, c\}$
2. n, $\{a, e, i, o, u\}$
3. $\tfrac{2}{3}$, N
4. 1492, N
5. Monday, {days of the week}
6. August, {months with 30 days}
7. ▨, $\{\oslash, \ominus, \oplus, \boxtimes\}$
8. $\times$, $\{\times, +, \square, \ast\}$

In Problems 9–18, replace the comma in each pair with the correct symbol $\subseteq$ or $\not\subseteq$. $A = \{1, 2, 3\}$, $B = \{1, 3, 2\}$, $C = \{1, 2\}$, $D = \{1, 3\}$, $E = \{2\}$.

9. A, B
10. C, A
11. E, D
12. 2, A
13. C, D
14. B, C
15. 3, A
16. A, E
17. $\{2, 3\}$, A
18. $\{2\}$, A

In Problems 19–26, list all subsets of the given set.

19. $\{a, b\}$
20. $\{x, y\}$
21. $\{1, 2, 3\}$
22. $\{d, e, f\}$
23. $\{a, b, c, d\}$
24. $\{w, x, y, z\}$
25. $\{a, b, c, d, e\}$
26. $\{1, 2, 3, 4, 5\}$

Justify your answers in Problems 27–34.

27. If $A \subseteq B$ and $x \in A$, must x be an element of B?
28. If $A \subseteq B$ and $y \in B$, must y be an element of A?
29. If $P \subseteq Q$ and $Q \subseteq P$, in what other way is P related to Q?
30. If $P \subseteq Q$ and $Q \subseteq R$, in what way is P related to R?
31. If $A \subseteq B$ and $C \subseteq B$, must A be a subset of C?
32. If $A \subseteq B$ and $C \subseteq B$, can A be a subset of C?
33. If $P = Q$ and $Q = R$, in what way is P related to R?
34. If $P = Q$ and $Q \subseteq R$, in what way is P related to R?

1.2 Operations on Sets

A binary operation is a process by means of which elements from two given sets are used to form a third set. The set so formed may or may not be different from either of the two given sets. In this section we discuss three binary operations on sets—**union, intersection**, and the **Cartesian product**.

Union

The **union** of two sets is a set containing every element that is a member of at least one of the given sets. ∪ is used to designate this operation. Thus $A \cup B$ is read, "the union of A and B."

Example If $A = \{a, b, c\}$ and $B = \{a, c, d, e\}$, then $A \cup B = \{a, b, c, d, e\}$.

Note that even if an element is a member of both sets, it is listed only once in the union. If one set is a subset of the other, the union of the two sets is equal to the set that contains the other as a subset.

Example If $A \subseteq B$, then $A \cup B = B$. Also note that $A \cup A = A$.

Venn Diagrams

Operations with sets are sometimes displayed pictorially by sketches called Venn diagrams. One method of constructing a Venn diagram is to use a rectangle to represent the universal set and circles or ovals to represent the sets involved

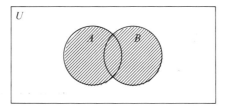

Figure 1.2-1

in the operation. The region representing the result of the operation is usually shaded in a manner that clearly distinguishes it from the remainder of the universal set. The union of sets A and B ($A \cup B$) is shown in Figure 1.2-1. Notice that A and B are entirely within U since they are both subsets of U.

Set-Builder Notation

In Section 1.1 we illustrated the list method of specifying a set. Another method is **set-builder notation**, which consists of describing the set symbolically by using a variable to represent the members of the set and stating the condition(s) for set membership. For example, the union of sets A and B can be symbolized as

$$A \cup B = \{x \mid x \in A \text{ or } x \in B\}.$$

The right-hand expression is read, "the set of all x such that x is a member of A or x is a member of B." Note that the word *or* is emphasized in the union of two sets.

Intersection

The **intersection** of two sets is the set of all elements common to both sets. It contains no elements that are not members of both sets. ∩ is used to designate the intersection. $A \cap B$ is read, "the intersection of A and B."

Example If $A = \{a, b, c\}$ and $B = \{a, c, d, e\}$, then $A \cap B = \{a, c\}$.

The intersection of two sets can be specified in set-builder notation as

$$A \cap B = \{x \mid x \in A \text{ and } x \in B\}.$$

Note that the word *and* is emphasized in the intersection of two sets.

If one set is a subset of another, then the intersection of the two sets is the subset.

Example If $A \subset B$, then $A \cap B = A$. Also note that $A \cap A = A$.

Venn diagrams for the intersection of two sets are shown in Figure 1.2-2. Two different techniques of shading are illustrated. In (a) only the intersection is shaded. In (b) each set is shaded in a different direction, so the intersection has both shadings and appears as the cross-hatched region.

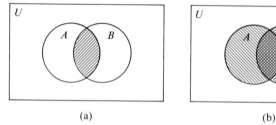

Figure 1.2-2

Two sets that have no common elements are called **disjoint sets**. The intersection of two disjoint sets is the empty set.

Example If $A = \{a, b, c\}$ and $B = \{x, y, z\}$, then $A \cap B = \emptyset$.

Complementation

If $A \subseteq U$, where U is the universal set, then the set of all elements in U that are not elements in A is called the **complement** of A and is designated by A'. Symbolically,

$$A' = \{x \mid x \in U \text{ and } x \notin A\}.$$

Example If $U = \{a, b, c, d, e, f\}$ and $A = \{b, d, f\}$, then $A' = \{a, c, e\}$.

Similarly, given two sets A and B, the set of all elements in A that are not elements in B is called the **relative complement** of B with respect to A and is designated by $A - B$. Symbolically,

$$A - B = \{x \mid x \in A \text{ and } x \notin B\}.$$

Example If $A = \{a, b, c, d, e, f\}$ and $B = \{a, c, g\}$, then $A - B = \{b, d, e, f\}$.

Ordered Pairs

If two elements are considered at the same time in such a way that one element must be taken first and the other must be taken second, we say that the two elements form an **ordered pair** and enclose them in parentheses. Thus, (a, b) is an ordered pair in which a is called the first component and b is called the second component.

Two ordered pairs are equal if and only if the first components are equal and the second components are equal. Thus $(a, b) = (c, d)$ if and only if $a = c$ and $b = d$. Also, $(a, b) \neq (b, a)$ unless $a = b$.

Cartesian Product

You have seen that the binary operations of union and intersection resulted in sets whose elements were elements of the original sets. We will now discuss a binary operation on sets that results in a set of ordered pairs.

The **Cartesian product** of two sets is the set of all possible ordered pairs such that the first component of each ordered pair is an element in the first set and each second component is an element in the second set. The Cartesian product of two sets A and B is designated by $A \times B$ (read "A cross B"). In set-builder notation, we have

$$A \times B = \{(a, b) \mid a \in A \text{ and } b \in B\}.$$

Example If $A = \{a, b, c\}$ and $B = \{x, y\}$, then
$$A \times B = \{(a, x), (a, y), (b, x), (b, y), (c, x), (c, y)\}.$$

When the two sets involved in a binary operation can be interchanged without affecting the result of the operation, we say that the operation is **commutative**. You can observe from the definitions of union and intersection that $A \cup B = B \cup A$ and $A \cap B = B \cap A$, since the order in which the elements are listed is not significant. Thus, both union and intersection are commutative operations. However, note that, in general, $A \times B \neq B \times A$. In other words, the Cartesian product is *not* commutative.

Example If the order of the two sets in the previous example is changed, we have
$$B \times A = \{(x, a), (y, a), (x, b), (y, b), (x, c), (y, c)\},$$
and from the definition of an ordered pair, we know that the ordered pairs listed here are not the same as those listed for $A \times B$.

Multiple Operations

What meaning can we give to expressions such as $A \cup B \cup C$, $A \cap B \cap C$, and $A \cup B \cap C$? Obviously, these do not represent binary operations, since more than two sets are involved. However, since the union of two sets is a set, we can use parentheses to indicate which operation is to be performed first. Thus, $(A \cup B) \cup C$ means that you first find the union of sets A and B and then find the union of this set with C. If we wish to find the union of B and C first, we write $A \cup (B \cup C)$.

In the exercises that follow, you will find that although

$$(A \cup B) \cup C = A \cup (B \cup C) \quad \text{and} \quad (A \cap B) \cap C = A \cap (B \cap C),$$

$(A \cup B) \cap C$ is not generally equal to $A \cup (B \cap C)$.

EXERCISE 1.2

ORAL

Give definitions for

1. The union of two sets.
2. The intersection of two sets.
3. The Cartesian product of two sets.
4. A commutative operation.

Read

5. $A \cup B = \{x \mid x \in A \text{ or } x \in B\}$.
6. $A \cap B = \{x \mid x \in A \text{ and } x \in B\}$.
7. $A \times B = \{(a, b) \mid a \in A \text{ and } b \in B\}$.

WRITTEN

In Problems 1–20, list the elements in each set, given that $A = \{a, b, c, d\}$, $B = \{e, f, g, h\}$, $C = \{a, c, e, g\}$, and $D = \{b, d, g, h\}$.

Examples	(a) $B \cup D$	(b) $B \cap C$
Solutions	(a) $\{b, d, e, f, g, h\}$	(b) $\{e, g\}$

1. $A \cup B$
2. $B \cup C$
3. $C \cup D$
4. $A \cup D$
5. $A \cap C$
6. $A \cap D$
7. $A \cap B$
8. $B \cap D$
9. $A \cap (B \cup C)$
10. $(A \cup C) \cup D$
11. $(A \cup B) \cup D$
12. $B \cup (C \cup D)$
13. $(A \cap B) \cap C$
14. $(B \cap C) \cap D$
15. $A \times B$
16. $B \times C$
17. $C \times D$
18. $B \times A$
19. $(A \cup B) \cap C$
20. $A \cup (B \cap C)$

In Problems 21–28, construct Venn diagrams to show the operations. A, B, and C are subsets of U.

21. $B \cup C$
22. $B \cap C$
23. $A \cup (B \cap C)$
24. $A \cap (B \cap C)$ where no pair of sets is disjoint.

25. $A \cap (B \cup C)$ where B and C are disjoint.
26. $A \cup (B \cap C)$ where $C \subseteq B$.   27. A' 28. $B - A$
29. If $x \in A$, must x be a member of $A \cup B$?
*30. If $x \in A$, must x be a member of $A \cap B$?
*31. If $A \cap B = \emptyset$, is there a set C such that $A \cap (B \cup C) \neq \emptyset$?
*32. Under what conditions is $A \times B = B \times A$ a true statement?
33. Explain why it is true that for any three sets A, B, and C,

$$(A \cup B) \cup C = A \cup (B \cup C).$$

*34. Explain why it is true that for any three sets A, B, and C,

$$(A \cap B) \cap C = A \cap (B \cap C).$$

1.3 Properties of Real Numbers

In your previous study of algebra, you were probably told that the set of real numbers can be placed in one-to-one correspondence with the points in a line. That is, each real number corresponds to exactly one point, and conversely, each point in the line corresponds to exactly one real number.

You also learned that several subsets of the real numbers, R, have special names. We repeat the definitions of these subsets and the symbols by which we identify them.

Natural numbers, N, represent the numbers of elements in non-empty sets.

$$N = \{1, 2, 3, \ldots\}$$

Whole numbers, W, represent the numbers of elements in sets, including the empty set.

$$W = \{0, 1, 2, \ldots\}$$

Integers, J, include the natural numbers and zero (as in the whole numbers) and the negatives of the natural numbers.

$$J = \{\ldots, -2, -1, 0, 1, 2, \ldots\}$$

Rational numbers, Q, are numbers that can be represented as the quotient of two integers.

$$Q = \left\{ \frac{a}{b} \,\middle|\, a, b \in J, b \neq 0 \right\}$$

Irrational numbers, H, are all the real numbers that are not rational numbers.

$$H = \{x \mid x \in R \text{ and } x \notin Q\}$$

Substitution Principle

In Section 1.1 two sets were defined to be equal if they contain exactly the same elements. Thus, the statement $A = B$ implies that two different names are used for the same set and that the names are interchangeable. The concept of different names or symbols for the same thing can be extended to the set of real numbers R. To do this we state some of the properties of the real numbers as *axioms*, statements that are assumed to be true. The fundamental axiom concerns the interchangeability of names for the same number and is called the **substitution axiom**.

> S-1 If $a, b \in R$ and $a = b$, then a may be replaced by b, or b by a, in any expression naming a real number without changing the truth or falsity of the statement.

The substitution axiom states one property of the equals relation and suggests others as follows.

AXIOMS OF EQUALITY (for all $a, b, c \in R$):

E-1 $a = a$. (Reflexive property)
E-2 If $a = b$, then $b = a$. (Symmetric property)
E-3 If $a = b$ and $b = c$, then $a = c$. (Transitive property)

Binary Operations on the Real Numbers

Binary operations on sets were discussed in the previous section. We now turn to binary operations on the elements of a set.

If a and b are any members of a set S, then a binary operation on a and b assigns to the ordered pair (a, b) some unique c which may or may not be a member of S. The set A of all c is called the set of results of the operation. If all c are also elements in S, then $A \subset S$, and S is said to be **closed** with respect to the operation. If the operation assigns the same c to both (a, b) and (b, a), then the operation is **commutative**. These relations will become clear as the axioms for binary operations on real numbers are stated.

We will wish to extend the binary operations involving real numbers to more than two numbers, just as we extended binary operations on sets to more than two sets. Symbols that represent extensions of binary operations must be defined so that the axioms of equality are applicable.

From this point on, definitions are stated more formally and are given reference numbers to indicate the section in which the definition is given and the sequence in which the definitions occur.

The binary operations of addition and multiplication are extended in R by the following.

DEFINITION 1.3-1 For all $a, b, c \in R$,

$$a + b + c = (a + b) + c,$$
$$a \cdot b \cdot c = (a \cdot b) \cdot c,$$
$$a \cdot b + c = (a \cdot b) + c,$$

and

$$a + b \cdot c = a + (b \cdot c).$$

The last two statements tell us that if a problem involves both addition and multiplication, the multiplication must be done first unless otherwise indicated by parentheses.

More About Axioms

In Section 1.2 we discussed the union of two sets and worked problems involving the union of two finite sets. In every case, the union of two such sets was a finite set. This leads us to assume that such a result is always true even though it is impossible to test all such unions. We state this assumption formally as an axiom.

For all A, B that are finite sets, $A \cup B$ is a finite set.

We now state the axioms concerning the properties of real numbers with respect to the binary operations of addition and multiplication.

AXIOMS OF ADDITION

A-1 For all $a, b \in R$, $a + b$ is a unique real number. (Law of closure for addition)

A-2 For all $a, b, c \in R$, (Associative law of addition)

$(a + b) + c = a + (b + c)$.

A-3 There exists a unique number $0 \in R$ such that for every $a \in R$, (Additive identity law)

$0 + a = a + 0 = a$.

A-4 For each $a \in R$, there exists a unique number $-a \in R$, called the negative of a such that (Additive inverse law)

$a + (-a) = 0$.

A-5 For all $a, b \in R$, (Commutative law of addition)

$a + b = b + a$.

AXIOMS OF MULTIPLICATION

M-1 For all $a, b \in R$, ab is a unique real number. (Law of closure for multiplication)

M-2 For all $a, b, c \in R$, (Associative law of multiplication)

$(ab)c = a(bc)$.

M-3 There exists a unique number $1 \in R$, $1 \neq 0$, such that for every $a \in R$, (Multiplicative identity law)

$a \cdot 1 = 1 \cdot a = a$.

M-4 For each $a \in R$, except 0, there exists a unique number $\dfrac{1}{a}$, called the multiplicative inverse or reciprocal of a, such that $a \cdot \dfrac{1}{a} = \dfrac{1}{a} \cdot a = 1$. (Multiplicative inverse law)

M-5 For all $a, b \in R$, $ab = ba$. (Commutative law of multiplication)

DISTRIBUTIVE LAW

D-1 For all $a, b, c \in R$, $a(b + c) = ab + ac$ and $(a + b)c = ac + bc$.

The existence of additive and multiplicative inverse elements permits us to define the arithmetic operations of subtraction and division in terms of addition and multiplication, respectively.

DEFINITION 1.3-2 For all $a, b \in R$, $a - b = a + (-b)$.

DEFINITION 1.3-3 For all $a, b \in R$, $\dfrac{a}{b} = a \cdot \dfrac{1}{b}$.

These two definitions are ordinarily called **theorems** since they can be proved to be true.

Since $ab \in R$, ab has a negative. If we wrote this number as $-ab$, we could interpret it as the product of $-a$ and b. To avoid this ambiguity, we state

DEFINITION 1.3-4 For all $a, b \in R$, $-ab = -(ab)$.

EXERCISE 1.3

ORAL OR WRITTEN

In Problems 1–20, each statement can be justified by one of the axioms of equality or by one of the properties of the real numbers. Name the axiom or property that justifies each of the following statements. Assume that all variables represent real numbers.

<u>Examples</u> (a) $x + y = x + y$ (b) $5 + z \in R$

<u>Solutions</u> (a) Reflexive property   (b) Closure for addition

1. $0 + (x + y) = x + y$
2. If $x - 7 = 3$, then $3 = x - 7$.
3. $x + 17 \in R$
4. If $x = y$ and $y = z + 7$, then $x = z + 7$.
5. $(x + y) \cdot \dfrac{1}{(x + y)} = 1$ $(x \neq -y)$

6. $(x + y)(p + q) = (x + y)p + (x + y)q$
7. $(x + 2y) + [-(x + 2y)] = 0$
8. If $(x + y) = z$ and $z = (x - 7)$, then $(x + y) = (x - 7)$.
9. $(xy)z \in R$
10. $\left(\dfrac{x}{y} + z\right) + \dfrac{p}{q} = \dfrac{x}{y} + \left(z + \dfrac{p}{q}\right)$ $(y, q \neq 0)$
11. $3x - (y + 5) = 3x - (5 + y)$
12. $\dfrac{x - 7}{4} = \dfrac{x - 7}{4}$
13. $[(x + 3)(x - 2)](x + 2) = (x + 3)[(x - 2)(x + 2)]$
14. $\dfrac{x}{y} \cdot \dfrac{y}{z} = \dfrac{y}{z} \cdot \dfrac{x}{y}$ $(y, z \neq 0)$
15. $x(x - 7y + 2) = (x - 7y + 2)x$
16. $(x + \sqrt{2}) \cdot \dfrac{1}{(x + \sqrt{2})} = 1$ $(x \neq -\sqrt{2})$
17. If $4 = y + 5$, then $y + 5 = 4$.
18. $(x + y) + 2 \in R$
19. $\sqrt{2}(x + 3) = \sqrt{2} \cdot x + \sqrt{2} \cdot 3$
20. $(x + y) + [-(x + y)] = 0$

For Problems 21-26, specify the definition which assures us that each expression represents a real number if all variables are real numbers.

21. $x + y + 2 = (x + y) + 2$
22. $3 \cdot x \cdot y = (3 \cdot x) \cdot y$
23. $\dfrac{5}{x} = 5 \cdot \dfrac{1}{x}$ $(x \neq 0)$
24. $3x - 7y = 3x + (-7y)$
25. $(x + y)(x + 2)(x - y) = [(x + y)(x + 2)](x - y)$
26. $(2x + 3) - (x + 4) = (2x + 3) + [-(x + 4)]$

1.4 Theorems and Deductive Proof

In the development of a mathematical system, we begin by defining the symbols we use and the operations involving the set. We also assume some properties of the elements with respect to the defined operations. Generally, we attempt to make use of as few definitions and assumptions as possible. Any statement concerning the properties of the elements that can be shown to be true as a consequence, or result, of definitions or axioms is called a **theorem**. After a theorem has been proved, it can be used to help prove other theorems.

The statement of a theorem, or proposition, consists of two parts: (1) an *if clause*, or **hypothesis**, which is assumed to be true, and (2) a *then clause*, or **conclusion**, which follows as a logical consequence of the hypothesis. In an informal statement the conclusion may be given before the hypothesis.

14 REAL NUMBERS

Examples (a) Underline{If it doesn't rain,} Underline{we will go to the beach.}
                     hypothesis   conclusion

(b) Underline{If $2 + 3 = 5$,} Underline{then $5 - 2 = 3$.}
       hypothesis   conclusion

(c) Underline{We will go to a movie} Underline{if I get my allowance.}
       conclusion   hypothesis

The following list of theorems includes some that you may remember from elementary algebra. Some have been given names that are used in this book. In Theorems 1.4-1 through 1.4-8, $a, b, c \in \mathbf{R}$.

THEOREM 1.4-1 If $a = b$, then $a + c = b + c$.   (Addition law)
THEOREM 1.4-2 If $a = b$, then $ac = bc$.   (Multiplication law)
THEOREM 1.4-3 For all a, $a \cdot 0 = 0$.   (Zero factor law)
THEOREM 1.4-4 If $a + c = b + c$, then $a = b$.   (Cancellation law of addition)
THEOREM 1.4-5 If $ac = bc$, and $c \neq 0$, then $a = b$.
                                             (Cancellation law of multiplication)
THEOREM 1.4-6 $-a + (-b) = -(a + b)$.
THEOREM 1.4-7 $a \cdot (-b) = -(a \cdot b)$.
THEOREM 1.4-8 $(-a) \cdot (-b) = a \cdot b$.

In Theorems 1.4-9 through 1.4-16, $a, b, c,$ and d are integers.

THEOREM 1.4-9 $\dfrac{a}{b} = \dfrac{c}{d}$ if and only if $a \cdot d = b \cdot c$ $(b, d \neq 0)$.

THEOREM 1.4-10 $\dfrac{a \cdot c}{b \cdot c} = \dfrac{a}{b}$ and $\dfrac{a}{b} = \dfrac{a \cdot c}{b \cdot c}$ $(b, c \neq 0)$.
                                            (The fundamental principle of fractions)

THEOREM 1.4-11 $\dfrac{a}{c} + \dfrac{b}{c} = \dfrac{a + b}{c}$ $(c \neq 0)$.

THEOREM 1.4-12 $\dfrac{a}{b} \cdot \dfrac{c}{d} = \dfrac{a \cdot c}{b \cdot d}$ $(b, d \neq 0)$.

THEOREM 1.4-13 $\dfrac{a}{b} \div \dfrac{c}{d} = \dfrac{a}{b} \cdot \dfrac{d}{c}$ $(b, c, d \neq 0)$.

THEOREM 1.4-14 $\dfrac{1}{a/b} = \dfrac{b}{a}$ $(a, b \neq 0)$.

THEOREM 1.4-15 $\dfrac{a}{b} = \dfrac{-a}{-b} = -\dfrac{-a}{b} = -\dfrac{a}{-b}$ $(b \neq 0)$.

THEOREM 1.4-16 $-\dfrac{a}{b} = \dfrac{-a}{b} = \dfrac{a}{-b} = -\dfrac{-a}{-b}$ $(b \neq 0)$.

Theorems 1.4-9 through 1.4-16, as stated here, apply specifically to operations in the set of rational numbers Q, which means that the replacement set for a, b, c, and d is the set of integers. It can be shown that each of these theorems is also true if the replacement set for the variables is R subject to the same restrictions as those given in the theorems. These proofs will not be given here. However, we will assume them to be true and will use them as the need arises.

Deductive Proof

The deductive method of proof involves starting with the hypothesis or some other statement known to be true, and then deriving from this another statement, or statements, which by the rules of logic can be shown to be true. One should never lose sight of the fact that the conclusion is the desired goal and each statement along the way must be justified by a definition, axiom, or previously proved theorem.

We have previously stated that the hypothesis is a statement assumed to be true. In proving mathematical theorems, the hypothesis must be true, because a false hypothesis leads to a false conclusion. It is also important that any replacements for variables meet the conditions of the hypothesis. For example, the theorem "If $a = b$ and $c \neq 0$, then $ac = bc$" is frequently used in the solution of equations. Consider the equation

$$\frac{x}{x-3} - 2 = \frac{3}{x-3}.$$

If both members of this equation are multiplied by $x - 3$, we obtain

$$x - 2x + 6 = 3,$$

from which

$$x = 3.$$

However, if $x = 3$, then $x - 3 = 0$, which is contrary to the hypothesis. Hence, we know that x cannot equal 3. In fact, it can be shown that the equation has no solution.

For many people, the proof of a theorem is one of the more difficult aspects of mathematics. On the other hand, a systematic logical analysis is the best approach to problem solving. Hence, it is worthwhile to analyze the proofs of some elementary theorems even though you have probably had some experience with such proofs. We begin with the proof of Theorem 1.4-1, the addition law. The theorem states

If $a, b, c \in R$ and $a = b$, then $a + c = b + c$.

From the wording of this theorem, it should be clear that $a, b, c \in R$ and $a = b$ is the hypothesis and that $a + c = b + c$ is the conclusion to be deduced.

PROOF:

STATEMENTS	JUSTIFICATIONS
1. $a, b, c \in R$ and $a = b$	1. Hypothesis

The conclusion must follow from the hypothesis, but we note that the conclusion involves sums and that no sums are evident in the hypothesis. Using a clue from the conclusion we write

2. $a + c \in R$ 2. Closure for addition

But the conclusion is a statement of equality of two sums. So

3. $a + c = a + c$ 3. Reflexive property of equality

Now since we have assumed $a = b$, we may replace a with b in the right-hand member to obtain

4. $a + c = b + c$ 4. Substitution

Since we have reached the desired conclusion, the proof is complete.

Frequently the connection between the hypothesis and the conclusion is vague. It is then much more difficult to get started. For example, consider the proof of

THEOREM 1.4-7   If $a, b \in R$, then $a(-b) = -ab$.

In analyzing this statement, we recognize that $-b$, $a(-b)$, and $-(ab)$ are real numbers, but the connection is not so obvious. The product $a(-b)$ is the key. Since $-b$ is not a part of the hypothesis it must be introduced.

PROOF:

     STATEMENTS   JUSTIFICATIONS

1. $a, b \in R$ 1. Hypothesis
2. $-b \in R$ and $b + (-b) = 0$ 2. Additive inverse law

Now the product $a(-b)$ can be formed.

3. $a[b + (-b)] = a \cdot 0$ 3. Theorem 1.4-2 (Multiplication law)
4. $a \cdot b + a(-b) = a \cdot 0$ 4. Distributive law
5. $ab + a(-b) = 0$ 5. Theorem 1.4-3 (Zero factor law)
6. $ab \in R$ 6. Closure for multiplication
7. $-ab \in R$ and $ab + (-ab) = 0$ 7. Additive inverse
8. $a(-b) = -ab$ 8. Additive inverse is unique

Since analysis and logical reasoning form an important part of mathematics, you should concentrate on trying to understand the reasoning involved in the above proofs and attempt to formulate proofs on your own.

EXERCISE 1.4

ORAL OR WRITTEN

Justify each statement in Problems 1–10 with one of the theorems 1.4-1 through 1.4-16. All variables represent real numbers. When a theorem has been given a name, the name can be used as your answer.

Examples (a) $(x + y) \cdot 0 = 0$

(b) $\dfrac{x}{4} + \dfrac{y}{4} = \dfrac{x + y}{4}$

Solutions (a) Theorem 1.4-3: $a \cdot 0 = 0$ or Zero factor law.

(b) Theorem 1.4-11: $\dfrac{a}{c} + \dfrac{b}{c} = \dfrac{a + b}{c}$

1. If $x + 4 = 7$, then $x + 4 + (-4) = 7 + (-4)$.
2. If $10x = 30$, then $10x(\tfrac{1}{10}) = 30(\tfrac{1}{10})$.
3. $\left(\dfrac{x + 5}{2}\right) \cdot 0 = 0$
4. $-2x + (-3y) = -(2x + 3y)$
5. $(-2x)(-3y) = (2x)(3y)$
6. If $\dfrac{x}{4} = \dfrac{y}{3}$, then $3x = 4y$.
7. $\dfrac{1}{\tfrac{2}{3}} = \dfrac{3}{2}$
8. $\dfrac{x + 2}{3} \cdot \dfrac{x - 1}{5} = \dfrac{(x + 2)(x - 1)}{3 \cdot 5}$
9. If $7(x + y) = 3(x - y)$, then $\dfrac{x + y}{3} = \dfrac{x - y}{7}$.
10. If $x + (y - 2) = y + 4$, then $x + (y - 2) + [-(y - 2)] = y + 4 + [-(y - 2)]$.

In Problems 11–16, justify each statement in the proof of the theorem with an axiom, definition, or previously proved theorem. The proofs outlined here are not necessarily complete in every detail. However, they are complete enough so that supporting reasons can be given for each statement.

11. *Theorem 1.4-2:* If $a, b, c \in R$ and $a = b$, then $ac = bc$.
    (a) $a, b, c \in R$ and $a = b$
    (b) $ac \in R$
    (c) $ac = ac$
    (d) $ac = bc$
12. *Theorem 1.4-3:* For all $a \in R$, $a \cdot 0 = 0$.
    (a) $a \in R$
    (b) $0 + 0 = 0$
    (c) $a(0 + 0) = a(0)$
    (d) $a \cdot 0 + a \cdot 0 = a \cdot 0$
    (e) $a \cdot 0 + a \cdot 0 + [-(a \cdot 0)] = a \cdot 0 + [-(a \cdot 0)]$
    (f) $a \cdot 0 + 0 = 0$
    (g) $a \cdot 0 = 0$
13. *Theorem 1.4-11:* If $a, b, c \in J$, $c \neq 0$, then $\dfrac{a}{c} + \dfrac{b}{c} = \dfrac{a + b}{c}$.
    (a) $\dfrac{a}{c} + \dfrac{b}{c} = a \cdot \dfrac{1}{c} + b \cdot \dfrac{1}{c}$
    (b) $a \cdot \dfrac{1}{c} + b \cdot \dfrac{1}{c} = (a + b) \cdot \dfrac{1}{c}$

(c) $(a+b) \cdot \dfrac{1}{c} = \dfrac{a+b}{c}$

(d) $\dfrac{a}{c} + \dfrac{b}{c} = \dfrac{a+b}{c}$

14. **Theorem 1.4-12:** If $\dfrac{a}{b}, \dfrac{c}{d} \in R$, then $\dfrac{a}{b} \cdot \dfrac{c}{d} = \dfrac{a \cdot c}{b \cdot d}$.

(a) $\dfrac{a}{b} \cdot \dfrac{c}{d} = \dfrac{a}{b} \cdot \dfrac{c}{d}$

(b) $= a \cdot \dfrac{1}{b} \cdot c \cdot \dfrac{1}{d}$

(c) $= a \cdot \dfrac{1}{b} \cdot c \cdot \dfrac{1}{d} \cdot b \cdot d \cdot \dfrac{1}{b \cdot d}$

(d) $= a \cdot c \cdot \dfrac{1}{b \cdot d} \cdot \left(\dfrac{1}{b} \cdot b\right)\left(\dfrac{1}{d} \cdot d\right)$

(e) $= a \cdot c \cdot \dfrac{1}{b \cdot d} \cdot 1 \cdot 1$

(f) $= a \cdot c \cdot \dfrac{1}{b \cdot d}$

(g) $\dfrac{a}{b} \cdot \dfrac{c}{d} = \dfrac{a \cdot c}{b \cdot d}$

15. **Theorem 1.4-5:** If $a, b, c \in R$, $c \neq 0$ and $ac = bc$, then $a = b$.

(a) $a, b, c \in R$ and $ac = bc$

(b) $\dfrac{1}{c} \in R$

(c) $ac \cdot \dfrac{1}{c} = bc \cdot \dfrac{1}{c}$

(d) $a\left(c \cdot \dfrac{1}{c}\right) = b\left(c \cdot \dfrac{1}{c}\right)$

(e) $a \cdot 1 = b \cdot 1$

(f) $a = b$

16. **Theorem 1.4-14:** If $\dfrac{a}{b}, \dfrac{b}{a} \in R$, then $\dfrac{1}{a/b} = \dfrac{b}{a}$.

(a) $\dfrac{a}{b}, \dfrac{b}{a} \in R$

(b) $\dfrac{1}{a/b} \in R$

(c) $\dfrac{a}{b} \cdot \dfrac{1}{a/b} = 1$

(d) $\dfrac{a}{b} \cdot \dfrac{b}{a} = \dfrac{a \cdot b}{b \cdot a}$

(e) $\dfrac{a \cdot b}{b \cdot a} = 1$

(f) $\dfrac{a}{b} \cdot \dfrac{1}{a/b} = \dfrac{a}{b} \cdot \dfrac{b}{a}$

(g) $\dfrac{1}{a/b} = \dfrac{b}{a}$

1.5 Order and Absolute Value

An important characterization of the set of real numbers is that they can be placed in one-to-one correspondence with the set of points in a line. This relationship can help you to understand other properties of real numbers.

The Number Line

Let us first demonstrate that the whole numbers form an ordered set. Draw a line and arbitrarily associate some point on the line with the number 0. This point is called the **origin**. Next, select some convenient line segment as a unit of measure and locate additional points to the right of the origin by repeated applications of this unit of measure. Since each point so located is a *whole* number of

Figure 1.5-1

units from the origin, we associate with each point the whole number corresponding to the number of units the point is from the origin. The point associated with a number is called the **graph** of the number, and the number is called the **coordinate** of the point. (See Figure 1.5-1.)

Order

Referring to the number line, you can compare any two numbers by noting the relative positions of their graphs. The number associated with the point to the left is **less than** (<) the number associated with the point to the right. Also, the

Figure 1.5-2

number associated with the point to the right is **greater than** (>) the other. In Figure 1.5-2, $a < b$, and $b > a$. Either statement implies the other. When the elements in a set of numbers are arranged so that any element is less than every element to the right and greater than every element to the left, the set is said to be arranged in increasing order.

Positive and Negative Numbers

In Section 1.3 you learned that every number $a \in R$ is assumed to have a unique additive inverse, or negative, denoted by $-a$, such that $a + (-a) = 0$. Since the origin separates the points on the number line into three disjoint subsets—that is, points to the left of the origin, the origin, and points to the right of the origin—and since the natural numbers are considered to be *positive* numbers, numbers associated with points to the left of the origin are called **negative numbers**. Thus, the set of real numbers can be separated into three disjoint subsets: R^-, the negative numbers; $\{0\}$; and R^+, the positive numbers. (See Figure 1.5-3.)

Figure 1.5-3

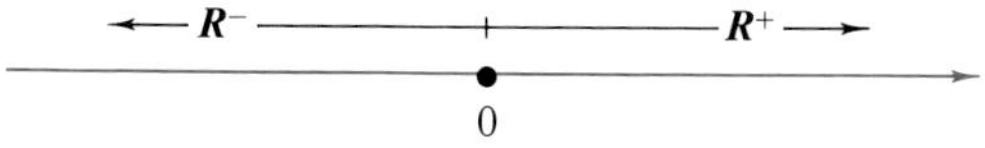

The same ordering principle that was used in graphing the whole numbers is used for the real numbers. That is,

For every $a \in \boldsymbol{R}^-$, $a < 0$.
For every $a \in \boldsymbol{R}^+$, $a > 0$.

Axioms of Order

The ordering principle stated above leads to the first axiom of order, called the trichotomy law:

O-1 For each $a \in \boldsymbol{R}$, one and only one of the following statements is true:

$$a < 0, \quad a = 0, \quad \text{or } a > 0.$$

Since $0 \in \boldsymbol{R}$, this axiom can be used to compare any two real numbers a and b. That is, a must be less than b, equal to b, or greater than b.

The second axiom is one of closure for the addition and multiplication operations with positive numbers.

O-2 If $a, b \in \boldsymbol{R}^+$, then $a + b \in \boldsymbol{R}^+$ and $ab \in \boldsymbol{R}^+$.

The Less Than Relation

The statement $a < b$ implies that the point on the number line associated with a is to the left of the point associated with b. Hence, to go from a to b means to move in a positive direction, and the distance from a to b can be represented by a positive number. This suggests the following.

DEFINITION 1.5-1 For every $a, b \in \boldsymbol{R}$, $a < b$ implies the existence of some positive number c such that $a + c = b$. Also, if $a + c = b$, then $a < b$.

The axioms of order and Definition 1.5-1 lead to some interesting consequences. Suppose, for example, that you wish to arrange the set of numbers $\{3, -2, 1\}$ in increasing order. If these numbers are associated with points on the number line, you note that $-2 < 1$, $1 < 3$, and $-2 < 3$. This arrangement may cause you to suspect that any set of three numbers, no two of which are equal, can be arranged in a similar manner. However, you cannot verify this by testing all cases, since the set of real numbers is infinite. Hence, the general case is stated as a theorem, whose proof applies to all individual cases.

THEOREM 1.5-1 For every $a, b, c \in \boldsymbol{R}$, if $a < b$ and $b < c$, then $a < c$.

PROOF:

STATEMENTS	JUSTIFICATIONS
1. $a < b$ and $b < c$	1. Hypothesis
2. d, e are positive real numbers such that $a + d = b$ and $b + e = c$	2. Definition 1.5-1

3. $(a + d) + e = c$	3. Substitution $[(a + d)$ for $b]$
4. $a + (d + e) = c$	4. Associative property of addition
5. $d + e \in R^+$	5. Axiom O-2
6. $a < c$	6. Definition 1.5-1

You may note that Theorem 1.5-1 is somewhat similar to Axiom E-3, the transitive law of equality. Are there other similarities between statements of equality and of order? Consider the statements $x = 3$ and $y < 3$. We know by Theorems 1.4-1 and 1.4-2 that $x + 2 = 3 + 2$ and that $2 \cdot x = 2 \cdot 3$, but are we certain that $y + 2 < 3 + 2$ or that $2 \cdot y < 2 \cdot 3$? These questions can be answered by the theorems that follow.

THEOREM 1.5-2 If $a, b, c \in R$ and $a < b$, then $a + c < b + c$.

PROOF:

STATEMENTS	JUSTIFICATIONS
1. $a < b$	1. Hypothesis
2. For some $d \in R^+$, $a + d = b$	2. Definition 1.5-1
3. $(a + d) + c = b + c$	3. Theorem 1.4-1 (Addition law)
4. $(a + c) + d = b + c$	4. Commutative and associative properties of addition
5. $a + c < b + c$	5. Definition 1.5-1

THEOREM 1.5-3 If $a, b, c \in R$, $a < b$ and $c > 0$, then $a \cdot c < b \cdot c$.

(The proof of Theorem 1.5-3 is similar to that of Theorem 1.5-2 and is presented as an exercise.)

In Theorem 1.5-3, the multiplier c was specified to be a positive number. The situation is somewhat different if c is a negative number. For example, $2 < 5$, but $-3 \cdot 2 > -3 \cdot 5$ since $-15 < -6$. This would indicate that when both members of an inequality are multiplied by the same negative number, the sense or direction of the inequality is changed. Before we can prove a theorem concerning this situation we must first consider another theorem.

THEOREM 1.5-4 For every $a \in R$, $a < 0$ if and only if $-a > 0$.

PROOF:

STATEMENTS	JUSTIFICATIONS

If:

1. $-a > 0$	1. Hypothesis
2. $-a + a > 0 + a$	2. Theorem 1.5-2

3. $0 > 0 + a$ 3. Additive inverse
4. $0 > a$ or $a < 0$ 4. Additive identity

Only if:

5. $a < 0$ 5. Hypothesis
6. $a + (-a) < 0 + (-a)$ 6. Theorem 1.5-1
7. $0 < 0 + (-a)$ 7. Additive inverse
8. $0 < -a$ or $-a > 0$ 8. Additive identity

We are now able to prove the multiplication property of order when the multiplier is a negative number.

THEOREM 1.5-5   If $a, b, c \in \mathbf{R}$, $a < b$, and $c < 0$, then $ac > bc$.

The proof of this theorem depends on Theorems 1.5-4 and 1.5-1 and is left as an exercise.

Here are some other useful theorems, proofs of which are not given. All variables $\in \mathbf{R}$ except where stated.

THEOREM 1.5-6 $a > b$ if and only if $a - b > 0$.

THEOREM 1.5-7   If $a \neq 0$, $a^2 > 0$.

THEOREM 1.5-8   If $a \neq 0$, $a > 0$ if and only if $\dfrac{1}{a} > 0$.

THEOREM 1.5-9   If $a, b \neq 0$ and $a < b$, then $\dfrac{1}{a} > \dfrac{1}{b}$.

THEOREM 1.5-10   For every a, b, c, if $a < b$ and $a = c$, then $c < b$.

Absolute Value

A measure of distance is considered to be a positive number because the distance is the same regardless of the direction in which it is measured. This concept leads to a property of real numbers called **absolute value**. For example, consider the two numbers 5 and -5. If we associate these numbers with points on the number line, the graph of 5 is 5 units to the right of the origin and the graph of -5 is 5 units to the left, but both points are the *same distance from the origin*. The measure of such a distance is called the absolute value of the number associated with the point. To indicate absolute value, the number symbol is placed between vertical bars. Thus, $|a|$ is read "the absolute value of a." Since a measure of distance cannot be negative, the absolute value of a number is non-negative. For example, $|5| = 5$ and $|-5| = -(-5) = 5$. This suggests the following.

DEFINITION 1.5-2   For every $a \in \mathbf{R}$,

$$|a| = \begin{cases} a & \text{if } a \geq 0 \\ -a & \text{if } a < 0 \end{cases} \quad \begin{array}{l} (a \text{ is a non-negative number}) \\ (a \text{ is a negative number}) \end{array}$$

1.5 Order and Absolute Value

If you compare the absolute values of two numbers, the number with the greater absolute value is associated with the point that is the greater distance from the origin.

Examples $|-15| > |4|$; $|-5| = |5|$; $|-3| < |7|$.

You must learn to distinguish between absolute value and relative value. Remember, *relative value depends upon the location of points to the right or left of the origin, and absolute value depends upon the distance from the origin.* Hence,

$$-15 < 4 \quad \text{but} \quad |-15| > |4|.$$

EXERCISE 1.5

ORAL

Read the symbols

1. $a < b$ 2. $a > b$ 3. $|a|$ 4. R^- 5. R^+

WRITTEN

1. Express the phrase "a is a positive number" symbolically in two ways.
2. Express the phrase "a is a negative number" symbolically in two ways.

In Problems 3–20, state an axiom, definition, or theorem that justifies each statement. All variables $\in R$ except where stated.

Examples (a) $3 + 5 \in R^+$ (b) $3 < 5$ because $3 + 2 = 5$
Solutions (a) Axiom O-2   (b) Definition 1.5-1

3. If $x \in R^+$, $5x \in R^+$.
4. If $x < y$, then $x + 5 < y + 5$.
5. If $x < y$, then $4x < 4y$.
6. If $-x > 0$, then $x < 0$.
7. If $x + y > 0$, then $\dfrac{1}{x+y} > 0$.
8. If $x - y > 0$, then $x > y$.
9. If $x - y \neq 0$, then $(x - y)^2 > 0$.
10. If $x + y < z$, then $x + y - 5 < z - 5$.
11. If $x + y < z - 2$ and $z - 2 < w$, then $x + y < w$.
12. If $x < y$ and $z \in R^-$, then $xz > yz$.
13. If $x < y$ and $y = z$, then $x < z$.
14. If $x, y \in R^+$ and $x < y$, then $\dfrac{1}{x} > \dfrac{1}{y}$.
15. If $x < y$ and $x + z = y$, then $z \in R^+$.
16. If $x + y < 0$, then $-(x + y) > 0$.
17. $x + y < 0$, $x + y = 0$, or $x + y > 0$.
18. If $x - y < w + z$ and $v \in R^-$ then $v(x - y) > v(w + z)$.

19. If $2x - 7 < 0$, then $2x - 7 + 7 < 0 + 7$.

20. If $\frac{x}{4} > 3$, then $4 \cdot \frac{x}{4} > 4 \cdot 3$.

The proofs of statements in Problems 21–24 require three steps. Write the steps and give the justifications.

Example If $x < 0$, then $x^2 > 0$.

Solution
(a) $x < 0$ Hypothesis
(b) $x \neq 0$ Axiom O-1
(c) $x^2 > 0$ Theorem 1.5-7

21. If $x < z$ and $x + y = z$, then $5y \in R^+$.

22. If $x < z$ and $x + y = z$, then $\frac{1}{y} > 0$.

23. If $x, y > 0$, then $\frac{1}{x+y} < \frac{1}{y}$.

24. If $x < z$ and $x + y = z$, then $xy < zy$.

In Problems 25 and 26, complete the proofs by supplying reasons for the statements.

∗25. *Theorem 1.5-3:* If $a, b, c \in R$, $a < b$, and $c > 0$, then $ac < bc$.
 (a) $a < b$
 (b) For some $d \in R^+$, $a + d = b$
 (c) $c(a + d) = cb$
 (d) $ca + cd = cb$
 (e) $cd \in R^+$
 (f) $ca < cb$
 (g) $ac < bc$

∗26. *Theorem 1.5-5:* If $a, b, c \in R$, $a < b$ and $c < 0$, then $ac > bc$.
 (a) Since $c < 0$, $-c > 0$
 (b) $-ca < -cb$
 (c) For some $d \in R^+$, $-ca + d = -cb$
 (d) $-ca + d + ca + cb = -cb + ca + cb$
 (e) $-ca + ca + cb + d = -cb + cb + ca$
 (f) $cb + d = ca$
 (g) $cb < ca$ or $ca > cb$
 (h) $ac > bc$

1.6 Review of Addition

In Section 1.4 you reviewed the axioms and some theorems concerning the operation of addition with real numbers. These statements made no distinction between positive and negative numbers. In this section you will review the operation of addition with such numbers.

There are many times when a collection of symbols denoting a binary operation can be replaced with a single numeral. For example, the sum $3 + 2$ can be replaced with the numeral 5. In such instances, the single numeral is called a **basic numeral** The phrase **basic fraction** is used to refer to a fraction in which the numerator and denominator are relatively prime integers. That is, they have no common integral factor and the fraction is in lowest terms.

Sums

Since this is primarily a review of material you have previously studied, we will not prove the theorems concerning the addition of real numbers but will merely restate those concerning sums that involve negative numbers, and we will limit the discussion to rational numbers.

In a sum involving negative numbers, both numbers may be negative, or one number may be negative and the other positive. When both numbers are negative, the sum is the negative of the sum of the absolute values. That is, the numbers may be added without the signs and then the minus sign is attached to the sum. When the numbers have different signs, the difference of the numbers without the sign is found, and then the sign of the number with the greater absolute value is attached to the sum.

Examples (a) $-3 + (-2) = -(3 + 2) = -5$

(b) $\dfrac{-2}{7} + \dfrac{-3}{7} = -\left(\dfrac{2}{7} + \dfrac{3}{7}\right) = \dfrac{-5}{7}$

(c) $13 + (-5) = 13 - 5 = 8$
(The answer is positive because $|13| > |-5|$.)

(d) $5 + (-13) = -(13 - 5) = -8$
(The answer is negative because $|-13| > |5|$.)

Differences

Recall that in subtraction the number to be subtracted is called the *subtrahend,* and the number from which it is subtracted is called the *minuend.* Now, since the difference of two real numbers $a - b = a + (-b)$, we can subtract one real number from another by *adding the negative of the subtrahend to the minuend.*

Examples (a) $3 - 8 = 3 + (-8) = -(8 - 3) = -5$
[Compare with Example (d) for addition.]

(b) $-2 - (-7) = -2 + [-(-7)]$
$= -2 + 7 = 7 - 2 = 5$
[Compare with Example (c) for addition.]

(c) $-5 - 9 = -5 + (-9) = -(5 + 9) = -14$
[Compare with Example (a) for addition.]

(d) $8 - (-4) = 8 + [-(-4)] = 8 + 4 = 12$
[Compare with the addition of natural numbers.]

Least Common Multiple

The statement $\frac{a}{c} + \frac{b}{c} = \frac{a+b}{c}$ (Theorem 1.4-11) justifies writing the sum of two rational numbers as a rational number when the denominators are the same. When the denominators are not the same, you must change each rational number to an equivalent number so that both rational numbers have the same, or common, denominator. Generally, we use a common denominator that is the least positive integer possible. This least positive integer is called the **least common multiple** of the denominators or the **least common denominator**.

The least common multiple of a set of positive integers is the smallest positive integer that is exactly divisible by every member of the set. You find a least common multiple by forming the product of all of the individual prime factors of the members of the set, using each factor the greatest number of times it appears as a factor of any member of the set.

Example   Find the least common multiple of $\{6, 8, 18, 27\}$.

Solution   First write each number as a product of its prime factors.

$6 = 2 \cdot 3; \quad 8 = 2 \cdot 2 \cdot 2; \quad 18 = 2 \cdot 3 \cdot 3; \quad \text{and } 27 = 3 \cdot 3 \cdot 3$

Observe that 2 and 3 are the only numbers that occur as factors and that each appears in any one number at most three times. Hence, the least common multiple is $2 \cdot 2 \cdot 2 \cdot 3 \cdot 3 \cdot 3 = 216$.

Once the least common denominator has been determined, Theorem 1.4-10 $\left[\frac{a}{b} = \frac{a \cdot c}{b \cdot c}\right]$ can be applied to convert each addend to a fraction so that both fractions have the same denominator. You can then write the basic numeral for the sum of the numerators over the common denominator by using Theorem 1.4-11 $\left[\frac{a}{c} + \frac{b}{c} = \frac{a+b}{c}\right]$.

It may be necessary for you to apply the fundamental principle of fractions $\left[\frac{a \cdot c}{b \cdot c} = \frac{a}{b}\right]$ to obtain the basic fraction.

Example   Find the basic fraction for $\frac{7}{24} + \frac{8}{15}$.

Solution   First find the least common denominator, which is the least common multiple of 24 and 15.

$24 = 2 \cdot 2 \cdot 2 \cdot 3; \quad 15 = 3 \cdot 5$

The least common denominator is $2 \cdot 2 \cdot 2 \cdot 3 \cdot 5 = 120$. Since $120 = 24 \cdot 5$ and $120 = 15 \cdot 8$, we have

$$\frac{7 \cdot 5}{24 \cdot 5} + \frac{8 \cdot 8}{15 \cdot 8} = \frac{35}{120} + \frac{64}{120}$$

$$= \frac{99}{120}$$

$$= \frac{33 \cdot 3}{40 \cdot 3}$$

$$= \frac{33}{40}.$$

Since $a - b = a + (-b)$ for any two real numbers, you can find the difference of two fractions by adding the negative of the second fraction (subtrahend) to the first (minuend).

Example Find the basic fraction for $\frac{7}{8} - \frac{3}{8}$.

Solution $\frac{7}{8} - \frac{3}{8} = \frac{7}{8} + \left(\frac{-3}{8}\right) = \frac{7-3}{8} = \frac{4}{8} = \frac{1}{2}.$

EXERCISE 1.6

WRITTEN

Find the basic numeral (or fraction) for each sum or difference in Problems 1–52.

1. $5 + 7$
2. $13 + 9$
3. $5 + 27$
4. $16 + 98$
5. $8 + (-3)$
6. $15 + (-7)$
7. $21 + (-18)$
8. $42 + (-37)$
9. $9 + (-14)$
10. $11 + (-23)$
11. $42 + (-66)$
12. $198 + (-301)$
13. $-8 + (-7)$
14. $-12 + (-5)$
15. $-38 + (-42)$
16. $-107 + (-275)$
17. $\frac{1}{2} + \frac{1}{3}$
18. $\frac{5}{7} + \frac{7}{5}$
19. $\frac{4}{15} + \frac{9}{25}$
20. $\frac{5}{12} + \frac{13}{16}$
21. $\frac{1}{4} + \left(\frac{-3}{2}\right)$
22. $\frac{3}{5} + \left(\frac{-4}{15}\right)$
23. $\frac{5}{18} + \left(\frac{-8}{27}\right)$
24. $\frac{7}{24} + \left(\frac{-17}{36}\right)$
25. $\frac{-2}{5} + \left(\frac{-4}{7}\right)$
26. $\frac{-7}{12} + \left(\frac{-4}{15}\right)$
27. $\frac{-7}{16} + \left(\frac{-11}{18}\right)$
28. $\frac{-13}{30} + \left(\frac{-19}{35}\right)$
29. $10 - 3$
30. $17 - 15$
31. $24 - 42$
32. $147 - 251$
33. $45 - (-15)$
34. $52 - (-61)$
35. $127 - (-84)$
36. $247 - (-392)$
37. $-18 - (-10)$
38. $-24 - (-19)$
39. $-47 - (-74)$
40. $-174 - (-248)$
41. $\frac{1}{2} - \frac{1}{3}$
42. $\frac{3}{4} - \frac{4}{5}$

28 REAL NUMBERS

43. $\dfrac{7}{15} - \dfrac{8}{25}$
44. $\dfrac{13}{16} - \dfrac{11}{18}$
45. $\dfrac{2}{3} - \left(\dfrac{-1}{2}\right)$
46. $\dfrac{7}{8} - \left(\dfrac{-11}{12}\right)$
47. $\dfrac{3}{14} - \left(\dfrac{-8}{21}\right)$
48. $\dfrac{11}{30} - \left(\dfrac{-17}{35}\right)$
49. $\dfrac{-7}{8} - \left(\dfrac{-2}{3}\right)$
50. $\dfrac{-5}{12} - \left(\dfrac{-8}{15}\right)$
51. $\dfrac{-9}{21} - \left(\dfrac{-8}{35}\right)$
52. $\dfrac{-14}{27} - \left(\dfrac{-34}{63}\right)$

1.7 Review of Multiplication

Products

Theorems 1.4-7 and 1.4-8 (p. 14) can be rephrased to help you find the basic numeral (or basic fraction) for the product of two rational numbers. Theorem 1.4-7 tells us that if two numbers have different signs, then the product is a negative number.

Examples   (a) $3 \cdot (-5) = -(3 \cdot 5) = -15$
(b) $-4 \cdot 7 = -(4 \cdot 7) = -28$
(c) $\dfrac{2}{3} \cdot \dfrac{-4}{5} = -\left(\dfrac{2}{3} \cdot \dfrac{4}{5}\right) = \dfrac{-8}{15}$

Theorem 1.4-8 tells us that the product of two negative numbers is a positive number.

Examples   (a) $-2 \cdot (-8) = 2 \cdot 8 = 16$
(b) $\dfrac{-3}{4} \cdot \dfrac{-7}{8} = \dfrac{3}{4} \cdot \dfrac{7}{8} = \dfrac{21}{28}$

Quotients

Since the quotient of two real numbers $\dfrac{a}{b} = a \cdot \dfrac{1}{b}$, Theorems 1.4-7 and 1.4-8 hold for quotients as well as products.

Example $\dfrac{4}{15} \div \dfrac{-3}{5} = \dfrac{4}{15} \cdot \dfrac{1}{-3/5} = \dfrac{4}{15} \cdot \dfrac{-5}{3} = \dfrac{-20}{45} = \dfrac{-4}{9}.$

EXERCISE 1.7

WRITTEN

Find the basic numeral (or fraction) for each product or quotient in Problems 1–52.

1. $5 \cdot 7$
2. $13 \cdot 9$
3. $5 \cdot 27$

4. $16 \cdot 98$
5. $8(-3)$
6. $15(-7)$
7. $21(-18)$
8. $42(-37)$
9. $-14(9)$
10. $-23(11)$
11. $-66(42)$
12. $-73(24)$
13. $-8(-7)$
14. $-12(-5)$
15. $-38(-42)$
16. $-31(-52)$
17. $\frac{1}{2} \cdot \frac{1}{3}$
18. $\frac{5}{7} \cdot \frac{7}{5}$
19. $\frac{4}{15} \cdot \frac{25}{9}$
20. $\frac{5}{12} \cdot \frac{13}{16}$
21. $\frac{1}{4}\left(\frac{-3}{2}\right)$
22. $\frac{3}{5}\left(\frac{-15}{4}\right)$
23. $\frac{-5}{18}\left(\frac{-27}{8}\right)$
24. $\frac{-7}{24}\left(\frac{36}{14}\right)$
25. $\frac{-2}{5}\left(\frac{-4}{7}\right)$
26. $\frac{-7}{12}\left(\frac{-15}{4}\right)$
27. $\frac{-7}{16}\left(\frac{-18}{11}\right)$
28. $\frac{-13}{20}\left(\frac{-35}{26}\right)$
29. $15 \div 3$
30. $45 \div 15$
31. $144 \div 24$
32. $225 \div 25$
33. $45 \div (-15)$
34. $91 \div (-13)$
35. $-225 \div 25$
36. $-275 \div 25$
37. $-75 \div (-15)$
38. $-121 \div (-11)$
39. $-147 \div (-21)$
40. $-111 \div (-37)$
41. $\frac{1}{2} \div \frac{2}{3}$
42. $\frac{3}{8} \div \frac{3}{4}$
43. $\frac{8}{15} \div \frac{4}{5}$
44. $\frac{5}{12} \div \frac{15}{16}$
45. $\frac{2}{3} \div \left(\frac{-1}{2}\right)$
46. $\frac{7}{8} \div \left(\frac{-7}{12}\right)$
47. $\frac{-7}{5} \div \frac{8}{25}$
48. $\frac{-13}{30} \div \frac{19}{35}$
49. $\frac{-7}{8} \div \left(\frac{-3}{2}\right)$
50. $\frac{-5}{12} \div \left(\frac{-15}{8}\right)$
51. $\frac{-9}{21} \div \left(\frac{-6}{35}\right)$
52. $\frac{-14}{27} \div \left(\frac{-42}{35}\right)$

1.8 Decimal and Radical Notation

In the introductory paragraph to Section 1.5, the set of real numbers was described as being in one-to-one correspondence with the set of points on a line. A further consideration of the number line and the use of decimal notation can help strengthen your understanding of this intuitive concept.

We begin by dividing each interval on the number line into 10 equal subintervals, each of which represents one-tenth of a unit. We may then place one digit to the right of the decimal point for each coordinate as shown in Figure 1.8-1.

Figure 1.8-1

30 REAL NUMBERS

We chose 10 equal parts because of the decimal system of numeration. If each subinterval is further subdivided into 10 equal parts, each of these new subintervals represents one-hundredth of a unit and we may place a second digit to the right of the decimal point for each coordinate as shown in Figure 1.8-2.

Figure 1.8-2

Infinite Decimals

The fact that this process can be continued indefinitely should help you to understand that the real number associated with any point on a number line can be represented by an infinite decimal numeral or, as we shall refer to it, simply an **infinite decimal**. Thus, a real number is a number that can be named by an infinite decimal. Some examples are:

$$2 = 2.000000\ldots$$
$$\tfrac{4}{3} = 1.333333\ldots$$
$$\tfrac{7}{8} = 0.875000\ldots$$
$$\sqrt{2} = 1.414214\ldots$$
$$\pi = 3.1415927\ldots$$

Terminating Decimals

Infinite decimals, such as $2.000000\ldots$ and $0.875000\ldots$, that have a finite number of nonzero digits followed by an infinite number of zeros are called **terminating decimals** and are usually written without the sequence of zeros.

Periodic Decimals

Some infinite decimals are **repeating** or **periodic**, while others are nonrepeating or nonperiodic. It is impossible to determine whether a particular infinite decimal is periodic or nonperiodic by an examination of a finite sequence of digits. However, we can specify that an infinite decimal is periodic or nonperiodic by using a special form of symbolism. In this text, we will specify a periodic decimal by placing a bar over the last shown repeating group of digits, and to indicate a nonperiodic decimal we will use three dots following the last digit given.

Examples (a) $2.3\overline{3}$, $0.12\overline{12}$, and $0.123\overline{434}$ are periodic.
(b) $3.14\ldots$, $0.101001\ldots$, and $1.414\ldots$ are nonperiodic.

Any rational number of the form $\dfrac{a}{b}$, $a, b \in J$, $b \neq 0$, can be written as an infinite decimal by application of the familiar division process of arithmetic.

1.8 Decimal and Radical Notation   **31**

Furthermore, you will find that the decimal numeral for any rational number is either terminating or periodic. Consequently, you can assume that any terminating or periodic decimal names a rational number and any nonperiodic infinite decimal names an irrational number.

Changing a Decimal to the Quotient of Two Integers

Any rational number named by a terminating or periodic decimal can be renamed as the quotient of two integers. In the case of a terminating decimal, the sequence of digits without using the decimal point names the integer in the numerator. The denominator will be 1, 10, 100, etc., depending upon the number of digits to the right of the decimal point. The resulting fraction is then reduced to lowest terms.

Example   Express 1.075 as the quotient of two integers.

Solution   We use 1075 as the numerator and 1000 as the denominator since there are three digits to the right of the decimal point. Then,

$$\frac{1075}{1000} = \frac{25 \cdot 43}{25 \cdot 40} = \frac{43}{40}$$

The following example illustrates one process of converting a periodic decimal into the quotient of two integers.

Example   Express $0.12\overline{12}$ as the quotient of two integers.

Solution   Assume that for some $x \in Q$,

$x = 0.12\overline{12}$.

Then, by the multiplication law,

$100x = 100 \cdot 0.12\overline{12} = 12.12\overline{12}$

$100x - x = 12.12\overline{12} - 0.12\overline{12}$

$99x = 12$

$x = \frac{12}{99} = \frac{4}{33}$.

In the foregoing example, we multiplied by 100 in order to have *exactly one repeating group to the left of the decimal point*. Thus, when the members of the original equation were subtracted from the respective members of the new equation, we obtained an integer as the new left-hand member. Sometimes it is necessary to multiply both members by two different powers of 10 before subtracting.

Example   Express $0.123\overline{3}$ as the quotient of two integers.

Solution   Assume that for some $x \in Q$,

$x = 0.123\overline{3}$.

In this case we must choose 1000 as a multiplier in order to have one repeating group to the left of the decimal point.

$$1000x = 1000 \cdot 0.123\overline{3} = 123.3\overline{3}.$$

The simplest way to eliminate the repeating part of the decimal is to subtract a multiple of x in which only the nonrepeating digits appear to the left of the decimal point. We can do this by using 100 as a multiplier. That is,

$$100x = 12.3\overline{3}.$$

Then, subtracting the respective members, we have

$$1000x - 100x = 123.3\overline{3} - 12.3\overline{3}$$
$$900x = 111$$
$$x = \tfrac{111}{900} = \tfrac{37}{300}.$$

Radical Notation

Radical notation is another method of symbolizing certain real numbers. Radical representation of square roots, cube roots, fourth roots, etc., with certain restrictions can be used for this purpose. Our discussion here is limited to the representation of square roots of non-negative numbers. Other radicals are discussed in Chapter 3.

DEFINITION 1.8-1 If $a \in R$ and $a > 0$, then $\sqrt{a}$ is the non-negative number such that

$$\sqrt{a} \cdot \sqrt{a} = a.$$

In the symbol $\sqrt{a}$, a is called the **radicand**. If a is a number that is the product of two equal rational numbers, $\sqrt{a}$ represents a rational number. If a is not the product of two equal rational numbers, then $\sqrt{a}$ represents an irrational number.

Examples (a) $\sqrt{4}$, $\sqrt{\tfrac{16}{25}}$, $\sqrt{0.25}$ are rational numbers.

(b) $\sqrt{2}$, $\sqrt{\tfrac{7}{8}}$, $\sqrt{0.5}$ are irrational numbers.

By Definition 1.8-1, $\sqrt{a^2}$ is a non-negative number such that $\sqrt{a^2} \cdot \sqrt{a^2} = a^2$. However, if $a \neq 0$, $a \cdot a = a^2$ and $-a(-a) = a^2$. Since $\sqrt{a^2}$ must be a non-negative number, we define $\sqrt{a^2}$ as

DEFINITION 1.8-2 $\sqrt{a^2} = |a|.$

The paragraph above indicates that any positive number actually has two square roots, which are negatives of each other. Of these two, the positive one is called the **principal square root**. If the negative square root is to be indicated, the radical symbol must be preceded by the minus (−) sign. In some cases, as will be seen in Chapter 4, we need to use both square roots.

Examples (a) $\sqrt{25} = 5$ (b) $-\sqrt{9} = -3$
(c) $7 = \sqrt{49}$ (d) $-8 = -\sqrt{64}$

The square roots of most of the numbers that you will encounter in a practical situation are irrational numbers. You have seen that the decimal representation of an irrational number is nonterminating and nonperiodic. Ordinarily, such numbers can be rounded off to three or four decimal places without seriously affecting the accuracy of most computations.

Since finding the square root of a number by arithmetic methods is somewhat tedious, we have included a table of squares and square roots of numbers as Appendix 2. To use the table, find the number in the left-hand column and read the square root under $\sqrt{N}$. For example, $\sqrt{2}$ is found to the right of 2 under $\sqrt{N}$. You can also use the table to find the square root of 10 times some numbers. $\sqrt{250}$ is found to the right of 25 under $\sqrt{10N}$.

Example Use the table of square roots to compute $\dfrac{\sqrt{2}\ \sqrt{370}}{\sqrt{45}}$.

Solution From the table,

$\sqrt{2} = 1.414, \quad \sqrt{370} = 19.235, \quad \text{and} \quad \sqrt{45} = 6.708.$

$$\dfrac{\sqrt{2}\ \sqrt{370}}{\sqrt{45}} = \dfrac{1.414(19.235)}{6.708}$$

$$= \dfrac{27.198}{6.708}$$

$$= 4.055$$

EXERCISE 1.8

WRITTEN

1. State three ways of naming rational numbers.
2. Define symbolically: $\sqrt{a}, \ a > 0; \ \sqrt{a^2}$.

In Problems 3–28, express each terminating and each periodic decimal as the quotient of two relatively prime integers.

Examples (a) 0.35 (b) $0.7\overline{7}$

Solutions (a) $0.35 = \dfrac{35}{100}$
$= \dfrac{7}{20}$

(b) Let $x = 0.7\overline{7}$. Then,

$$10x = 7.7\overline{7}$$
$$10x - x = 7.7\overline{7} - 0.7\overline{7}$$
$$9x = 7$$
$$x = \tfrac{7}{9}$$

3. 0.125
4. 7.95
5. 1.21
6. 2.375
7. 0.675
8. 4.312
9. -5.42
10. -0.025
11. 3.84
12. 6.98
13. -0.035
14. -4.60
15. $0.9\overline{9}$
16. $0.2\overline{2}$
17. $0.5\overline{5}$
18. $0.35\overline{35}$
19. $-0.23\overline{23}$
20. $-0.45\overline{45}$
21. $1.14\overline{14}$
22. $3.75\overline{75}$
23. $0.123\overline{123}$
24. $1.25\overline{25}$
25. $0.124\overline{24}$ (HINT: Subtract $10x$ from $1000x$.)
26. $0.203\overline{03}$
27. $0.112\overline{12}$
28. $2.312\overline{12}$

In Problems 29–32, list the elements in each set that name (a) whole numbers, (b) integers, (c) rational numbers, and (d) irrational numbers.

Example $\{-\sqrt{9}, -3.25, -2.12\ldots, -\sqrt{2}, -1.1\overline{1}, 0.23, 0.2\overline{3}\ldots, \sqrt{17}, \sqrt{49}\}$.

Solution (a) $\{\sqrt{49}\}$ (b) $\{-\sqrt{9}, \sqrt{49}\}$
(c) $\{-\sqrt{9}, -3.25, -1.1\overline{1}, 0.23, \sqrt{49}\}$
(d) $\{-2.12\ldots, -\sqrt{2}, 0.2\overline{3}\ldots, \sqrt{17}\}$

29. $\{-\sqrt{23}, -5.75, -\sqrt{6.75}, -3.45\ldots, 0, 1.25\overline{5}, \sqrt{2.25}, \sqrt{3}, 2.01\ldots, \tfrac{14}{3}\}$
30. $\{-3.3\overline{3}, \sqrt{6.76}, -\sqrt{16}, \tfrac{2}{3}, \sqrt{1.21}, \sqrt{2.5}, 3.05, 5\sqrt{2}, \sqrt{81}\}$
31. $\left\{\dfrac{-32}{13}, \dfrac{-8}{4}, -1.22, -\sqrt{1.44}, 0.03\overline{3}, \sqrt{1.69}, 2\sqrt{0.49}, 25\sqrt{0.01}\right\}$
32. $\left\{\dfrac{-45}{15}, -2\sqrt{1.96}, -\sqrt{0.25}, \dfrac{3-7}{2}, \dfrac{2 \cdot 6}{3}, \sqrt{-49}, 3\sqrt{6.25}, \tfrac{13}{2}\sqrt{1.44}\right\}$

In Problems 33–44, write each expression without a radical sign.

33. $\sqrt{49}$
34. $-\sqrt{121}$
35. $-\sqrt{1.44}$
36. $\sqrt{6.25}$
37. $-\sqrt{\tfrac{25}{49}}$
38. $\sqrt{\tfrac{81}{121}}$
39. $\sqrt{x^2}$
40. $\sqrt{(2y)^2}$
41. $\sqrt{(-2)^2}$
42. $-\sqrt{(-3)^2}$
43. $\sqrt{\left(\dfrac{-2}{3}\right)^2}$
44. $-\sqrt{\left(\dfrac{-4}{7}\right)^2}$

Use the table of square roots for the computations in Problems 45–54.

45. $\sqrt{2} + \sqrt{3}$
46. $\sqrt{7} + \sqrt{5}$
47. $\sqrt{320} - \sqrt{170}$
48. $\sqrt{280} - \sqrt{130}$
49. $\dfrac{\sqrt{3}\sqrt{5}}{\sqrt{6}}$
50. $\dfrac{\sqrt{10}\sqrt{15}}{\sqrt{7}}$
51. $\dfrac{\sqrt{27} + \sqrt{32}}{\sqrt{8}}$
52. $\dfrac{\sqrt{48}}{\sqrt{2}} + \sqrt{5}\sqrt{8}$
53. $\sqrt{480}(\sqrt{8} - \sqrt{6})$
54. $\sqrt{520}(\sqrt{13} + \sqrt{15})$

Summary

The numbers in the brackets to the right indicate the section in which the concept is discussed. This system is used in all subsequent chapter summaries.

1. A **set** is a collection of items that are called the **members** or **elements** of the set. [1.1]
2. A **variable** is a symbol used to represent an unspecified element in a set. [1.1]
3. Two sets are **equivalent** if they contain the same number of elements, and they are **equal** if they contain exactly the same elements. [1.1]
4. One set is a **subset** of a second set if every member of the first set is also a member of the second set. If the second set contains at least one element that is not in the first set, then the first set is a **proper subset** of the second. [1.1]
5. The **union** of two sets is the set containing every element that belongs to at least one of the sets. Symbolically,
$$A \cup B = \{x \mid x \in A \text{ or } x \in B\}.$$ [1.2]
6. The **intersection** of two sets is the set that contains all elements common to both sets and no other elements. Symbolically,
$$A \cap B = \{x \mid x \in A \text{ and } x \in B\}.$$ [1.2]
7. The **Cartesian product** of two sets is the set of all possible ordered pairs such that the first component of each ordered pair is a member of the first set and the second component is a member of the second set. Symbolically,
$$A \times B = \{(a, b) \mid a \in A \text{ and } b \in B\}.$$ [1.2]
8. The **properties of equality** of real numbers are the **substitution, reflexive, symmetric,** and **transitive** properties. [1.3]
9. If a and b are any two elements of a set S, a **binary operation** on a and b assigns to the ordered pair (a, b) some unique c which may or may not be an element of S. [1.3]
10. The eleven properties of real numbers with respect to the binary operations of addition and multiplication were discussed. [1.3]
11. A **theorem** is a statement concerning the properties of the elements of a set that can be shown to be true as a consequence of axioms, definitions, or previously proved theorems. [1.4]
12. The **real numbers** form a set that can be separated into three disjoint subsets: **negative numbers, zero,** and **positive numbers.** [1.5]
13. The **axioms of order** state that a particular real number belongs to exactly one of the subsets mentioned above, and that both the sum and the product of two positive numbers are positive numbers. [1.5]
14. The **absolute value** of a real number is always non-negative. [1.5]

15. The sum of two negative numbers is the negative of the sum of their absolute values. [1.6]
16. The sum of two numbers having different signs is the difference of their absolute values preceded by the sign of the number with the greater absolute value. [1.6]
17. The difference of two numbers is the sum of the minuend and the negative of the subtrahend. [1.6]
18. The **least common multiple** of a set of positive integers is the smallest positive integer that is exactly divisible by each member of the set. [1.6]
19. The **least common denominator** of a set of fractions is the least common multiple of the denominators. [1.6]
20. The product (or the quotient) of two real numbers is positive if both numbers have the same sign, and negative if the numbers have different signs. [1.7]
21. Every real number can be represented as an **infinite decimal**. Rational numbers can be represented as either **terminating** decimals or **periodic** decimals. Irrational numbers can be represented as **nonperiodic infinite decimals**. [1.8]
22. A **square root** of a non-negative real number is one of the two equal factors whose product is the number. [1.8]
23. $\sqrt{a^2} = |a|$. [1.8]

REVIEW EXERCISES

SECTION 1.1

1. $\{a, b, c\}$ is a set specified by the _____ or roster method.
2. If $A = B$ and $A = \{a, b, c, d\}$, then $B = $ _____.
3. Since $\{a, b, c\}$ and $\{1, 2, 3\}$ have the same number of elements, they are _____ sets.
4. $x \in A$ means that ___ is a member of A.
5. In $x \in A$, x is called a _____, and A is the _____ _____ for x.
6. The sets $\{a, b, c\}$ and $\{1, 2, 3\}$ can be placed in _____ _____ _____ correspondence with each other.
7. If every element in A is also a member of B, then A is said to be a _____ of B. This relationship written in symbolic form is _____.

SECTION 1.2

8. The process that is used to combine two sets or elements to form a third set or element is called a _____ operation.
9. $\{1, 2, 3\} \cup \{3, 4, 5\}$, called the _____ of the two sets, equals $\{$ _____ $\}$.
10. $\{1, 2, 3, 4\} \cap \{2, 4, 6, 8\}$, called the _____ of the two sets, equals $\{$ _____ $\}$.
11. The symbolic form in which some set is specified as $\{x |$ some condition on $x\}$ is called _____ _____.
12. $\{1, 2\} \times \{3, 4\}$, called the _____ _____ of the two sets, equals $\{$ _____ $\}$.
13. Each element in $A \times B$, where A and B are sets, is called a(n) _____ pair.
14. If a binary operation assigns the same element c to both (a, b) and (b, a), then the operation is _____.

SECTION 1.3

15. The five subsets of the real numbers to which we have given names are ___ ___, ___ ___, ___, ___ ___, and ___ ___.
16. The statement $a = a$ is called the ___ ___ of equality.
17. The symbolic form of the symmetric property of equality is: If $a = b$, then ___.
18. If $a = b$ and $b = c$, then ___. This is called the ___ property of equality.
19. If the result of a binary operation on any pair of elements in a set is also a member of the set, the set is ___ with respect to the operation.
20. By definition, $a + b + c =$ ___, and ___ $= (a \cdot b) \cdot c$.
21. List the symbolic forms of the eleven properties of the real numbers with respect to addition and multiplication.

SECTION 1.4

22. A statement concerning properties of real numbers with respect to a binary operation that can be proved to be true as a consequence of axioms or definitions is called a(n) ___.
23. The statement of a theorem contains an "if clause," called the ___, and a "then clause," called the ___.
24. The method of proving a theorem in which we begin with a true statement and then derive from this a sequence of other true statements until we reach the desired conclusion is called the ___ method of proof.

SECTION 1.5

25. The point on the number line corresponding to zero is called the ___.
26. The point on the number line corresponding to a specific number is called the ___ of the number.
27. The number corresponding to a specific point on the number line is called the ___ of the point.
28. If $a, b, c \in R$ as shown by

    $\xrightarrow{\quad \overset{a}{\bullet} \quad \overset{b}{\bullet} \quad \overset{c}{\bullet} \quad}$

    then a ___ b and c ___ b.
29. For each $a \in R$, one and only one of the following is true. Either $a <$ ___, $a =$ ___, or $a >$ ___.
30. If a and b are positive numbers, then $a + b$ and ab are ___ ___.
31. $a < b$ implies the existence of some ___ ___ c such that $a + c = b$.
32. If $a < b$ and $b < c$, then ___.
33. If $a < b$ and $c > 0$, then ac ___ bc.
34. If $a < b$ and $c < 0$, then ac ___ bc.
35. The measure of the distance from the origin to the point associated with a number is called the ___ ___ of the number.
36. If $a < 0$, then $-a$ ___ 0.

SECTION 1.6

37. If $a, b < 0$, then $a + b =$ ___.
38. If $a > 0$, $b < 0$, and $|a| > |b|$, then $a + b =$ ___.
39. If $a > 0$, $b < 0$, and $|a| < |b|$, then $a + b =$ ___.
40. $a - b = a +$ ___.

SECTION 1.7

41. The product (quotient) of two negative numbers is a _____ number.
42. The product (quotient) of a positive number and a negative number is a _____ number.
43. $7(-4) = $ _____.   **44.** $132 \div (-11) = $ _____.   **45.** $-1\frac{3}{4}\left(-2\frac{1}{4}\right) = $ _____.

SECTION 1.8

46. Any real number can be represented as a(n) _____ decimal.
47. Any rational number can be represented either as a _____ decimal or as a _____ decimal.
48. A number in the form of a nonperiodic infinite decimal is a(n) _____ number.
49. If $b \in Q$ and $a = b \cdot b$, then $\sqrt{a}$ is a _____ number.
50. $\sqrt{(-3)^2} = $ _____.

First-Degree Open Sentences with One Variable

A **statement** is a declarative sentence that can be judged to be true or false. The sentence "It is raining" is a statement because we can easily determine whether it is true or false. Sentences such as

$$2 + 3 = 7 \quad \text{and} \quad 4 + 5 = 9$$

are statements for the same reason.

A sentence whose truth or falsity depends upon the selection of a particular subject of the sentence is called an **open sentence**. For example,

$$\text{He is six feet tall.} \tag{1}$$

is an open sentence because the truth or falsity depends upon whose name we use to replace the pronoun *he*. The sentence

$$x + 2 = 7 \tag{2}$$

is an open sentence, because it is true or false depending upon the number selected as a replacement for x.

The word *he* in the first sentence and the letter x in the second sentence are **variables**, because they are used to represent unspecified elements of particular sets. The sets whose elements are represented by these variables are called the **replacement sets** of the variables. Thus, the replacement set for *he* is the set of all names of males, and the replacement set for x is a set of numbers. The only condition is that for any replacement the sentence must be either true or false.

We are primarily concerned with open sentences involving numbers. In this discussion, we specify the replacement set for all variables as the set of real numbers. Furthermore, since we must be able to determine whether or not a statement is true, *we limit the replacement set to those real numbers for which the sentence is defined.* We call such replacements **permissible replacements**.

Examples (a) 3 is not a permissible replacement for x in

$$\frac{x}{x-3} = 5$$

because $3 - 3 = 0$, and division by 0 is undefined.

(b) The permissible replacements for y in

$$y + 2 = \sqrt{9 - y^2}$$

must be between -3 and 3, inclusive, because the square of any number outside of this interval is greater than 9 and the square root of a negative number is not a real number.

Open sentences may be either statements of equality—**equations**—or statements of order—**inequalities**. The equals sign ($=$) or the order symbol ($>$ or $<$) separates the sentence into two parts that we call the **members** of the sentence. For the present, we will be concerned with open sentences that involve only one variable.

Not all elements of the replacement set for a variable make a given statement true. Those that do are called **solutions** of the open sentence.

Examples (a) 2 is a solution of $3x + 5 = 11$, because when x is replaced with 2, $3(2) + 5 = 11$ is a true statement.
(b) -5 is a solution of $2x + 8 < 3$, because $2(-5) + 8 = -2$ and $-2 < 3$.
(c) 3 is not a solution of $2x - 7 = 4$, because $2(3) - 7 = 4$ is not a true statement.
(d) 4 is not a solution of $3x - 11 < 1$, because $3(4) - 11 = 1$ and 1 is not less than 1.

While the selection of a logical replacement set for a variable is important, in this chapter we are primarily concerned with finding the solutions of open sentences. The set of all solutions of an open sentence is called the **solution set**.

2.1 Equivalent Equations and Inequalities

You are sometimes able to determine the solution set of an open sentence by inspection. For example, it should be evident that the only replacement for x which makes the equation $x + 3 = 7$ a true statement is 4. Hence, the solution set of the equation is $\{4\}$.

More frequently, you will find that you are unable to determine the solution set of an open sentence by inspection. In such a case, you can form a new equation whose solution set is evident by application of one or more axioms, definitions, or theorems. Whenever an axiom, definition, or theorem is used to form a new open sentence, and both open sentences have the same solution set, they are called **equivalent open sentences**.

DEFINITION 2.1-1 Equivalent open sentences are open sentences which have the same solution set.

Use of Axioms in Generating Equivalent Equations

The following illustrates how some of the axioms can be used to generate equivalent equations from the equation

$$3x + 4 + 5x - 2 = 10.$$

By using the commutative property of addition we can write

$$3x + 5x + 4 - 2 = 10.$$

Then, by using the associative property of addition and the distributive law, we obtain

$$8x + 2 = 10.$$

It is obvious that $\{1\}$ is the solution set of this equation. If we replace x with 1 in the first equation, we obtain

$$3(1) + 4 + 5(1) - 2 = 10,$$

which is a true statement. Thus, the equations are equivalent.

In all of the above applications, the substitution axiom has been used. In general, an equivalent open sentence is generated whenever an expression is replaced with another expression naming the same number.

Use of Theorems in Generating Equivalent Open Sentences

Many of the theorems stated earlier can be used for transforming open sentences. Those most commonly used in this text are Theorems 1.4-1 and 1.4-2 for transformations of equations and Theorems 1.5-2, 1.5-3, and 1.5-5 for transformations of inequalities. These theorems are restated below in abbreviated form for your convenience. $a, b, c \in \mathbf{R}$.

THEOREM 1.4-1   If $a = b$, then $a + c = b + c$.

Example $5x + 2 = 2x + 11$ is transformed into the equivalent equation $3x = 9$ by adding $-2 - 2x$ to each member of the first equation in order to isolate the variable in one member and eliminate it from the other.

THEOREM 1.4-2   If $a = b$, and $c \neq 0$, then $a \cdot c = b \cdot c$.

Example $\frac{3}{4}x = 15$ is transformed into the equivalent equation $x = 20$ by multiplying both members of the first equation by $\frac{4}{3}$, the multiplicative inverse of $\frac{3}{4}$.

THEOREM 1.5-2   If $a < b$, then $a + c < b + c$.

Example $x - 3 < 7 - x$ is transformed into the equivalent inequality $2x < 10$ by adding $3 + x$ to both members of the first inequality.

THEOREM 1.5-3   If $a < b$ and $c > 0$, then $ac < bc$.

Example $2x < 7$ is transformed into the equivalent inequality $x < \frac{7}{2}$ by multiplying both members of the first inequality by $\frac{1}{2}$.

THEOREM 1.5-5   If $a < b$ and $c < 0$, then $ac > bc$.

Example $-(x/3) < 5$ is transformed into the equivalent inequality $x > -15$ by multiplying both members of the first inequality by -3.

Since $a < b$ means that $b > a$, Theorems 1.5-2, 1.5-3, and 1.5-5 can be applied to inequalities of either sense.

Eliminating Fractions from an Open Sentence

When one or both members of an open sentence contain fractions, an open sentence without fractions can be formed by multiplying both members of the open sentence by the least common denominator (L.C.D.) of the fractions. However, if the transformation is to generate an equivalent open sentence, it must be specified in all cases that the multiplier names a nonzero number. You must also keep in mind that if the multiplier is an expression involving the variable, the resulting equation may not be equivalent to the first. If the open sentence is an inequality, it is necessary to make a distinction between the cases where the multiplier names a positive number and where it names a negative number.

Examples   Generate equivalent open sentences free of fractions.

(a) $\dfrac{1}{2} + \dfrac{1}{x} = \dfrac{5}{2}.$ (b) $\dfrac{2}{3} - \dfrac{3}{x+1} < 2.$

Solutions   (a) The L.C.D. is $2x$. If $x \neq 0$.

$$2x\left(\frac{1}{2} + \frac{1}{x}\right) = 2x\left(\frac{5}{2}\right),$$

and

$x + 2 = 5x$

is equivalent to the original open sentence.

(b) The L.C.D. is $3(x + 1)$. If $x \neq -1$ and $x + 1 > 0$,

$$3(x+1)\left(\frac{2}{3} - \frac{3}{x+1}\right) < 3(x+1)2,$$

and

$2x + 2 - 9 < 6x + 6$

is equivalent for all $x > -1$.

     If $x \neq -1$ and $x + 1 < 0$,

$$3(x+1)\left(\frac{2}{3} - \frac{3}{x+1}\right) > 3(x+1)2,$$

and

$$2x + 2 - 9 > 6x + 6$$

is equivalent to the original open sentence.

Many open sentences contain more than one variable. At this time we will not discuss solution sets of such open sentences. However, the techniques of generating equivalent open sentences can be used to "solve" an open sentence for one variable in terms of the others.

Example Write an equation equivalent to $2x + 3y = 5 + 2z$ that expresses y in terms of x and z.

Solution Add $-2x$ to both members to obtain

$$3y = 5 - 2x + 2z.$$

Multiply both members by $\frac{1}{3}$; then

$$y = \tfrac{1}{3}(5 - 2x + 2z),$$

which is the desired form.

EXERCISE 2.1

ORAL

In Problems 1–14, state the axiom or theorem which justifies the transformation of the first open sentence into the second.

Examples (a) $3x + 2 = x - 5$; $3x + 2 + (-x - 2) = x - 5 + (-x - 2)$
(b) $x - 7 + 4x - 3 = 2$; $x + 4x - 7 - 3 = 2$
(c) $3 - 2x < 5$; $-6 + 4x > -10$

Solutions (a) Theorem 1.4-1: If $a = b$, then $a + c = b + c$.
(b) Commutative property of addition.
(c) Theorem 1.5-5: If $a < b$ and $c < 0$, then $a \cdot c > b \cdot c$.

1. $x = 5$; $x + 4 = 9$
2. $x + 1 = 6$; $3(x + 1) = 18$
3. $4x + 2x + 3 = 21$; $(4 + 2)x + 3 = 21$
4. $\frac{1}{x} + 3 = 7$; $x\left(\frac{1}{x} + 3\right) = 7x$ $(x \neq 0)$
5. $2x - 5 = 7 - x$; $2x - 5 + x + 5 = 7 - x + x + 5$
6. $2x - 5 + x + 5 = 7 - x + x + 5$; $2x + x - 5 + 5 = 7 + 5 - x + x$
7. $2x + x - 5 + 5 = 7 + 5 - x + x$; $2x + x + 0 = 7 + 5 + 0$
8. $\dfrac{2}{x + 2} = 8$; $(x + 2)\dfrac{2}{x + 2} = (x + 2)8$ $(x \neq -2)$
9. $x < 3$; $3 \cdot x < 3 \cdot 3$

10. $2x + 5 > 9$; $2x + 5 - 5 > 9 - 5$
11. $5 - 2x < 4$; $\frac{-1}{2}(5 - 2x) > \frac{-1}{2} \cdot 4$
12. $2x - 3 < x + 4$; $2x - 3 - x + 3 < x + 4 - x + 3$
13. $5x - x > 12$; $(5 - 1)x > 12$
14. $3 - \frac{2x}{3} > 4 - x$; $-3\left(3 - \frac{2x}{3}\right) < -3(4 - x)$

WRITTEN

Transform each open sentence in Problems 1–12 into an equivalent open sentence free of fractions by multiplying both members by the L.C.D. of the terms that are fractions. State any restrictions on the variable necessary to guarantee equivalence. Assume that all denominators in the inequalities represent positive numbers.

Examples   (a) $\frac{2}{3} = \frac{5}{2x + 1}$ (b) $\frac{1}{2} + \frac{3}{x - 2} > \frac{5}{x - 2}$

Solutions   (a) The L.C.D. is $3(2x + 1)$. If $x \neq -\frac{1}{2}$,

$$3(2x + 1)\frac{2}{3} = 3(2x + 1)\frac{5}{2x + 1}$$

$$4x + 2 = 15.$$

(b) The L.C.D. is $2(x - 2)$. If $x > 2$, $x - 2$ is positive, then

$$2(x - 2)\left(\frac{1}{2} + \frac{3}{x - 2}\right) > 2(x - 2)\frac{5}{x - 2}$$

$$x - 2 + 6 > 10.$$

1. $\frac{2x - 3}{x} = \frac{3}{5}$
2. $\frac{3y + 7}{y} = \frac{4}{3}$
3. $\frac{2}{x - 1} = \frac{5}{x + 2}$
4. $\frac{3}{2x + 1} = \frac{5}{x - 3}$
5. $\frac{1}{x} - 7 < \frac{3}{4}$
6. $\frac{3}{z} + 5 > \frac{2}{3}$
7. $\frac{4}{x - 3} + 5 > 2$
8. $6 - \frac{3}{x + 2} < 5$
9. $\frac{3}{x - 7} > \frac{4}{x + 2}$
10. $\frac{5}{x + 3} < \frac{7}{x + 2}$
11. $1 > 5 - \frac{x}{x + 1}$
12. $\frac{x}{x - 2} - 1 < 2$

For each open sentence in Problems 13–28, write an equivalent open sentence that expresses, in simplest form, the specified variable in terms of the other variables. In this group of problems all variables denote positive real numbers.

Examples   (a) $d = rt$; t (b) $x + y < z$; y

Solutions   (a) $t = \frac{d}{r}$ (b) $y < z - x$

13. $3x - a = x + a$; x
14. $ax = a^2b - ax$; x
15. $a^2x + b = ax$; x $(a \neq 1)$
16. $\frac{ax}{b} - c = a$; x

17. $a(a - x) = b(b - x)$; x $(a \neq b)$
18. $y(b + 3) = 2b$; y
19. $x < 2y + 3$; y
20. $2x - y < 3z$; x
21. $4x - y < 3$; y
22. $\frac{x}{2} + y > 2 - y$; x
23. $\frac{x}{y + 2} > 1$; y
24. $\frac{x}{1 + x} > y$; x $(x \neq 1)$

Example $\frac{x}{2} - \frac{y + 1}{3} = 2$; y

Solution Multiply by 6, the L.C.D.

$$3x - 2(y + 1) = 12$$
$$3x - 2y - 2 = 12$$
$$-2y = 14 - 3x$$
$$y = \frac{14 - 3x}{-2}$$
$$y = \frac{3x - 14}{2}$$

25. $\frac{x}{5} + \frac{x - y}{2} = 4$; x
26. $\frac{x}{6} + \frac{x + y}{2} = 2$; y
27. $\frac{x}{a} + \frac{x}{b} = c$; x
28. $\frac{1}{x} + \frac{1}{y} = \frac{1}{z}$; y $(x \neq z)$

2.2 Solutions of First-Degree Open Sentences

An open sentence involving one variable in which the variable has no exponent other than 1 is called an **open sentence of the first degree**. However, some open sentences that are not of the first degree are equivalent to open sentences of the first degree. For example, the equation $1/x + 3 = 5$ is not of the first degree since $1/x = x^{-1}$, as will be shown in Chapter 3. However, if $x \neq 0$, Theorem 1.4-2 can be applied to generate the equivalent equation $1 + 3x = 5x$, which is of the first degree.

Permissible Replacements

The illustration above indicates that not all real numbers are suitable replacements for the variable in every open sentence. Hence, as was stated on page 39, the replacement set for the variable will include only those real numbers for which both members of the open sentence are defined. Such a replacement for the variable is called a **permissible replacement**.

Types of Open Sentences

Some equations are true for every member of the replacement set, and some are true for one or more replacements but not for others. Those of the first type are called **identities**, and those of the second type are called **conditional equations**.

46 FIRST-DEGREE OPEN SENTENCES WITH ONE VARIABLE

Inequalities can be classified in a similar manner. An inequality that is true for all permissible replacements is an **absolute inequality**. If it is true for some replacements but not for others it is a **conditional inequality**.

The techniques for generating equivalent equations can be used to determine whether or not a given equation is an identity. If both members can be transformed into the same, or identical, expression, the equation is an identity; otherwise it is conditional. However, it must be kept in mind that only transformations that can be justified by an axiom, definition, or theorem may be used.

Examples Determine whether or not the following equations are identities.

(a) $\dfrac{x}{x+1} + 2 = \dfrac{11}{4}$ (b) $\dfrac{(x+2)(x-3)}{x+2} = x - 3$

Solutions (a) Both members of the equation are defined for every real number except -1. Hence, if $x \neq -1$, we can multiply both members of the equation by $4(x + 1)$.

$$4(x+1)\left(\dfrac{x}{x+1} + 2\right) = 4(x+1)\dfrac{11}{4}$$

$$4x + 8x + 8 = 11x + 11$$

$$12x + 8 = 11x + 11$$

Since the two members are not identical, the equation is conditional.

(b) Both members of the equation are defined for every real number except -2. Hence, if $x \neq -2$, we can apply the fundamental principle of fractions to the left-hand member to obtain

$$x - 3 = x - 3.$$

Since the two expressions are identical, the equation is an identity.

Solution Sets of Equations

In Section 2.1 you learned that if the solution set of an open sentence is not evident by inspection, you may be able to generate an equivalent open sentence whose solution set is evident. When it is necessary for you to find the solution set of a particular equation, you should continue the process of generating equivalent equations until you can determine the solution set by inspection. If you have taken care to insure that each equation is equivalent to the previous equation, the solution set of the final equation is the solution set for each equation including the given equation.

Example Find the solution set of the equation

$$3x - 7 - x + 4 = 5 - 2x.$$

Solution If you cannot determine the solution set by inspection, you can use the commutative, associative, and distributive laws and the substitution

axiom to obtain

$$2x - 3 = 5 - 2x,$$

which is equivalent to the given equation. If the solution set still is not evident, adding $2x + 3$ to both members gives you the equivalent equation

$$4x = 8.$$

At this point you should recognize that 2 is the only replacement for x that makes the equation a true statement and that $\{2\}$ is the solution set. If the solution set is obvious to you, nothing is to be gained by writing the trivial equation $x = 2$.

Solution Sets of Inequalities

The solution set of an inequality is, in general, an infinite set. There are exceptions to this statement, particularly when the replacement set for the variable is a subset of J, W, or N. However, we shall continue to use R as the replacement set for all variables. To specify the replacement set for a particular inequality, the usual technique is to generate an equivalent inequality for which you can determine by inspection that any selected number either is or is not a solution. This inequality is then used in set-builder notation to specify the solution set.

Example Specify and graph on a number line the solution set of the inequality

$$5 - 4x < 2x - 7.$$

Solution We first apply Theorem 1.5-2 by adding $4x + 7$ to both members of the inequality to obtain

$$12 < 6x.$$

We now apply Theorem 1.5-3 (If $a < b$ and $c > 0$, then $a \cdot c < b \cdot c$), multiplying both members by $\frac{1}{6}$ to obtain the equivalent inequality

$$2 < x,$$

which implies

$$x > 2.$$

It should now be evident that any real number greater than 2 is a solution, and that any real number less than or equal to 2 is not a solution. Hence, we specify the solution set over R to be

$$\{x \mid x > 2\}.$$

The graph of the solution set is shown in Figure 2.2-1. An open dot is placed at 2 to indicate that 2 is not a member of the solution set.

Figure 2.2-1

If the variable appears in the denominator of a fraction in an inequality, not only must we restrict its replacement set to those values for which both members are defined, but we must also consider as separate cases the values for which the denominator is a positive number and the values for which it is a negative number. Recall

THEOREM 1.5-3   If $a < b$, and $c > 0$, then $ac < bc$

and

THEOREM 1.5-5   If $a < b$, and $c < 0$, then $ac > bc$.

Thus, for every value of the variable for which the denominator is a negative number, we must change the sense of the inequality if we multiply by the denominator.

Example   Specify and graph on a number line the solution set of $1/x > 2$.

Solution   Since x appears in the denominator, we must consider $x > 0$ and $x < 0$ as separate cases.

PART I: If $x > 0$, then by Theorem 1.5-3,

$$x \cdot \frac{1}{x} > x \cdot 2$$

$$1 > 2x$$

$$\tfrac{1}{2} > x \quad \text{or} \quad x < \tfrac{1}{2}.$$

We now have two conditions on x, both of which must be met. That is, $x > 0$ and $x < \tfrac{1}{2}$. Hence, for this part, we specify the solution set as

$$\{x \mid x > 0\} \cap \{x \mid x < \tfrac{1}{2}\} = \{x \mid 0 < x < \tfrac{1}{2}\}.$$

PART II: If $x < 0$, then by Theorem 1.5-5,

$$x \cdot \frac{1}{x} < x \cdot 2$$

$$1 < 2x$$

$$\tfrac{1}{2} < x \quad \text{or} \quad x > \tfrac{1}{2}.$$

However, if $x < 0$, the statement $x > \tfrac{1}{2}$ is impossible, because there is no number that can meet both conditions. Thus, we specify the solution set for this part as

$$\{x \mid x < 0\} \cap \{x \mid x > \tfrac{1}{2}\} = \varnothing.$$

The solution set of the inequality is the union of the solution sets of the two parts. That is,

$$\{x \mid 0 < x < \tfrac{1}{2}\} \cup \varnothing = \{x \mid 0 < x < \tfrac{1}{2}\}.$$

The graph is shown in Figure 2.2-2.

Figure 2.2-2

EXERCISE 2.2

ORAL

Define:

1. First-degree open sentence in one variable.
2. Permissible replacement for the variable.
3. Conditional equation.
4. Identity.
5. Conditional inequality.
6. Absolute inequality.

WRITTEN

Find the solution set over **R** for each open sentence. If an open sentence is either an identity or an absolute inequality, state this as your answer. Also state any necessary restrictions on the variable so that each member will be defined. Graph the solution set of each conditional inequality.

Examples   (a) $2x + 3 - x = x + 3$ (b) $2x + 3 > 5$

Solutions   (a) $2x + 3 - x = x + 3$
$2x - x + 3 = x + 3$
$x + 3 = x + 3$
An identity.

(b) $2x + 3 > 5$
$2x + 3 - 3 > 5 - 3$
$2x > 2$
$x > 1$
The solution set is $\{x \mid x > 1\}$ (see Figure 2.2-3).

Figure 2.2-3

1. $x - 3 = 5x + 7$
2. $3x - 6 = 18 + 7x$
3. $5 - 3y = 2 - 3y$
4. $3 - 4x = 6 - 8x$
5. $4z + 5 = 10 + 8z$
6. $3z - 4 = 1 - 3z$
7. $2(4 + y) = 8 + 3y$
8. $5 - 4x = 5 - 5x$
9. $3 - 4x < -7 - 6x$
10. $2 - 4y > 5 - 3y$
11. $5x - 4 > 3x + 4$
12. $6x + 3 < 5x + 2$
13. $x + 2 < x + 5$
14. $2x + 4 > 3 + 2x$

Example $\dfrac{3}{x+1} + 5 = \dfrac{18}{x+1}$

Solution If $x \neq -1$, then $x + 1 \neq 0$, and Theorem 1.4-2 can be applied to obtain

$$(x + 1)\frac{3}{x + 1} + 5(x + 1) = (x + 1)\frac{18}{x + 1}$$

$$3 + 5x + 5 = 18$$

$$5x = 10,$$

for which the solution set is $\{2\}$.

15. $\dfrac{x}{x + 2} = \dfrac{3}{5}$

16. $\dfrac{x + 1}{x - 7} = 9$

17. $\dfrac{2}{x + 1} + \dfrac{3}{5} = 1$

18. $\dfrac{4}{3 - x} + 5 = \dfrac{9}{2}$

19. $\dfrac{x + 7}{x - 2} + \dfrac{2}{3} = \dfrac{32}{3}$

20. $\dfrac{4 - x}{1 + x} + \dfrac{3}{2} = 3$

21. $\dfrac{x + 2}{x - 1} + \dfrac{2}{3} = \dfrac{3x - 2}{x - 1} - \dfrac{1}{3}$

22. $\dfrac{3x + 7}{x + 4} - \dfrac{1}{2} = \dfrac{2x + 1}{x + 4} + \dfrac{3}{2}$

23. $\dfrac{x + 3}{x} + \dfrac{2}{x} = \dfrac{5}{x} + 1$

24. $\dfrac{x - 2}{x} + \dfrac{4}{x} = 3 - \dfrac{8}{x}$

Example $\dfrac{2x - 6}{3} \geq 0$

Solution $3\left(\dfrac{2x - 6}{3}\right) \geq 3 \cdot 0$ (Theorem 1.5-3)

$2x - 6 \geq 0$

$2x \geq 6$ (Theorem 1.5-2)

$x \geq 3$ (Theorem 1.5-3)

The solution set is $\{x \mid x \geq 3\}$ (see Figure 2.2-4).

Figure 2.2-4

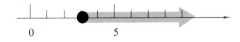

25. $\dfrac{2x - 3}{2} \leq 5$

26. $\dfrac{5x - 7x}{3} > 4$

27. $\dfrac{x - 3x}{5} \leq 6$

28. $\dfrac{2x - 5x}{2} < 7$

29. $\dfrac{x - 6x}{2} < -20$

30. $\dfrac{x + 4}{3} \leq 6 + x$

31. $\dfrac{x}{2} + 1 < \dfrac{x}{3} - x$

32. $\dfrac{1}{2}(x + 2) \geq \dfrac{2x}{3}$

33. $2(x + 2) \leq \dfrac{3x}{4} - 1$

34. $\dfrac{2(x - 1)}{3} + \dfrac{3(x + 1)}{4} < 0$

35. $\dfrac{3(2x - 1)}{4} - \dfrac{4x + 3}{2} \geq 0$

36. $\dfrac{3(3x + 2)}{5} - \dfrac{2(2x - 1)}{3} \leq 0$

Example $\dfrac{x^2}{x - 1} < x + 2$

2.3 Open Sentences with Absolute Value

Solution $x - 1$ cannot equal zero because if it did, $x^2/(x - 1)$ would be undefined. Hence, $x - 1 > 0$, or $x - 1 < 0$. If $x - 1 > 0$, then $x > 1$ and

$$(x - 1)\frac{x^2}{x - 1} < (x - 1)(x + 2)$$

$$x^2 < x^2 + x - 2$$

$$0 < x - 2.$$

Therefore, $2 < x$, which means $x > 2$. Then, since $2 > 1$, any value of x greater than 2 is also greater than 1. Hence, the solution set for this part is $\{x \mid x > 2\}$.

If $x - 1 < 0$, then $x < 1$ and

$$(x - 1)\frac{x^2}{x - 1} > (x - 1)(x + 2)$$

$$x^2 > x^2 + x - 2$$

$$0 > x - 2.$$

In this case, $2 > x$, which means $x < 2$. Now x must be less than 1, and since $1 < 2$, the solution set for this part is $\{x \mid x < 1\}$.

Finally, since $x - 1$ may be either greater than zero *or* less than zero, the complete solution set is the union of the two parts.

$$\{x \mid x < 1\} \cup \{x \mid x > 2\}$$

The graph is shown in Figure 2.2-5.

Figure 2.2-5

37. $\dfrac{1}{x} > 3$ 38. $\dfrac{1}{x} < 2$

39. $\dfrac{2}{x + 1} > 5$ 40. $\dfrac{4}{x - 2} > 1$

41. $\dfrac{x^2}{x + 3} < x - 4$ 42. $\dfrac{x^2}{x - 2} > x + 1$

43. $\dfrac{x^2}{x + 2} > x + 2$ 44. $\dfrac{2x^2}{x - 1} < 2x - 1$

2.3 Open Sentences with Absolute Value

The definition

$$|a| = \begin{cases} a & \text{if } a \geq 0 \\ -a & \text{if } a < 0 \end{cases}$$

states that the absolute value of any expression is non-negative. Thus, any nonzero number and its negative have the same absolute value. This leads to two possibilities if the expression inside the absolute value sign contains a variable. For example, if $|x| = 3$, then x can be either 3 or -3.

Equations

If an equation includes an absolute value symbol containing the variable, then we must consider both the possibility that the expression within the absolute value sign represents a positive number and the possibility that it represents a negative number. This means that we must actually solve two separate equations.

Examples   Find the solution sets of

    (a) $|2x + 5| = 7$ (b) $|4 - 2x| = 3$

Solutions   (a) Since both $|7|$ and $|-7|$ equal 7,

$$2x + 5 = 7 \quad \text{or} \quad 2x + 5 = -7.$$

Solving these equations separately, we obtain $\{1\}$ as the solution set of the first and $\{-6\}$ as the solution set of the second. Then, since either 1 *or* -6 is a solution, we have $\{-6, 1\}$ as the solution set of the original equation.

(b) Since both $|3|$ and $|-3|$ equal 3,

$$4 - 2x = 3 \quad \text{or} \quad 4 - 2x = -3.$$

If we solve these equations separately and combine the results, we obtain $\{\frac{1}{2}, \frac{7}{2}\}$.

Inequalities

An inequality involving absolute value can be separated into two inequalities in a manner similar to that above. However, the fact that the expression within the absolute value sign represents either a positive or a negative number places a condition on the variable that must be taken into consideration when the solution set is specified. The following examples illustrate this concept.

Examples   Specify and graph on a number line the solution sets of

    (a) $|x - 2| > 3$ (b) $|x - 2| < 3$

Solutions   (a) If $x - 2 \geq 0$ $(x \geq 2)$, then $x - 2 > 3$, which gives $x > 5$. However, x must satisfy both this condition *and* the condition $x \geq 2$. Hence, the solution set of this inequality is

$$\{x | x \geq 2\} \cap \{x | x > 5\} = \{x | x > 5\}.$$

If $x - 2 < 0$ $(x < 2)$, then $-(x - 2)$ is positive and $-(x - 2) > 3$. Then, multiplying both members by -1, we have $x - 2 < -3$, from which $x < -1$. Again, both of two conditions must be met, $x < 2$ and $x < -1$. Hence, the solution set of this inequality is

$$\{x | x < -1\} \cap \{x | x < 2\} = \{x | x < -1\}.$$

Finally, since $x - 2$ can be either positive *or* negative, the solution

set of the original inequality is

$\{x | x < -1\} \cup \{x | x > 5\}$.

The graph is shown in Figure 2.3-1.

Figure 2.3-1

(b) If $x - 2 \geq 0$ ($x \geq 2$), then $x - 2 < 3$, or $x < 5$. Since both conditions must be met, the solution set is

$\{x | x \geq 2\} \cap \{x | x < 5\} = \{x | 2 \leq x < 5\}$.

If $x - 2 < 0$ ($x < 2$), then $-(x - 2) < 3$, which gives $x - 2 > -3$. Thus, $x > -1$. The set of values that meets both of these conditions is

$\{x | x > -1\} \cap \{x | x < 2\} = \{x | -1 < x < 2\}$.

Finally, the solution set of the original inequality is

$\{x | -1 < x < 2\} \cup \{x | 2 \leq x < 5\} = \{x | -1 < x < 5\}$.

(See Figure 2.3-2.)

Figure 2.3-2

When the intersection of two sets, such as those in Example (b) above, is written as a single set, it is essential that the transitive property of order (Theorem 1.5-1) be followed. In general, if the absolute value of an expression involving a variable is *less than* some positive constant, the absolute value sign can be removed and the statement written in transitive form. For example,

$$|x - a| < k \quad (k > 0)$$

and

$$-k < x - a < k$$

are equivalent statements.

EXERCISE 2.3

WRITTEN

Find the solution set of each open sentence in Problems 1–18. Graph the solution set of each inequality.

Examples (a) $|x - 6| = 3$ (b) $|2x - 1| \leq 5$

Solutions (a) If $x - 6 > 0$, then

$x - 6 = 3$

$x = 9$. (Solution continued on next page.)

If $x - 6 < 0$, then
$$-(x - 6) = 3$$
$$-x + 6 = 3$$
$$x = 3.$$

The solution set is $\{3, 9\}$.

(b) If $2x - 1 \geq 0$, then
$$2x - 1 \leq 5$$
$$2x \leq 6$$
$$x \leq 3$$

If $2x - 1 < 0$, then
$$-(2x - 1) \leq 5$$
$$2x - 1 \geq -5$$
$$2x \geq -4$$
$$x \geq -2.$$

The solution set is $\{x \mid -2 \leq x \leq 3\}$. (See Figure 2.3-3.)

Figure 2.3-3

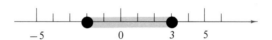

1. $|x| = 6$
2. $|x - 1| = 4$
3. $|x - \frac{3}{4}| = \frac{1}{2}$
4. $|x - \frac{2}{3}| = \frac{1}{3}$
5. $|3x + 7| = 1$
6. $|2x + 5| = 2$
7. $\left|1 + \frac{3x}{2}\right| = \frac{1}{2}$
8. $\left|1 - \frac{x}{2}\right| = \frac{3}{4}$
9. $|\frac{1}{3} - 4x| = \frac{2}{3}$
10. $|2x + \frac{1}{2}| = \frac{1}{4}$
11. $|x| < 2$
12. $|x| \geq 5$
13. $|x + 3| < 4$
14. $2|x + 1| \leq 8$
15. $|5 - 2x| > 3$
16. $3|4 - 2x| > 6$
17. $\left|\frac{x}{2} - 3\right| < 5$
18. $\left|4 - \frac{3x}{2}\right| > 8$

Example Find a value for k so that the equation $4x - 3 = k(x + 1)$ has $\{6\}$ as its solution set.

Solution If 6 is a solution of the equation, then $4(6) - 3 = k(6 + 1)$ is a true statement.
$$24 - 3 = k(6 + 1)$$
$$21 = 7k$$
$$k = 3$$

19. Find a value for k so that the equation $5x - 6 = 2x + k$ will have $\{3\}$ as its solution set.
20. For what value of k will the equation $3x - 1 = k$ be equivalent to $2x + 5 = 1$?
21. Explain why the solution set of $|2x - 3| < -2$ is the empty set.
22. Specify the replacement set for k so that the solution set of $|3x + 2| < -k$ will be nonempty.

Specify the solution set of each inequality.

23. $|2x - 1| > x + 2$
24. $|x - 3| > 2x - 5$
25. $|3x - 4| < x - 5$
26. $|2x + 3| < 4 - x$
27. $|3 - x| > x + 2$
28. $|5 - x| < x - 3$

2.4 Word Problems

Solutions to word problems involving numerical quantities can sometimes be found by determining the solution set of the algebraic statement of the problem. Such a statement may be an equation or inequality in one variable, or it may consist of a system of equations or inequalities. In this section we use one equation or one inequality in one variable as the algebraic statement of each problem.

The technique outlined below can be helpful to you in solving word problems.

(1) Read the statement of the problem carefully to determine what quantity or quantities the problem requires you to find. Choose some variable to represent a number associated with the desired quantity. If more than one quantity is to be found, represent each of them in terms of the same variable.

(2) Translate the word statement into an algebraic statement using the expressions above and signs of operation.

(3) Find the solution set of the algebraic statement.

(4) Interpret the solution set in terms of the conditions stated in the problem. (This is essential, because not all members of the solution set may meet the conditions of the problem.)

Example A man has $2.40 in nickels and dimes. If there are twice as many nickels as dimes, how many of each kind of coin does he have?

Solution (1) If we let x represent the number of dimes, then $2x$ represents the number of nickels.

(2) Since the value of a dime is 10 cents and the value of a nickel is 5 cents, the total value of x dimes and $2x$ nickels in cents can be expressed by

Value of dimes + Value of nickels = Total value

$$10(x) + 5(2x) = 240.$$

(3) $20x = 240$

The solution set is $\{12\}$.

(4) If there are 12 dimes, there are 24 nickels, and since 12 dimes plus 24 nickels equals $2.40, we have the solution.

If the algebraic statement of a problem is an inequality rather than an equation, the solution will usually be an interval of values.

56 FIRST-DEGREE OPEN SENTENCES WITH ONE VARIABLE

Example How many ounces of water must be added to 24 ounces of a solution which is 13% alcohol in order to make a solution which is less than 3.2% alcohol?

Solution
(1) Let x represent the number of ounces of water to be added.
(2) The total quantity of alcohol in the final mixture must be less than 3.2%. Since only water is to be added, the number of ounces of alcohol must be 13% of the original 24 ounces, or 0.13(24). The number of ounces in the final mixture is $24 + x$, so we have

$$\frac{0.13(24)}{24 + x} < 0.032.$$

(3) We first multiply both members of the inequality by 100 to remove the decimal point on the left.

$$\frac{13(24)}{24 + x} < 3.2$$

Since we know that x must represent a non-negative number, we can multiply both members by $24 + x$ to obtain

$$312 < 76.8 + 3.2x,$$

which is equivalent to

$$3.2x > 235.2,$$

whose solution set is $\{x \mid x > 73.5\}$.
(4) It would be necessary to add more than 73.5 ounces of water.

EXERCISE 2.4

WRITTEN

For each problem,

(1) Write a statement or statements to express the required quantity or quantities in terms of one variable.
(2) Translate the word statement into an algebraic statement.
(3) Find the solution set of the algebraic statement.
(4) Interpret this solution set in terms of the stated problems.

Example $350 is to be divided between Jones and Smith so that Jones will receive $30 more than Smith. How much does each receive?

Solution
(1) Let x represent the number of dollars Smith receives and $x + 30$ represent the number of dollars Jones receives.
(2) Since the sum of the amounts each is to receive is $350, the equation is

$$x + x + 30 = 350.$$

(3) Combine like terms and add -30 to both members to obtain the equivalent equation

$$2x = 320,$$

for which the solution set is $\{160\}$.
(4) Smith received $160 and Jones $190.

(CHECK: $160 + $190 = $350)

1. The sum of two consecutive odd integers is 72. Find the integers.
2. Find two consecutive odd positive integers whose squares differ by 64.
3. The length of a rectangle is three times its width. If the length is increased by 5 feet and the width is decreased by 2 feet, the area is decreased by 21 square feet. Find the dimensions of the rectangle.
4. One dimension of a rectangle is 3 feet greater than the other. If each dimension is increased by 2 feet, the perimeter becomes 62 feet. Find the dimensions.
5. John, Jerry, and Paul plan a camping trip that will cost $90. John is to furnish the car, so Jerry must pay $10 more than John. Jerry is to provide the boat, so Paul will pay $10 more than Jerry. How much will each pay?
6. How much will each of three people receive if $180 is to be divided so that the second receives $\frac{2}{3}$ as much as the first and the third receives $\frac{1}{2}$ as much as the second?

Example A man purchased a number of 8¢ and 11¢ stamps for $5.85. If the number of 11¢ stamps was $\frac{3}{4}$ the number of 8¢ stamps, how many of each kind did he buy?

Solution (1) Let x represent the number of 8¢ stamps, and $\frac{3}{4}x$ or $\frac{3x}{4}$ represent the number of 11¢ stamps.
(2) The total value of the stamps can be expressed as

$$8x + 11\left(\frac{3x}{4}\right) = 585.$$

(3) $32x + 33x = 2340$
$\qquad 65x = 2340$

The solution set is $\{36\}$.
(4) The man purchased 36 eight-cent stamps and 27 eleven-cent stamps.

(CHECK: $0.08 \times 36 + $0.11 \times 27 = $2.88 + $2.97 = $5.85)

7. A collection of coins consisting of dimes and quarters has a value of $11.60. How many coins of each kind are there in the collection if there are 32 more dimes than quarters?
8. A man has 23 coins, all dimes and nickels, with a total value of $1.80. How many dimes and how many nickels does he have?
9. 2500 people paid a total of $2456.25 in admissions to a football game. If the price of admission was $1.50 for adults and 75¢ for children, how many adult tickets and how many children's tickets were sold?
10. A collection of coins with a value of $24.50 consists of three times as many quarters as half dollars, twice as many dimes as quarters, and twice as many nickels as dimes. How many of each kind of coin are there in the collection?

58 FIRST-DEGREE OPEN SENTENCES WITH ONE VARIABLE

Example Figure 2.4-1 shows a lever in a position of equilibrium. If the weight of the lever is disregarded, then the sum of the moments of force on one side of the support point, or fulcrum, is equal to the sum of the moments of force on the other side. [A moment of force in this case is the product of the magnitude of the force (weight) and its distance from the fulcrum.]

$$w_1 \cdot d_1 = w_2 \cdot d_2 + w_3 \cdot d_3$$

If $w_1 = 10$ pounds, $w_2 = 5$ pounds, $w_3 = 8$ pounds, $d_1 = 15$ feet, and $d_2 = 6$ feet, what must be the measure of d_3 so that the lever will be in equilibrium?

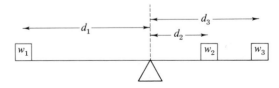

Figure 2.4-1

Solution (1) Let d_3 represent a measure of the required distance.
(2) By substituting the given values in the equation above, we obtain

$$10(15) = 5(6) + 8(d_3).$$

(3) $150 = 30 + 8d_3$
   $120 = 8d_3$
   The solution set is $\{15\}$.
(4) The required distance is 15 feet.

11. A weight of 300 pounds is placed on a lever 10 feet from the fulcrum. How far from the fulcrum on the other side must a weight of 250 pounds be placed in order to give equilibrium?
12. Two weights are placed on a lever on opposite sides of the fulcrum at distances of 3 and 4 feet, respectively. Find the size of each weight if their total is 350 pounds.
13. A 30-pound weight is placed 5 feet from the fulcrum on a lever, and a 50-pound weight is placed 7 feet from the fulcrum on the other side. Where should a 40-pound weight be placed to achieve equilibrium?
14. How many pounds of force must be applied on one end of a 9-foot lever to just balance a 250-pound rock on the other end, if the fulcrum is placed 2 feet from the rock?

Example How long will it take workers A and B to complete a job by working together if A can do the job alone in 6 days and B can do it alone in 4 days?

Solution If A does the job in 6 days, he can do $\frac{1}{6}$ of the job in 1 day. B can do $\frac{1}{4}$ of the job in 1 day.

(1) Let x represent the number of days it would take both A and B to complete the job. Then $1/x$ represents that part of the job they could do in 1 day working together.
(2) The quantity of work the two men can do together must equal the sum of the quantities they can do separately. Hence,

$$\frac{1}{x} = \frac{1}{6} + \frac{1}{4}.$$

(3) Multiply by $12x$ to obtain

$$12 = 2x + 3x,$$

for which the solution set is $\{\frac{12}{5}\}$.

(4) It would take A and B $\frac{12}{5}$ or $2\frac{2}{5}$ days to complete the job by working together.

15. A reservoir can be filled through one pipe in 5 days or through a second pipe in 3 days. How long would it take to fill the reservoir if both pipes were used?
16. One typist can address a certain number of envelopes in 6 hours, and another typist can address the same number of envelopes in 9 hours. How long will it take to address this number of envelopes if the typists work together?
17. If 1000 articles of a given type can be manufactured by one machine in 8 hours, by a second machine in 6 hours, and by a third machine in 4 hours, how long would it take to manufacture 1000 articles if all three machines were working?
18. A tank can be filled through one pipe in 5 hours and emptied through another pipe in 10 hours. If the tank is empty, how many hours will it take for the tank to be filled if both pipes are open?

Example An airplane flew 950 miles in $2\frac{1}{2}$ hours against a head wind of 30 miles per hour. How fast could the airplane fly in still air?

Solution (1) This problem involves the formula $d = rt$. If we let r represent the normal rate in miles per hour, then $r - 30$ will represent the rate on the flight against the head wind.

(2) Substituting in the formula gives the equation

$$\tfrac{5}{2}(r - 30) = 950.$$

(3) $5r - 150 = 1900$

$$5r = 2050$$

The solution set is $\{410\}$.

(4) The plane can fly 410 miles per hour in still air.

19. With a wind velocity of 30 miles per hour, it takes an airplane as long to fly 840 miles with the wind as it does to fly 660 miles against the wind. What would the speed of the airplane be if no wind were blowing?
20. Two men start at 8:00 A.M. from the same place and travel in opposite directions at average speeds of 45 and 60 miles per hour, respectively. At what time will they be 245 miles apart?
21. In a 1500-meter race between two men, the winner's time is 3 minutes 40 seconds and his lead at the finish is 70 meters. Assuming his rate to be constant, how many seconds would it take the loser to run 1500 meters?
22. Two rivers flow at constant rates of 6 and 4 miles per hour, respectively. A man in a motor boat can travel 36 miles downstream on the first river in the same time it takes him to travel 16 miles upstream on the second river. If the speed of the motor boat is constant, how long would it take him to go 24 miles in still water?

Example An algebra student has made scores of 75, 62, 69, and 67, respectively, on four tests. What score must he make on a fifth test to receive a grade of C (70-80)?

Solution (1) Let x represent a score on the fifth test.
(2) The average is the sum of the scores divided by the number of tests, and this average must be between 70 and 80. Thus, we have the inequality

$$70 \leq \frac{75 + 62 + 69 + 67 + x}{5} < 80.$$

(3) $350 \leq 273 + x < 400$

The solution set is $\{x \mid 77 \leq x < 127\}$.

(4) The student must make a score of 77 or higher in order to receive a C grade.

23. What score must a student make on a test in order to receive a grade of B (80–90) if he has made scores of 75, 82, 61, 86, and 78 on five previous tests?
24. A bowler made scores of 145, 133, 152, 160, and 148 in five successive games. What score must he make in his next game so that his average will be between 150 and 160?
25. A certain alloy weighs 8 grams per cubic centimeter. If a block of this alloy with a base 10 centimeters square weighs 5000 ± 40 grams, what is the height of the block?
26. The Fahrenheit and Celsius temperature scales are related by the equation $°C = \frac{5}{9}(°F - 32)$. Find the interval on the Fahrenheit scale so that the Celsius temperature will lie between $-10°$ and $20°$.

Summary

1. **Equivalent open sentences** are sentences that have the same solution set. [2.1]
2. Equivalent open sentences can be **generated** from a given open sentence by application of the axioms and certain theorems. [2.1]
3. An **equation free of fractions** can be formed by multiplying both members of the given equation by the least common denominator (L.C.D.). If the L.C.D. contains the variable, the resulting equation may not be equivalent to the original equation. [2.1]
4. If both members of an **inequality** are multiplied by an expression that contains the variable, the cases where the expression represents a positive number and those where it represents a negative number must both be considered. [2.1]
5. **Conditional equations** and **conditional inequalities** are open sentences that are true for some permissible replacements of the variable but not for others. If the sentence is true for every permissible replacement, it is either an **identity** or an **absolute inequality**. [2.2]
6. The solution set of an open sentence is specified by generating equivalent open sentences, progressively simpler in form, until the solution set is obvious by inspection. [2.2]
7. If there are **two sets of conditions** on the replacements of the variable, the solution set of an inequality is the **intersection** of the two sets if the conditions must be met at the same time, and the **union** of the two sets if the conditions are not met at the same time. [2.2]

8. The solution set of $|ax + b| = k$ is

$$\{x \mid ax + b = k\} \cup \{x \mid ax + b = -k\}. \quad [2.3]$$

9. The solution set of $|ax + b| > k$, $k > 0$, is

$$\{x \mid ax + b > k\} \cup \{x \mid ax + b < -k\}. \quad [2.3]$$

10. The solution set of $|ax + b| < k$, $k > 0$, is

$$\{x \mid ax + b < k\} \cap \{x \mid ax + b > -k\}, \text{ or}$$

$$\{x \mid -k < ax + b < k\}. \quad [2.3]$$

11. Problems involving numerical quantities can sometimes be solved by finding the solution set of an **algebraic statement** of the problem. This statement may be an equation or an inequality. [2.4]

12. One method of analysis that can be helpful in solving word problems was discussed. [2.4]

REVIEW EXERCISES

SECTION 2.1

In Problems 1–6, give the name of the axiom or the number of the theorem that justifies the equivalence of each pair of open sentences.

1. $x - 4 + 2x + 3 = 5$; $x + 2x - 4 + 3 = 5$
2. $x + 2x - 4 + 3 = 5$; $(1 + 2)x - 1 = 5$
3. $3x - 1 = 5$; $3x - 1 + 1 = 5 + 1$
4. $3x = 6$; $x = 2$
5. $\dfrac{x}{x+2} > 5, x > -2$; $x > 5(x + 2)$
6. $\dfrac{x}{x+2} > 5, x < -2$; $x < 5(x + 2)$
7. Find an equation equivalent to $\dfrac{x}{1+x} = y$ that expresses x in terms of y.
8. Find an inequality equivalent to $\dfrac{x+y}{x-y} < 3$ that expresses y in terms of x. State the condition on x and y necessary for equivalence. (Two solutions)

SECTION 2.2

In Problems 9–12, find the solution set of the given equation. If the equation is an identity, state this as your answer.

9. $3x - 7 = 3 - 2x$
10. $4x + 1 - 3x + 7 = 5x + 8$
11. $\dfrac{x+2}{x} - \dfrac{3}{x} = 1 - \dfrac{1}{x}$ $\quad (x = 0)$
12. $\dfrac{x}{x-3} - \dfrac{5}{x-3} = 4$ $\quad (x = 3)$

62 FIRST-DEGREE OPEN SENTENCES WITH ONE VARIABLE

Specify the solution set of each inequality in Problems 13–16. If the statement is an absolute inequality, state this as your answer.

13. $3x + 2 < 4x + 1$
14. $x + 7 > 2x - 1$
15. $\dfrac{x+1}{x} > 1$ $\quad (x > 0)$
16. $\tfrac{3}{4}(2x - 1) - \tfrac{1}{2}(4x + 3) > 0$

SECTION 2.3

Find the solution set of each open sentence.

17. $|x - 6| = 3$
18. $|2x + 5| = 15$
19. $|1 - 4x| = 2$
20. $|2x - 1| > 7$
21. $|2x - 5| < 3$
22. $|x + 1| > x$

SECTION 2.4

Use the four-step method of analysis to find the solution for each problem.

23. The difference between twice a certain number and two-thirds of the number is 32. Find the number.
24. The value of a number of dimes and three less that number of nickels is $1.80. Find the number of dimes and the number of nickels.
25. At a certain manufacturing plant it costs $1000 to set up the machinery to manufacture an article. The cost of materials and labor is $2.50 per article. How many articles must be made if the total cost per article is to be less than $3.00?

Exponents, Roots, and Radicals

You learned in elementary algebra that factors often occur more than once in a given product. For example, x is a double factor of $x \cdot x$, and 2 occurs as a factor three times in the product $2 \cdot 2 \cdot 2$. You also learned that such products are called **powers** and that powers can be represented by special symbols. Thus, $a \cdot a = a^2$, and $2 \cdot 2 \cdot 2 = 2^3$. More generally, any power can be represented by a symbol of the form
$$a^n,$$
where a represents the repeated factor (called the **base** of the power), and n denotes a natural number (called the **exponent** of the power) that is the number of times the base occurs as a factor in the product. The power a^n is defined formally as follows:

DEFINITION 3.0-1 If $a \in R$ and $n \in N$, then
$$a^n = \overbrace{a \cdot a \cdots a}^{n \text{ factors}}.$$

Powers

Many problems in algebra involve operations with powers. If these situations are to be handled satisfactorily, you must be familiar with the properties of powers. These properties (stated as theorems without proof) are repeated here for reference.

If $a, b \in R$ and $m, n \in N$, then

THEOREM 3.0-1 $a^m \cdot a^n = a^{m+n}.$

THEOREM 3.0-2 $\dfrac{a^m}{a^n} = \begin{cases} a^{m-n} & \text{if } m > n \\ \dfrac{1}{a^{n-m}} & \text{if } m < n \end{cases} \quad (a \neq 0).$

63

64 EXPONENTS, ROOTS, AND RADICALS

THEOREM 3.0-3 $(a^m)^n = a^{mn}$.

THEOREM 3.0-4 $(ab)^n = a^n b^n$.

THEOREM 3.0-5 $\left(\dfrac{a}{b}\right)^n = \dfrac{a^n}{b^n} \qquad (b \neq 0)$.

In this chapter we discuss the use of other real numbers as exponents and some of the properties of powers that have such exponents.

3.1 Integers as Exponents

If we assume that powers of variables with integers as exponents represent real numbers, then such powers must have the same properties as those listed in Theorems 3.0-1 through 3.0-5. Since we have agreed that the natural numbers and the positive integers have the same properties, it follows that powers with positive integral exponents possess the listed properties, but what about zero and the negative integers as exponents of powers?

Zero as an Exponent

If Theorems 3.0-1 and 3.0-2 are to hold for powers with exponent 0, then

$$a^n \cdot a^0 = a^{n+0} = a^n \qquad (a \neq 0),$$

and

$$\frac{a^n}{a^n} = a^{n-n} = a^0 \qquad (a \neq 0).$$

In either case it is clear that a^0 represents the multiplicative identity 1, except when $a = 0$.

This discussion suggests the following.

DEFINITION 3.1-1 If $a \in \mathbf{R}$, $a \neq 0$, then $a^0 = 1$.

Thus, in any expression, 1 may be substituted for a power with a nonzero base and 0 as its exponent. Consequently, it can be shown that Theorems 3.0-3, 3.0-4, and 3.0-5 also hold for such powers.

Examples (a) $x^0 = 1 \quad (x \neq 0)$ (b) $(12{,}345)^0 = 1$ (c) $(\sqrt{3})^0 = 1$

Negative Integers as Exponents

If n is a positive integer, can we give meaning to the symbol a^{-n}? If Theorem 3.0-1 is to be valid for a^{-n}, then

$$a^n \cdot a^{-n} = a^{n+(-n)} = a^0 = 1.$$

3.1 Integers as Exponents

However, if the product of two numbers is 1, the two numbers are multiplicative inverses of each other, and since the multiplicative inverse of any nonzero number is unique, a^{-n} and $1/a^n$ must name the same number. We may now give meaning to a power with an exponent that is a negative integer as follows.

DEFINITION 3.1-2 If n is a positive integer and a is any nonzero real number, then
$$a^{-n} = \frac{1}{a^n}.$$

Thus, any power with a negative integer as an exponent represents the same number as the reciprocal of the power of the same base whose exponent is the additive inverse of the original exponent.

Examples Write each expression without using negative exponents.

(a) x^{-3} (b) $\dfrac{1}{y^{-2}}$

Solutions (a) $x^{-3} = \dfrac{1}{x^3}$ by Definition 3.1-2.

(b) $\dfrac{1}{y^{-2}} = \dfrac{1}{1/y^2}$ by Definition 3.1-2.

$= y^2$ by Theorem 1.4-14 $\left(\dfrac{1}{a/b} = \dfrac{b}{a}\right)$.

EXERCISE 3.1

WRITTEN

For all problems in this exercise, assume that a, b, x, y, and z represent nonzero real numbers and that m and n are integers. In Problems 1–12 state the theorem or definition that justifies each statement.

Examples (a) $2^4 2^n = 2^{4+n}$ (b) $(x^2)^5 = x^{10}$

Solutions (a) Theorem 3.0-1: $a^m \cdot a^n = a^{m+n}$

(b) Theorem 3.0-3: $(a^m)^n = a^{mn}$

1. $\left(\dfrac{2}{y}\right)^3 = \dfrac{8}{y^3}$

2. $\left(\dfrac{y^2}{-2}\right)^4 = \dfrac{(y^2)^4}{(-2)^4}$

3. $(xy)^4 = x^4 y^4$

4. $(-3x^2)^2 = (-3)^2(x^2)^2$

5. $\dfrac{z^2}{z^5} = z^{-3}$

6. $\dfrac{y^2}{y^m} = y^{2-m}$

7. $\left(\dfrac{xy}{-2}\right)^4 = \dfrac{(xy)^4}{(-2)^4}$

8. $(-2x^2)^3 = (-2)^3(x^2)^3$

$\dfrac{(xy)^4}{(-2)^4} = \dfrac{x^4 y^4}{16}$

$(-2)^3(x^2)^3 = -8x^6$

EXPONENTS, ROOTS, AND RADICALS

9. $\left(\dfrac{x^4}{x}\right)^2 = (x^3)^2$

   $(x^3)^2 = x^6$

10. $\left(\dfrac{2}{x}\right)^3 \left(\dfrac{2}{x}\right)^4 = \left(\dfrac{2}{x}\right)^7$

    $\left(\dfrac{2}{x}\right)^7 = \dfrac{128}{x^7}$

11. $(x^2 y^{-2})^2 = x^4 y^{-4}$

    $x^4 y^{-4} = x^4 \cdot \dfrac{1}{y^4}$

12. $(x^{-2} y^3)^{-1} = x^2 y^{-3}$

    $x^2 y^{-3} = x \cdot \dfrac{1}{y^3}$

For each expression in Problems 13–30, write an equivalent expression in which each variable is written only once and all exponents are positive integers.

Examples (a) $(x^2 y^2)(x^2 y^3)$ (b) $\dfrac{(5x)^2}{(3x^2)^3}$

Solutions (a) $x^4 y^5$ (b) $\dfrac{25}{27 x^4}$

13. $\dfrac{18 x^3 y^4}{6 x^2 y}$
14. $\dfrac{18 y^5 z^6}{9 y^3 z^2}$
15. $(3x^3)(2x^4)$
16. $(-3x^5)^3$
17. $\dfrac{27 x^3 y}{18 x^2 y^3}$
18. $\dfrac{32 y^2 t}{12 y^5 t^4}$
19. $(2x^0 y^{-1} z^{-3})(4 y^2 z^3)$
20. $(2x^2 t)(3x^{-3} t^{-2})$

Examples (a) $x^{n-1} x^{2-n}$ (b) $\dfrac{x^{n+2} y^{m-1}}{x^{n+1} y^{1-m}}$

Solutions (a) $x^{(n-1)+(2-n)} = x$

(b) $x^{(n+2)-(n+1)} y^{(m-1)-(1-m)} = xy^{2m-2}$

21. $x^{3+n} x^{1-2n}$
22. $y^{m+5} y^{2m-3}$
23. $(x^{n+1} x^{2n+3})^2$
24. $(y^{m+2})^n$
25. $\dfrac{x^{n+1}}{x^{1-n}}$
26. $\dfrac{x^{n+2} y^{m-1}}{x^{n-1} y^{2m-1}}$
27. $\dfrac{(x^{n+2} y^{n+1})^2}{x^{2n} y^2}$
28. $\dfrac{(x^{2n-3} y^{n-2})^2}{x^{n-8} y^{3n-7}}$
29. $\dfrac{(a^{3n+1} t^{2n+3})^3}{a^{6n} t^6}$
30. $\dfrac{(a^{n-2} t^{n+1})^4}{(a^{n+1} t^{n-1})^3}$

For each expression in Problems 31–40, use negative exponents, if necessary, to write an equivalent expression in which no variable appears in a denominator.

Examples (a) $\dfrac{1}{x^2 y^2}$ (b) $\dfrac{4xy^3}{2x^2 y^6}$

Solutions (a) $x^{-2} y^{-2}$ (b) $2x^{-1} y^{-3}$

31. $\dfrac{2x^2}{y^{-3}}$
32. $\dfrac{x^3}{y^{-2}}$
33. $\dfrac{x^3}{y^2}$
34. $\dfrac{2y}{x^2 y^3}$
35. $\dfrac{3x^2}{y z^3}$
36. $\dfrac{x^2 y^{-1}}{x y^{-2} z^{-3}}$
37. $\dfrac{2 x^0 y}{2^{-1} x^{-1} y^{-3}}$

38. $\dfrac{8x^{-3}y^{-1}}{4x^{-1}y^{-2}z^{0}}$ 39. $\left(\dfrac{x^{-1}}{y^{-2}}\right)^{3}$ 40. $\left(\dfrac{x^{-2}y^{3}z^{-1}}{x^{-3}z^{-2}}\right)^{3}$

3.2 Other Rational Numbers as Exponents

You saw in the previous section that we can define powers like a^0 and a^{-3} in such a way that Theorems 3.0-1 through 3.0-5 are true for powers with integers as exponents. The question now arises, can powers like $a^{1/2}$ and $a^{2/3}$ be defined in a similar way?

If Theorem 3.0-3 (p. 64) is to hold for a power of the form $a^{1/n}$, where $a \in R$ and $n \in N$, then

$$(a^{1/n})^n = a^{(1/n)n} = a^1 = a.$$

This statement suggests the following.

DEFINITION 3.2-1   The symbol $a^{1/n}$, $a \in R$, $n \in N$, names a number such that

$$\underbrace{a^{1/n} \cdot a^{1/n} \cdots a^{1/n}}_{n \text{ factors}} = a.$$

Thus, $a^{1/n}$ is one of the n equal factors of a. We may also say that $a^{1/n}$ is an **nth root** of a. For example, $a^{1/2} = \sqrt{a}$, $a^{1/3} = \sqrt[3]{a}$, etc.

Definition 3.2-1, as stated, does not specify that $a^{1/n}$ is a real number. However, if $a^{1/n}$ does represent a real number, then a must be a non-negative number if n is an even integer. Recall that $\sqrt{a}$ is a real number if and only if $a \geq 0$.

Roots of Real Numbers

There are several important facts that you must consider when nth roots of real numbers are discussed. These facts, which are ordinarily proved as theorems in more advanced courses, are listed in the following paragraphs.

(1) Every $a \in R$ has n distinct nth roots, some of which may not be real numbers. For example, 16 has 4 fourth roots, two of which are real numbers. -16 has 4 fourth roots, none of which is a real number. Numbers with roots that are not real numbers will be discussed in Chapter 11.

(2) If $a > 0$ and n is a *positive even integer,* there are two real numbers, one positive and one negative, that are nth roots of a. The positive root is called the **principal nth root** and is denoted by $a^{1/n}$ or $\sqrt[n]{a}$. The negative root is denoted by $-a^{1/n}$ or $-\sqrt[n]{a}$.

(3) If $a < 0$ and n is a positive even integer, a has no real nth roots.

(4) If n is a positive odd integer, a has one and only one real nth root, which is the principal root. This root is positive if $a > 0$ and negative if $a < 0$.

(5) For any positive integer n, the nth root of zero is zero.

Exponents of the Form m/n

Our discussion can now be extended to include any rational number m/n as an exponent. However, to keep the discussion from becoming too complicated, we specify that $m \in J$ and $n \in N$. We also restrict a to be a non-negative real number.

The definition of a quotient

$$\left(\frac{a}{b} = a \cdot \frac{1}{b}\right)$$

permits you to write m/n as

$$m \cdot \frac{1}{n} \quad \text{or as} \quad \frac{1}{n} \cdot m.$$

Consequently, $a^{m/n}$ is defined as follows:

DEFINITION 3.2-2 For all $a \in R$, $a > 0$, $m \in J$, and $n \in N$,

$$a^{m/n} = (a^{1/n})^m.$$

It can also be shown, but we shall not do so, that subject to the restrictions noted in Definition 3.2-2 the following is true.

THEOREM 3.2-1 If $a > 0$, then $(a^{1/n})^m = (a^m)^{1/n}$.

The example below illustrates the necessity for the restriction on a.

Example If Theorem 3.2-1 did not require that a be a positive number, it could be used to write

$$(-1)^{5/3} = [(-1)^{1/3}]^5 = [(-1)^5]^{1/3} = -1.$$

However, if 5/3 is replaced with 10/6, we have

$$(-1)^{10/6} = [(-1)^{1/6}]^{10} = [(-1)^{10}]^{1/6},$$

which is not a true statement since $[(-1)^{1/6}]^{10}$ is not always a real number* and $[(-1)^{10}]^{1/6} = 1^{1/6} = 1$.

It can be shown, but we will not do so, that Theorems 3.0-1 through 3.0-5 are true for all powers with non-negative bases and rational numbers as exponents.

EXERCISE 3.2

WRITTEN

Find the value of each power in Problems 1–10.

---

*There are 6 sixth roots of -1, none of which is a real number. If a proper choice of these roots is made, $[(-1)^{1/6}]^{10} = -1$.

Examples (a) $16^{3/4}$ (b) $27^{-4/3}$

Solutions (a) $16^{3/4} = (16^{1/4})^3 = 2^3 = 8$
(b) $27^{-4/3} = (\frac{1}{27})^{4/3} = [(\frac{1}{27})^{1/3}]^4 = (\frac{1}{3})^4 = \frac{1}{81}$

1. $4^{3/2}$ 2. $9^{5/2}$ 3. $16^{-3/2}$ 4. $25^{-3/2}$ 5. $32^{3/5}$
6. $64^{5/6}$ 7. $64^{-4/3}$ 8. $81^{-3/4}$ 9. $125^{-2/3}$ 10. $128^{4/7}$

In the problems that follow, assume that all variables represent positive real numbers.

Which theorem, 3.0-1 through 3.0-5, can be used to justify each statement in Problems 11–20?

Examples (a) $(x^{3/2})^{2/3} = x$
(b) (1) $(y^{1/2}y^{3/4})^{2/5} = (y^{5/4})^{2/5}$
(2) $(y^{5/4})^{2/5} = y^{1/2}$

Solutions (a) Theorem 3.0-3: $(a^m)^n = a^{mn}$
(b) (1) Theorem 3.0-1: $a^m \cdot a^n = a^{m+n}$
(2) Theorem 3.0-3: $(a^m)^n = a^{mn}$

11. $\left(\frac{yz}{x}\right)^{1/2} = \frac{(yz)^{1/2}}{x^{1/2}}$

12. $\left(\frac{x^3y}{z^2}\right)^{1/3} = \frac{(x^3y)^{1/3}}{(z^2)^{1/3}}$

13. $(x^{-2}y^4)^{1/2} = x^{-1}y^2$

14. $(x^3y^6)^{-1/3} = x^{-1}y^{-2}$

15. $\frac{x^{2/3}}{x^{1/6}} = x^{1/2}$

16. $\frac{x^{-1/2}}{x^{-1/3}} = x^{-1/6}$

17. (a) $\left(\frac{x^{3/2}}{x^{-1/2}}\right)^{1/2} = (x^2)^{1/2}$
    (b) $(x^2)^{1/2} = x$

18. (a) $(x^{2/3}y^{1/2})^6 = (x^{2/3})^6(y^{1/2})^6$
    (b) $(x^{2/3})^6(y^{1/2})^6 = x^4y^3$

19. (a) $\left(\frac{x^{2/3}y^{1/4}}{3z^{1/6}}\right)^3 = \frac{(x^{2/3}y^{1/4})^3}{(3z^{1/6})^3}$
    (b) $\frac{(x^{2/3}y^{1/4})^3}{(3z^{1/6})^3} = \frac{(x^{2/3})^3(y^{1/4})^3}{3^3(z^{1/6})^3}$
    (c) $\frac{(x^{2/3})^3(y^{1/4})^3}{3^3(z^{1/6})^3} = \frac{x^2y^{3/4}}{27z^{1/2}}$

20. (a) $\left(\frac{x^{3/4}y}{x}\right)^{2/3} = (x^{-1/4}y)^{2/3}$
    (b) $(x^{-1/4}y)^{2/3} = (x^{-1/4})^{2/3}y^{2/3}$
    (c) $(x^{-1/4})^{2/3}y^{2/3} = x^{-1/6}y^{2/3}$

Write each expression in Problems 21–32 as a product or a quotient in which each variable appears only once.

21. $x^{1/2}x^{3/2}$
22. $x^{1/2}x^{1/3}$
23. $\frac{x^{3/2}y^{3/4}}{x^{1/2}y^{1/2}}$

24. $\frac{x^{1/2}y^{2/3}}{x^{1/3}y^{1/2}}$
25. $\frac{(x^{1/2}y^{1/2})^3}{xy}$
26. $\frac{(x^{1/3}y^{1/3})^2}{(xy)^{1/3}}$

27. $(x^{1/2}y^{1/3})(x^0y^{2/3})$
28. $(x^{3/4}y^{2/3})(x^{1/3}y^{1/4})$
29. $(x^{1/5}y^{2/3})^2(xy)^{1/3}$

30. $(x^2y^3)^{1/4}(x^{1/2}y^{1/3})^2$
31. $\left(\frac{x^2}{y^3}\right)^{1/4}\left(\frac{y^2}{x^3}\right)^{1/2}$
32. $\left(\frac{x^{1/2}}{y^{1/3}}\right)^3\left(\frac{y^{1/3}}{x^{1/2}}\right)^2$

In Problems 33–56, write each expression as a product or a quotient of powers in which each variable appears only once and all exponents are positive.

70 EXPONENTS, ROOTS, AND RADICALS

Examples (a) $x^{-1/4} \cdot x^{1/2}$ (b) $\left(\dfrac{x^{3/2}}{x}\right)^{2/3}$ (c) $\left(\dfrac{x^{1/2}}{y^{2/3}}\right)^{-6}$

Solutions (a) $x^{(-1/4+1/2)} = x^{1/4}$ (b) $\dfrac{x}{x^{2/3}} = x^{1/3}$ (c) $\dfrac{x^{-3}}{y^{-4}} = \dfrac{y^4}{x^3}$

33. $x^{-2}y^{1/2}$
34. $x^{-1}y^{-2}$
35. $x^0 y^{1/2} z^{-2/3}$
36. $x^{-1}y^{-3/4}z^0$
37. $\dfrac{2x^{-3}y^{-1/2}}{2^{-2}x^3 y^{-2/3}}$
38. $\dfrac{2^{-3}x^{-1}y^{-2/3}}{2^{-4}x^{-2}y}$
39. $\dfrac{3^{-3}x^{-2}y^{-3/4}}{3^{-2}x^2 y^{-5/4}}$
40. $\dfrac{2^{-5}x^{-6}y^{-3/2}}{2^{-4}x^{-4}y^{-1/2}}$
41. $\left(\dfrac{x^{-2}}{y^3}\right)^{-2/5}$
42. $\left(\dfrac{x^{-1}}{y^{-3}}\right)^{2/3}$
43. $\left(\dfrac{x^0}{y^2}\right)^{-3/4}$
44. $\left(\dfrac{y^4}{x^{-3}}\right)^{-2/3}$
45. $(x^{-2}y)^{-2/3}$
46. $(y^2 z^{-3})^{2/5}$
47. $(x^{-3}y^2)^{-3/4}$
48. $(x^{-1}y^{-2})^{-3/2}$
49. $x^{(n-1)/2} \cdot x^{-n/2}$
50. $x^{-2n}x^{2/n}$
51. $\left(\dfrac{x^{1-n}}{x^{2-n}}\right)^{-2/n}$
52. $\left(\dfrac{x^{-n}y^{n-1}}{y^n}\right)^{-2/n}$
53. $\left(\dfrac{x^7}{x^{n-1}}\right)^{-1/n}$
54. $\left(\dfrac{x^n y^{2n-1}}{x^{n+1}y^{2n}}\right)^{1/n}$
55. $\left[\dfrac{x^{2n}y^{n-1}}{x^{2n-3}y^{n+4}}\right]^{2/(n+1)}$
56. $\left[\dfrac{x^{4-n}y^{3+n}}{x^{-2-n}y^{2+n}}\right]^{-1/(n+1)}$

3.3 Sums and Differences Involving Powers

So far in our discussion of powers, we have considered expressions involving simple products and quotients with one or more variables. Before this discussion is extended to include sums and differences or more complicated products and quotients, you should recall a few more facts from elementary algebra.

Algebraic Expressions

Collections of symbols such as

$$3x + 2, \quad 5x^2 - 7x + 3, \quad 2x^2 y - 3xy^2, \quad \text{and} \quad \dfrac{2x+1}{y}$$

are called *algebraic expressions,* or simply expressions, and represent real numbers for permissible real number replacements of the variables.

The parts of an expression formed by products or quotients are called *terms.* Two or more terms that have identical variable factors are called **similar terms** or **like terms**.

Examples (a) $3, 2x^2, 4xy/z, 2 + x/3$ are expressions composed of a single term each.

(b) $3x + 5y - 7/z$ is an expression of three terms.

(c) $2x^2 + 4x^2 y - 3x^2 + y^2 - 2xy^2$ is an expression of five terms. $2x^2$ and $-3x^2$ are similar terms, because they have identical variable factors. $4x^2 y$ and $-2xy^2$ are not similar terms, because the variable factors are not identical.

Sums of Expressions

A sum involving two or more expressions can be written equivalently as an expression with no similar terms by applying the commutative and associative properties of addition and the distributive law.

Example Write

$$(2x^2 + 5x - 7) + (3x + 2) + (-x^2 + x + 4)$$

as an expression containing no similar terms.

Solution The commutative property of addition can be used to write the given expression as

$$2x^2 - x^2 + 5x + 3x + x - 7 + 2 + 4,$$

whose terms are collected as follows by the associative property of addition:

$$(2x^2 - x^2) + (5x + 3x + x) + (-7 + 2 + 4).$$

The distributive law is applied to each of the sums within the first two sets of parentheses to obtain

$$(2 - 1)x^2 + (5 + 3 + 1)x + (-7 + 2 + 4).$$

The sums within the parentheses are then simplified to obtain the desired form,

$$x^2 + 9x - 1.$$

Differences of Expressions

The difference of two real numbers was defined in Section 1.3 as follows:

DEFINITION 1.3-3 If $a, b \in R$, then $a - b = a + (-b)$.

If b represents an expression containing more than one term, then $-b$ represents the negative of the expression. Since $b + (-b) = 0$, the negative of an expression is an expression such that the sum of the expression and its negative is zero. Theorem 1.4-6 (p. 14) can be reversed to show that $-(a + b) = -a - b$. This leads us to conclude that the negative of an expression is the expression in which each term is the negative of the corresponding term in the original.

Examples Write the negative of each expression.

(a) $x^2 - 3x + 2$ (b) $3x^{1/2} + x^{-3}$

(c) $4x^2y^{-1} - 3xy^0 + 6x^{1/2}y^{1/2}$

Solutions (a) $-x^2 + 3x - 2$ (b) $-3x^{1/2} - x^{-3}$

(c) $-4x^2y^{-1} + 3xy^0 - 6x^{1/2}y^{1/2}$

The foregoing discussion suggests a technique that you can use to simplify differences involving expressions.

Example Write the difference

$$(3x^{3/2} - 2x^{1/2} + 7) - (x^{3/2} - 5x^{1/2} + 4)$$

as an expression containing no similar terms.

Solution $(3x^{3/2} - 2x^{1/2} + 7) - (x^{3/2} - 5x^{1/2} + 4)$
$$= (3x^{3/2} - 2x^{1/2} + 7) + (-x^{3/2} + 5x^{1/2} - 4)$$
$$= (3x^{3/2} - x^{3/2}) + (-2x^{1/2} + 5x^{1/2}) + (7 - 4)$$
$$= (3 - 1)x^{3/2} + (-2 + 5)x^{1/2} + (7 - 4)$$
$$= 2x^{3/2} + 3x^{1/2} + 3$$

EXERCISE 3.3

WRITTEN

Write each sum or difference as a single expression containing no similar terms. Assume that each expression represents a real number.

Examples (a) $(x^2y^2 + 3xy - 5) + (2x^2y^2 - xy + 7)$
(b) $(5x^{1/2}y^{-1} + 2x^{-1/2}y + 4xy^2) + (3x^{1/2}y^{-1} + 4x^{-1/2}y - 8xy^2)$

Solutions (a) $(x^2y^2 + 2x^2y^2) + (3xy - xy) + (-5 + 7)$
$$= (1 + 2)x^2y^2 + (3 - 1)xy + (-5 + 7)$$
$$= 3x^2y^2 + 2xy + 2$$

(b) $(5x^{1/2}y^{-1} + 3x^{1/2}y^{-1}) + (2x^{-1/2}y + 4x^{-1/2}y) + (4xy^2 - 8xy^2)$
$$= (5 + 3)x^{1/2}y^{-1} + (2 + 4)x^{-1/2}y + (4 - 8)xy^2$$
$$= 8x^{1/2}y^{-1} + 6x^{-1/2}y - 4xy^2$$

1. $(x^2 - 2x) + (x^2 + 2x) + (2x^2 - x + 2)$
2. $(x^2 - 2x - 1) + (2x^2 - x + 3)$
3. $(2y^2 + 4y - 5) + (2 + y - y^2)$
4. $(4y^2 + 3y + 5) + (8 - 6y - 3y^2)$
5. $(3x^2y^2 + 2xy - 1) + (4x^2y^2 - 3xy + 3)$
6. $(7x^2y + 2x - 3y) + (x^2y + 2x)$
7. $(2y^{-2} - y^{-1} + 1) + (3y^{-2} + 2y^{-1} + 7)$
8. $(4x^{-2} - 2x^{-1} + 1) + (3x^{-2} - 1)$
9. $(2x^{3/2} - 7x^{1/2} + 1) + (5x^{3/2} + x^{1/2} - 5)$
10. $(x^{1/2}y^{-2} + 2x^{-1/2}y^{-1} + 1) + (3 - x^{-1/2}y^{-1} - x^{1/2}y^{-2})$

Examples (a) $(3x^2 - 2xy + 5y^2) - (2x^2 + 2xy - y^2)$
(b) $(x^{-2}y^{1/2} - x^{-1}y + y^{3/2}) - (2x^{-2}y^{1/2} + 3x^{-1}y - y^{3/2})$

Solutions (a) $(3x^2 - 2xy + 5y^2) + [-(2x^2 + 2xy - y^2)]$
$$= 3x^2 - 2xy + 5y^2 - 2x^2 - 2xy + y^2$$
$$= 3x^2 - 2x^2 - 2xy - 2xy + 5y^2 + y^2$$
$$= (3 - 2)x^2 + (-2 - 2)xy + (5 + 1)y^2$$
$$= x^2 - 4xy + 6y^2$$

(b) $(x^{-2}y^{1/2} - x^{-1}y + y^{3/2}) + [-(2x^{-2}y^{1/2} + 3x^{-1}y - y^{3/2})]$
$= x^{-2}y^{1/2} - x^{-1}y + y^{3/2} - 2x^{-2}y^{1/2} - 3x^{-1}y + y^{3/2}$
$= x^{-2}y^{1/2} - 2x^{-2}y^{1/2} - x^{-1}y - 3x^{-1}y + y^{3/2} + y^{3/2}$
$= (1-2)x^{-2}y^{1/2} + (-1-3)x^{-1}y + (1+1)y^{3/2}$
$= -x^{-2}y^{1/2} - 4x^{-1}y + 2y^{3/2}$

11. $(2x^2 - 3xy + 5y^2) - (3x^2 + 2xy - y^2)$
12. $(2x^2 - 3x + 1) - (2x^2 + 3x - 1)$
13. $(7y^2 + 2y - 3) - (y^2 + 2y + 5)$
14. $(xy^2 + xy - x^2y) - (2x^2y - 3xy + xy^2)$
15. $(x^{-2} - 3x^{-1} + 1) - (2x^{-2} - 2x^{-1} + 5)$
16. $(2x^{-3} + 5x^{-2} + 3x^{-1} + 5) - (x^{-3} + 4x^{-2} - 2x^{-1} - 4)$
17. $(3x^{3/2} - 4x^{1/2} + 7) - (8 + 3x^{1/2} - 5x^{3/2})$
18. $(5x^{3/2} + 6x^{1/2} - 4x^{-1/2}) - (2x^{-1/2} - 4x^{1/2} + 2x^{3/2})$
19. $(x^{1/2}y^{-2} + x^{1/2}y^{-1} - xy) - (2x^{1/2}y^{-2} - 3x^{1/2}y^{-1} + 4xy)$
20. $(2x^{3/2}y^{1/2} + 3x^{1/2}y - x) - (2x^{1/2}y + x - 2x^{3/2}y^{1/2})$

Example $(x^3 - 3z^{1/2}) + (2x^2 - x + 2z^{1/2}) - (x^3 + x^2 + x - z^{1/2})$

Solution $(x^3 - 3z^{1/2}) + (2x^2 - x + 2z^{1/2}) + [-(x^3 + x^2 + x - z^{1/2})]$
$= x^3 - 3z^{1/2} + 2x^2 - x + 2z^{1/2} - x^3 - x^2 - x + z^{1/2}$
$= x^3 - x^3 + 2x^2 - x^2 - x - x - 3z^{1/2} + 2z^{1/2} + z^{1/2}$
$= (1-1)x^3 + (2-1)x^2 + (-1-1)x + (-3+2+1)z^{1/2}$
$= x^2 - 2x$

21. $(3x^2 - x - 1) + (2x^2 - 3x + 2) - (-x^2 + 4x - 3)$
22. $(5x^2 + 7x + 3) + (2x - 5 - 3x^2) - (1 + x^2 - 3x)$
23. $(2x^2y^2 + 4xy + 8) + (xy - 5 - x^2y^2) - (4 + 2xy - 3x^2y^2)$
24. $(4xy^2 - 7xy + 2x) + (3xy^2 + 4xy - 3x) - (xy^2 - 3xy - 5x)$
25. $(2x^{3/2} + 5x^{1/2} - 8) + (4x^{3/2} - 7x^{1/2} + 4) - (3x^{3/2} - 4x^{1/2} + 7)$
26. $(6x^{3/2} - 4x^{1/2} + 3) + (5x^{3/2} + 2x^{1/2} - 7) - (9x^{3/2} + 2x^{1/2} + 3)$
27. $(3y^{-2} + 2y^{-1} - 1) + (6y^{-1} + 5 - 8y^{-2}) - (8 + 5y^{-1} - 6y^{-2})$
28. $(8y^{-3} + 7y^{-2} - 5y^{-1}) + (4y^{-3} - 3y^{-2} + 6y^{-1}) - (9y^{-3} + 4y^{-2} - 3y^{-1})$
29. $(4x^{-2}y^{3/2} - 6x^{-1}y^{1/2} + 4) + (2x^{-2}y^{3/2} + 2x^{-1}y^{1/2} - 3) - (3x^{-2}y^{3/2} - 2x^{-1}y^{1/2} + 5)$
30. $(7x^{1/2}y^{-1} + 4x^{3/2}y^{-2} - 8) + (4 - 5x^{1/2}y^{-1}) - (2x^{3/2}y^{-2} - 5)$

3.4 Further Applications of the Distributive Law

Products involving expressions in which one or more of the factors is a sum can be written equivalently as a sum by applying the distributive law and Theorem 3.0-1 (p. 63).

Examples Write each product as a sum.

(a) $x^2(2x^2 + 5x)$
(b) $x^{1/2}(x + x^{-1/2})$
(c) $(x^{1/2} + y^{1/2})^2$
(d) $(x - y)^{1/2}[(x - y) + (x - y)^{-1}]$

Solutions (a) $x^2(2x^2 + 5x) = x^2(2x^2) + x^2(5x)$
$$= 2x^2 \cdot x^2 + 5x \cdot x^2$$
$$= 2x^{2+2} + 5x^{1+2}$$
$$= 2x^4 + 5x^3$$

(b) $x^{1/2}(x + x^{-1/2}) = x^{1/2} \cdot x + x^{1/2} \cdot x^{-1/2}$
$$= x^{(1/2)+1} + x^{(1/2)+(-1/2)}$$
$$= x^{3/2} + x^0$$
$$= x^{3/2} + 1$$

(c) $(x^{1/2} + y^{1/2})^2 = (x^{1/2} + y^{1/2})(x^{1/2} + y^{1/2})$
$$= (x^{1/2} + y^{1/2})x^{1/2} + (x^{1/2} + y^{1/2})y^{1/2}$$
$$= x^{1/2} \cdot x^{1/2} + y^{1/2} \cdot x^{1/2} + x^{1/2} \cdot y^{1/2} + y^{1/2} \cdot y^{1/2}$$
$$= x^{(1/2)+(1/2)} + x^{1/2}y^{1/2} + x^{1/2}y^{1/2} + y^{(1/2)+(1/2)}$$
$$= x + 2x^{1/2}y^{1/2} + y$$

(d) $(x - y)^{1/2}[(x - y) + (x - y)^{-1}]$
$$= (x - y)^{1/2}(x - y) + (x - y)^{1/2}(x - y)^{-1}$$
$$= (x - y)^{(1/2)+1} + (x - y)^{(1/2)-1}$$
$$= (x - y)^{3/2} + (x - y)^{-1/2}$$

Since you can represent any number as a sum of two other numbers, you can write a power equivalently so that the exponent is a sum of two rational numbers. Thus, an expression involving a sum of two or more powers having the same base can be written equivalently so that each exponent is a sum and each sum contains a common addend. As a result, by Theorem 3.0-1, you can represent each of the powers as a product so that each has a common factor. Then, if the base is such that all powers represent real numbers, the distributive law can be applied to write the original sum as a product. You will find this technique useful in simplifying certain complicated expressions.

Examples Write each sum as a product containing the given factor.

(a) $3x^3 + 4x^2 - 5x;\quad x$ (b) $x^{2/3} + x^{5/3};\quad x^{2/3}$

Solutions (a) $3x^3 + 4x^2 - 5x = 3 \cdot x^{2+1} + 4 \cdot x^{1+1} - 5 \cdot x$
$$= 3x^2 \cdot x + 4x \cdot x - 5 \cdot x$$
$$= (3x^2 + 4x - 5)x$$

(b) $x^{2/3} + x^{5/3} = x^{(2/3)+0} + x^{(2/3)+(3/3)}$
$$= x^{2/3} \cdot x^0 + x^{2/3} \cdot x^{3/3} \qquad (x \neq 0)$$
$$= x^{2/3}(x^0 + x^{3/3})$$
$$= x^{2/3}(1 + x)$$

EXERCISE 3.4

WRITTEN

In this exercise, assume that the values for the variables are such that all expressions represent real numbers.

In Problems 1–34, write each product as a sum by using the distributive law and Theorem 3.0-1. Do as much of the work mentally as you can.

Examples (a) $x^{3/2}(x^{3/2} + x^{1/2})$ (b) $(x - 1)[(x - 1)^{3/2} + (x - 1)^{-3/2}]$

Solutions (a) $x^{3/2}(x^{3/2} + x^{1/2}) = x^{3/2} \cdot x^{3/2} + x^{3/2} \cdot x^{1/2}$
$$= x^{(3/2)+(3/2)} + x^{(3/2)+(1/2)}$$
$$= x^3 + x^2$$

(b) $(x - 1)[(x - 1)^{3/2} + (x - 1)^{-3/2}]$
$$= (x - 1)(x - 1)^{3/2} + (x - 1)(x - 1)^{-3/2}$$
$$= (x - 1)^{1+(3/2)} + (x - 1)^{1-(3/2)}$$
$$= (x - 1)^{5/2} + (x - 1)^{-1/2}$$

1. $x(3x^2 + x)$
2. $y^3(y^2 + y)$
3. $x^{-2}(4x^3 + x^2)$
4. $y^{-1}(y^4 + 2y^5)$
5. $x^{1/2}(x^2 + x^{-1})$
6. $y^{-1}(y^2 + y)$
7. $x^{-2/3}(x^{2/3} - x^0)$
8. $z^{-3/2}(z^3 - z^{1/2})$
9. $y^{1/3}(y^{2/3} - y^{-1/3})$
10. $x^{2/5}(x^{3/5} + x^{-3/5})$
11. $(x + 2)^{1/2}[(x + 2)^2 + (x + 2)^{-1}]$
12. $(x - y)^{1/3}[(x - y)^{2/3} + (x - y)^{1/3}]$ $x-y + x-y^{2/3}$
13. $(y + z)^{2/3}[(y + z)^{1/3} - (y + z)^{-2/3}]$
14. $(x + 2y)^{1/2}[(x + 2y)^{1/2} - (x + 2y)^{-1/2}]$
15. $x^n(x^{n+1} + x^{-n})$
16. $y^m(y^{2m} + y^2)$
17. $z^{n/2}(z^n - z^{-n})$
18. $x^{n+1}(x^{n+1} + x^{n-1})$
19. $(x - 3)^{n-2}[(x - 3)^{n+1} + (x - 3)^{n/2}]$
20. $(2x - y)^{(n+2)/3}[(2x - y)^{n/3} - (2x - y)^{-2/3}]$

Examples (a) $(x^2 - 5)(x^2 + 2)$ (b) $(x^{1/2} + 2)(x^{1/2} - 3)$

Solutions (a) $(x^2 - 5)(x^2 + 2) = (x^2 - 5)x^2 + (x^2 - 5)2$
$$= x^2 \cdot x^2 - 5 \cdot x^2 + x^2 \cdot 2 - 5 \cdot 2$$
$$= x^{2+2} + (-5 + 2)x^2 - 5 \cdot 2$$
$$= x^4 - 3x^2 - 10$$

(b) $(x^{1/2} + 2)(x^{1/2} - 3) = (x^{1/2} + 2)x^{1/2} + (x^{1/2} + 2)(-3)$
$$= x^{1/2} \cdot x^{1/2} + 2 \cdot x^{1/2} + x^{1/2}(-3) + 2(-3)$$
$$= x^{(1/2)+(1/2)} + (2 - 3)x^{1/2} + 2(-3)$$
$$= x - x^{1/2} - 6$$

76 EXPONENTS, ROOTS, AND RADICALS

21. $(x + 3)(x - 7)$
22. $(2x + 4)(3x + 2)$
23. $(2x + 3)(x^2 - 4)$
24. $(2x^2 - 7)(x^3 + 2)$
25. $(2x - 1)(x^2 + x - 2)$
26. $(3x + 2)(9x^2 - 6x + 4)$
27. $(x^{-1} + 4)(x^{-1} + 2)$
28. $(2x^{-2} + 3)(2x^{-2} - 3)$
29. $(3x^{-2} - x^{-1})(x^{-1} + 1)$
30. $(5x^{-3} - 2x^{-2})(3x^{-1} + 4)$
31. $(x^{1/3} + 1)(x^{2/3} - x^{-1/3})$
32. $(x^{1/2} - y^{1/2})(x^{1/2} - y^{1/2})$
33. $(x^{1/3} + 1)(x^{2/3} - x^{1/3} + 1)$
34. $(x^{1/3} - 1)(x^{2/3} + x^{1/3} + 1)$

In Problems 35–54, find the second factor so that each sum is represented as a product.

Examples (a) $3x^3 + 7x^2 + 4x = x(?)$
(b) $x^{-2} - 4x^{-1} + 8x = x^{-2}(?)$
(c) $x^{1/2} + x^{-1/2} = x^{-1/2}(?)$
(d) $(x - 5)^{1/3} - (x - 5)^{4/3} = (x - 5)^{1/3}(?)$

Solutions (a) $3x^3 + 7x^2 + 4x = 3x^{2+1} + 7x^{1+1} + 4x$
$= 3x^2 \cdot x + 7x \cdot x + 4 \cdot x$
$= x(3x^2 + 7x + 4)$

(b) $x^{-2} - 4x^{-1} + 8x = x^{-2} - 4x^{1-2} + 8x^{3-2}$
$= 1 \cdot x^{-2} - 4x \cdot x^{-2} + 8x^3 \cdot x^{-2}$
$= x^{-2}(1 - 4x + 8x^3)$

(c) $x^{1/2} + x^{-1/2} = x^{1-(1/2)} + x^{-1/2}$
$= x \cdot x^{-1/2} + 1 \cdot x^{-1/2}$
$= x^{-1/2}(x + 1)$

(d) $(x - 5)^{1/3} - (x - 5)^{4/3} = (x - 5)^{1/3} - (x - 5)^{1+(1/3)}$
$= 1(x - 5)^{1/3} - (x - 5)(x - 5)^{1/3}$
$= (x - 5)^{1/3}[1 - (x - 5)]$
$= (x - 5)^{1/3}(6 - x)$

35. $5x^4 + 15x^3 - 20x^2 = 5x^2(?)$
36. $2x^5 - 5x^4 + 3x^3 = x^3(?)$
37. $3x^3y - 4x^2y^2 + 7xy^3 = xy(?)$
38. $4x^4y^2 + 6x^3y^3 - 8x^2y^4 = 2x^2y^2(?)$
39. $x^{-3} + x^{-2} - x = x^{-2}(?)$
40. $3x^{-5} + 2x^{-3} + x^2 = x^{-3}(?)$
41. $x^{3/2} + x^{1/2} = x^{1/2}(?)$
42. $x + x^{1/2} = x(?)$
43. $y - y^{2/3} = y^{1/3}(?)$
44. $y^{2/3} - y^{1/3} = y(?)$
45. $x^{-3/2} - x^{-1/2} = x^{-1/2}(?)$
46. $y^{1/2} + y^{-1/2} = y^{-1/2}(?)$
47. $x^{4/5} + x^{-1/5} = x^{-1/5}(?)$
48. $y^{3/5} - y^{-6/5} = y^{-6/5}(?)$

49. $(x + 2)^2 + (x + 2) = (x + 2)(?)$
50. $(4x - y)^3 + 5(4x - y)^2 = (4x - y)^2(?)$
51. $(x - 1)^{-2} - (x - 1)^{-1} = (x - 1)^{-1}(?)$
52. $(x + 1)^{-1} + (x + 1)^{-2} = (x + 1)^{-2}(?)$
53. $(x - 2)^{-1/3} + (x - 2)^{2/3} = (x - 2)^{-1/3}(?)$
54. $(2x - 3)^{1/4} + (2x - 3)^{-3/4} = (2x - 3)^{-3/4}(?)$

3.5 Changing the Form of Radical Expressions

Recall from Section 3.2 that $a^{1/n}$ and $\sqrt[n]{a}$ represent the same number. This means that we can think of a radical as a power whose base is the radicand and whose exponent is a rational number with the index of the radical as the denominator. This means that radical expressions are subject to the same conditions as powers when any operations are performed.

Let us first consider products and quotients of radicals that have the same index. Since $\sqrt{16} = 4$ and $\sqrt{25} = 5$, $\sqrt{16} \cdot \sqrt{25} = 4 \cdot 5 = 20$. However, since $16 \cdot 25 = 400$, $\sqrt{16 \cdot 25} = \sqrt{400} = 20$. Thus, $\sqrt{16} \cdot \sqrt{25} = \sqrt{16 \cdot 25}$. Similarly, $\dfrac{\sqrt{16}}{\sqrt{25}} = \dfrac{4}{5}$ and $\sqrt{\dfrac{16}{25}} = \dfrac{4}{5}$, from which $\dfrac{\sqrt{16}}{\sqrt{25}} = \sqrt{\dfrac{16}{25}}$. These illustrations lead to the following theorem.

THEOREM 3.5-1 If $a, b \in \mathbf{R}$, $a, b > 0$, and $n \in \mathbf{N}$, then

$$\sqrt[n]{a} \cdot \sqrt[n]{b} = \sqrt[n]{ab} \quad \text{and} \quad \dfrac{\sqrt[n]{a}}{\sqrt[n]{b}} = \sqrt[n]{\dfrac{a}{b}}.$$

PROOF:

Since $\sqrt[n]{a} = a^{1/n}$ and $\sqrt[n]{b} = b^{1/n}$,

$$\sqrt[n]{a} \cdot \sqrt[n]{b} = a^{1/n} \cdot b^{1/n}.$$

Then by Theorem 3.0-4,

$$a^{1/n} \cdot b^{1/n} = (ab)^{1/n} = \sqrt[n]{ab}.$$

Hence, $\sqrt[n]{a}\sqrt[n]{b} = \sqrt[n]{ab}$.

The proof of $\dfrac{\sqrt[n]{a}}{\sqrt[n]{b}} = \sqrt[n]{\dfrac{a}{b}}$ is similar.

Examples (a) $\sqrt{3}\sqrt{5} = \sqrt{15}$ (b) $\dfrac{\sqrt{5}}{\sqrt{3}} = \sqrt{\dfrac{5}{3}}$

(c) $\sqrt[4]{50} = \sqrt[4]{2}\sqrt[4]{25}$ (d) $\sqrt[3]{\dfrac{15}{4}} = \dfrac{\sqrt[3]{15}}{\sqrt[3]{4}}$

Simplest Radical Form

Theorem 3.5-1 is the basis for writing radicals in what we call the **simplest radical form**. Such an expression is said to be in simplest radical form if

78 EXPONENTS, ROOTS, AND RADICALS

(1) The radicand does not contain a factor that is a power of an integer or a variable having an exponent equal to or greater than the index of the radical.

(2) The radicand does not contain a fraction.

(3) There is no radical in the denominator of a fraction.

(4) The radicand is not a power whose exponent contains a factor that is an exact divisor of the index of the radical.

Examples (a) $\sqrt{50}$ is not in simplest radical form, because it contains the factor 25, which equals 5^2.

(b) $\sqrt[3]{x^4}$ is not in simplest radical form, because the exponent of x is greater than the index.

(c) $\sqrt{\tfrac{1}{2}}$ is not in simplest radical form, because the radicand is a fraction.

(d) $\dfrac{5}{\sqrt{3}}$ is not in simplest radical form, because there is a radical in the denominator.

(e) $\sqrt[8]{x^6}$ is not in simplest radical form, because the exponent of the radicand and the index contain the common factor 2.

The following examples illustrate techniques for simplifying radicals.

Example 1 Simplify $\sqrt{98}$.

Solution $\sqrt{98} = \sqrt{49}\sqrt{2}$ (Theorem 3.5-1: $\sqrt{ab} = \sqrt{a}\sqrt{b}$)
$= 7\sqrt{2}$

Example 2 Simplify $\sqrt{\tfrac{2}{3}}$.

Solution $\sqrt{\dfrac{2}{3}} = \dfrac{\sqrt{2}}{\sqrt{3}}$ $\left(\text{Theorem 3.5-1: } \sqrt{\dfrac{a}{b}} = \dfrac{\sqrt{a}}{\sqrt{b}}\right)$

$= \dfrac{\sqrt{2}}{\sqrt{3}} \cdot \dfrac{\sqrt{3}}{\sqrt{3}}$ (Theorem 1.4-12, fundamental principle of fractions)

$= \dfrac{\sqrt{6}}{\sqrt{3}\sqrt{3}}$ (Theorem 3.5-1)

$= \dfrac{\sqrt{6}}{3}$ (Definition 1.6-1: $\sqrt{a}\sqrt{a} = a$)

Example 3 Simplify $\dfrac{\sqrt{x}}{\sqrt{y}}$ $(x, y > 0)$.

Solution $\dfrac{\sqrt{x}}{\sqrt{y}} = \dfrac{\sqrt{x}}{\sqrt{y}} \cdot \dfrac{\sqrt{y}}{\sqrt{y}}$ $\left(\text{Theorem 1.4-12: } \dfrac{a}{b} = \dfrac{ac}{bc}\right)$

$= \dfrac{\sqrt{xy}}{y}$ (Theorem 3.5-1)
(Definition 1.6-1: $\sqrt{a}\sqrt{a} = a$)

3.5 Changing the Form of Radical Expressions

Example 4 Simplify $\sqrt[4]{36x^2}$ $(x > 0)$.

Solution

$$\begin{aligned}
\sqrt[4]{36x^2} &= (36x^2)^{1/4} & & (a^{1/n} = \sqrt[n]{a}) \\
&= (6^2 x^2)^{1/4} \\
&= 6^{2/4} x^{2/4} & & \text{(Theorem 3.04: } (ab)^n = a^n b^n \text{)} \\
&= 6^{1/2} x^{1/2} & & \text{(Theorem 1.4-12: } ac/bc = a/b \text{)} \\
&= \sqrt{6} \sqrt{x} \\
&= \sqrt{6x} & & \text{(Theorem 3.5-1)}
\end{aligned}$$

Rationalizing Denominators

The fractional expressions in Examples 2 and 3 above were simplified by using the fundamental principle of fractions to *rationalize* the denominators. To rationalize the denominator of a fraction, we multiply both the numerator and the denominator by an expression, called a **rationalizing factor**, whose product with the denominator is a rational expression. In this context, the phrase "rational expression" refers to a radical in which the radicand is a power whose exponent is the same natural number as (or a multiple of) the index of the radical. If the denominator is a single radical, the rationalizing factor is a radical with the same index.

Examples Find a rationalizing factor, and then write each expression equivalently so that the denominator is rationalized.

(a) $\dfrac{3}{\sqrt{x}}$ $(x > 0)$ (b) $\dfrac{\sqrt[3]{3}}{\sqrt[3]{4}}$

Solutions (a) A rationalizing factor is $\sqrt{x}$, because $\sqrt{x} \cdot \sqrt{x} = x$. Then,

$$\frac{3}{\sqrt{x}} \cdot \frac{\sqrt{x}}{\sqrt{x}} = \frac{3\sqrt{x}}{x}.$$

(b) A rationalizing factor is $\sqrt[3]{2}$, because $\sqrt[3]{4} = \sqrt[3]{2^2}$, so that $\sqrt[3]{4} \cdot \sqrt[3]{2} = \sqrt[3]{8} = \sqrt[3]{2^3}$.

Then,

$$\frac{\sqrt[3]{3}}{\sqrt[3]{4}} \cdot \frac{\sqrt[3]{2}}{\sqrt[3]{2}} = \frac{\sqrt[3]{6}}{\sqrt[3]{8}} = \frac{\sqrt[3]{6}}{2}.$$

Rationalizing Numerators

Sometimes it is necessary to rationalize the numerator of a fraction. The technique for this operation is essentially the same as that for rationalizing the denominator. Look for a rationalizing factor whose product with the numerator is a rational expression as described in the paragraph above.

Examples Find a rationalizing factor, and then write each expression equivalently so that the numerator is rationalized.

(a) $\dfrac{\sqrt{x^3}}{\sqrt{2}}$ (b) $\dfrac{\sqrt[4]{27}}{\sqrt[4]{2}}$

Solutions (a) A rationalizing factor is $\sqrt{x}$, because $\sqrt{x} \cdot \sqrt{x^3} = \sqrt{x^4} = x^2$. Then,

$$\dfrac{\sqrt{x^3}}{\sqrt{2}} \cdot \dfrac{\sqrt{x}}{\sqrt{x}} = \dfrac{\sqrt{x^4}}{\sqrt{2x}} = \dfrac{x^2}{\sqrt{2x}}.$$

(b) Since $\sqrt[4]{27} = \sqrt[4]{3^3}$, a rationalizing factor is $\sqrt[4]{3}$. Then,

$$\dfrac{\sqrt[4]{27}}{\sqrt[4]{2}} \cdot \dfrac{\sqrt[4]{3}}{\sqrt[4]{3}} = \dfrac{\sqrt[4]{81}}{\sqrt[4]{6}} = \dfrac{3}{\sqrt[4]{6}}.$$

It should be pointed out that when the numerator of such an expression is rationalized, the resulting expression is not generally in simplest radical form.

EXERCISE 3.5

ORAL

1. Define rationalizing factor.
2. Explain the conditions for a radical expression to be in simplest radical form.

WRITTEN

In Problems 1–16, write each radical in simplest radical form. All variables $\in R^+$.

Examples (a) $\sqrt{72}$ (b) $\sqrt{27x^3}$ (c) $\sqrt[3]{128x^4y^2}$

Solutions (a) $\sqrt{72} = \sqrt{36}\sqrt{2}$
$= 6\sqrt{2}$

(b) $\sqrt{27x^3} = \sqrt{9x^2}\sqrt{3x}$
$= 3x\sqrt{3x}$

(c) $\sqrt[3]{128x^4y^2} = \sqrt[3]{64x^3}\sqrt[3]{2xy^2}$
$= 4x\sqrt[3]{2xy^2}$

1. $\sqrt{8}$
2. $\sqrt{18}$
3. $\sqrt[3]{x^4}$
4. $\sqrt[4]{x^5}$
5. $\sqrt{98}$
6. $\sqrt{75}$
7. $\sqrt[3]{24}$
8. $\sqrt[3]{625}$
9. $\sqrt{4x^5}$
10. $\sqrt{18x^3y^2}$
11. $\sqrt[3]{32x^5}$
12. $\sqrt[5]{64xy^6}$
13. $\sqrt[3]{3x^5y^5}$
14. $\sqrt[5]{x^{11}y}$
15. $\sqrt[3]{-32y^4}$
16. $\sqrt[3]{-8x^6}$

In Problems 17–28, find a rationalizing factor expression such that the product of the factor and the radical is rational. All variables $\in R^+$.

Examples (a) $\sqrt{32}$ (b) $\sqrt{8x}$ (c) $\sqrt[3]{25x^2}$

3.5 Changing the Form of Radical Expressions   81

Solutions (a) $\sqrt{2}$, because $\sqrt{2}\sqrt{32} = \sqrt{64} = 8$.
(b) $\sqrt{2x}$, because $\sqrt{2x}\sqrt{8x} = \sqrt{16x^2} = 4x$.
(c) $\sqrt[3]{5x}$, because $\sqrt[3]{5x}\sqrt[3]{25x^2} = \sqrt[3]{125x^3} = 5x$.

17. $\sqrt{27}$ 18. $\sqrt{75}$ 19. $\sqrt{x^5 y}$ 20. $\sqrt{xy^2}$

21. $\sqrt[3]{9}$ 22. $\sqrt[3]{64}$ 23. $\sqrt[3]{4x^2 y}$ 24. $\sqrt[3]{16x^3 y^2}$

25. $\dfrac{1}{\sqrt{5}}$ 26. $\dfrac{3}{2\sqrt{3}}$ 27. $\dfrac{5}{\sqrt[3]{16}}$ 28. $\dfrac{7}{3\sqrt[3]{100}}$

In Problems 29–48, write each expression in simplest radical form. All variables $\in R^+$.

Examples (a) $\dfrac{2}{\sqrt{5}}$ (b) $\dfrac{\sqrt{3}}{\sqrt{7}}$ (c) $\sqrt[3]{\dfrac{5}{16}}$

Solutions (a) $\dfrac{2}{\sqrt{5}} \cdot \dfrac{\sqrt{5}}{\sqrt{5}} = \dfrac{2\sqrt{5}}{5}$

(b) $\dfrac{\sqrt{3}}{\sqrt{7}} \cdot \dfrac{\sqrt{7}}{\sqrt{7}} = \dfrac{\sqrt{21}}{7}$

(c) $\sqrt[3]{\dfrac{5}{16}} = \dfrac{\sqrt[3]{5}}{\sqrt[3]{16}} \cdot \dfrac{\sqrt[3]{4}}{\sqrt[3]{4}} = \dfrac{\sqrt[3]{20}}{\sqrt[3]{64}} = \dfrac{\sqrt[3]{20}}{4}$

29. $\dfrac{1}{\sqrt{2}}$ 30. $\dfrac{2}{\sqrt{y}}$ 31. $\dfrac{x}{\sqrt{x}}$ 32. $\dfrac{-x}{\sqrt{2y}}$

33. $\sqrt{\tfrac{7}{8}}$ 34. $\sqrt[3]{\tfrac{3}{5}}$ 35. $\dfrac{2}{\sqrt[3]{3x^2}}$ 36. $\dfrac{5}{\sqrt[3]{72y^2}}$

37. $\dfrac{2}{\sqrt[5]{16}}$ 38. $\sqrt[4]{\dfrac{8}{27}}$ 39. $\dfrac{1}{\sqrt{x+1}}$ 40. $\dfrac{x-2}{\sqrt{x-2}}$ $(x > 2)$

Examples (a) $\sqrt[4]{25x^2}$ (b) $\sqrt[9]{8y^6}$

Solutions (a) $\sqrt[4]{25x^2} = (25x^2)^{1/4}$ (b) $\sqrt[9]{8y^6} = (8y^6)^{1/9}$
$= (5^2 x^2)^{1/4}$ $= (2^3 y^6)^{1/9}$
$= 5^{2/4} x^{2/4}$ $= 2^{3/9} y^{6/9}$
$= 5^{1/2} x^{1/2}$ $= 2^{1/3} y^{2/3}$
$= (5x)^{1/2}$ $= (2y^2)^{1/3}$
$= \sqrt{5x}$ $= \sqrt[3]{2y^2}$

41. $\sqrt[4]{3^2}$ 42. $\sqrt[6]{2^2}$ 43. $\sqrt[4]{16x^2}$ 44. $\sqrt[6]{8y^3}$
45. $\sqrt[4]{36x^6}$ 46. $\sqrt[9]{64y^6}$ 47. $\sqrt[4]{25y^{10}}$ 48. $\sqrt[8]{16x^8 y^{12}}$

In Problems 49–58, write each expression equivalently by rationalizing the numerator.

Examples (a) $\dfrac{\sqrt{7}}{\sqrt{5}}$ (b) $\dfrac{\sqrt[3]{4}}{\sqrt[3]{2}}$ (c) $\dfrac{\sqrt[4]{8x^3}}{\sqrt[3]{y}}$

82 EXPONENTS, ROOTS, AND RADICALS

Solutions (a) A rationalizing factor is $\sqrt{7}$. Then,

$$\frac{\sqrt{7}}{\sqrt{5}} = \frac{\sqrt{7}}{\sqrt{5}} \cdot \frac{\sqrt{7}}{\sqrt{7}} = \frac{7}{\sqrt{35}}.$$

(b) A rationalizing factor is $\sqrt[3]{2}$. Then,

$$\frac{\sqrt[3]{4}}{\sqrt[3]{2}} = \frac{\sqrt[3]{4}}{\sqrt[3]{2}} \cdot \frac{\sqrt[3]{2}}{\sqrt[3]{2}}$$

$$= \frac{\sqrt[3]{8}}{\sqrt[3]{4}}$$

$$= \frac{2}{\sqrt[3]{4}}.$$

(c) A rationalizing factor is $\sqrt[4]{2x}$. Then,

$$\frac{\sqrt[4]{8x^3}}{\sqrt[4]{y}} = \frac{\sqrt[4]{8x^3}}{\sqrt[4]{y}} \cdot \frac{\sqrt[4]{2x}}{\sqrt[4]{2x}}$$

$$= \frac{\sqrt[4]{16x^4}}{\sqrt[4]{2xy}}$$

$$= \frac{2x}{\sqrt[4]{2xy}}.$$

49. $\dfrac{\sqrt{3}}{\sqrt{11}}$ 50. $\dfrac{\sqrt{5}}{\sqrt{6}}$ 51. $\dfrac{\sqrt{3x}}{\sqrt{2y}}$ 52. $\dfrac{\sqrt{6x}}{\sqrt{7y}}$ 53. $\dfrac{\sqrt[3]{9}}{\sqrt[3]{4}}$

54. $\dfrac{\sqrt[3]{16}}{\sqrt[3]{5}}$ 55. $\dfrac{\sqrt[3]{4x^2}}{\sqrt[3]{3y}}$ 56. $\dfrac{\sqrt[3]{25y^2}}{\sqrt[3]{2x}}$ 57. $\dfrac{\sqrt[3]{27x^3}}{\sqrt[3]{5y}}$ 58. $\dfrac{\sqrt[3]{125y^2}}{\sqrt[3]{4x^2}}$

3.6 Products and Quotients Involving Radicals

Products involving radicals can be simplified by applying Theorem 3.5-1 ($\sqrt{a} \cdot \sqrt{b} = \sqrt{ab}$) and the distributive law, *provided the radicals involved have the same index.*

Examples Write the products in simplest radical form.

(a) $\sqrt{3xy} \cdot \sqrt{6y}$ (b) $\sqrt[3]{4}(\sqrt[3]{6} + \sqrt[3]{10})$ (c) $(2 + \sqrt{3})(\sqrt{2} - 3)$

Solutions (a) $\sqrt{3xy}\sqrt{6y} = \sqrt{18xy^2}$

$$= \sqrt{9y^2}\sqrt{2x}$$

$$= 3y\sqrt{2x}$$

(b) $\sqrt[3]{4}(\sqrt[3]{6} + \sqrt[3]{10}) = \sqrt[3]{4}\sqrt[3]{6} + \sqrt[3]{4}\sqrt[3]{10}$

$$= \sqrt[3]{24} + \sqrt[3]{40}$$

$$= \sqrt[3]{8}\sqrt[3]{3} + \sqrt[3]{8}\sqrt[3]{5}$$

$$= 2\sqrt[3]{3} + 2\sqrt[3]{5}$$

(c) $(2 + \sqrt{3})(\sqrt{2} - 3) = (2 + \sqrt{3})\sqrt{2} + (2 + \sqrt{3})(-3)$
$= 2\sqrt{2} + \sqrt{6} - 6 - 3\sqrt{3}$

Quotients

An expression of the form $\sqrt[n]{x} / \sqrt[n]{y}$ can be written directly as a single radical if y is an exact divisor of x. Otherwise, the technique of rationalizing the denominator (Section 3.5) must be used.

Examples Write the quotients in simplest radical form.

(a) $\dfrac{\sqrt{33}}{\sqrt{3}}$ (b) $\dfrac{\sqrt{35xy^2}}{\sqrt{7xy}}$ (c) $\dfrac{\sqrt[3]{5x^2}}{\sqrt[3]{2}}$

Solutions (a) $\dfrac{\sqrt{33}}{\sqrt{3}} = \sqrt{\dfrac{33}{3}}$
$= \sqrt{11}$

(b) $\dfrac{\sqrt{35xy^2}}{\sqrt{7xy}} = \sqrt{\dfrac{35xy^2}{7xy}}$
$= \sqrt{5y}$

(c) $\dfrac{\sqrt[3]{5x^2}}{\sqrt[3]{2}} = \dfrac{\sqrt[3]{5x^2} \cdot \sqrt[3]{4}}{\sqrt[3]{2} \cdot \sqrt[3]{4}}$
$= \dfrac{\sqrt[3]{20x^2}}{2}$

Binomial Expressions

Recall that
$$(a + b)(a - b) = a^2 - b^2.$$

Similarly, the product
$$(\sqrt{a} + \sqrt{b})(\sqrt{a} - \sqrt{b}) = (\sqrt{a})^2 - (\sqrt{b})^2$$
$$= a - b.$$

Thus, if a and b are both non-negative rational numbers, $a - b$ is a rational number, and the product $(\sqrt{a} + \sqrt{b})(\sqrt{a} - \sqrt{b})$ is a rational number. Two numbers such as these factors are called **conjugates** of each other.

If the denominator of a fraction contains two terms, one (or both) of which is the square root of a non-negative rational number, the rationalizing factor will be the conjugate of the denominator.

Examples Express each fraction in simplest radical form.

(a) $\dfrac{2 + \sqrt{3}}{3 + \sqrt{3}}$ (b) $\dfrac{x + \sqrt{y}}{2x - \sqrt{y}}$

Solutions (a) The conjugate of $3 + \sqrt{3}$ is $3 - \sqrt{3}$. Then,

$$\frac{2 + \sqrt{3}}{3 + \sqrt{3}} = \frac{2 + \sqrt{3}}{3 + \sqrt{3}} \cdot \frac{3 - \sqrt{3}}{3 - \sqrt{3}}$$

$$= \frac{(2 + \sqrt{3})(3 - \sqrt{3})}{9 - 3}$$

$$= \frac{6 + \sqrt{3} - 3}{6}$$

$$= \frac{3 + \sqrt{3}}{6}.$$

NOTE: Even though we have not discussed sums of radicals, you should notice that the distributive law can be applied to $-2\sqrt{3} + 3\sqrt{3}$ to obtain $(-2 + 3)\sqrt{3} = \sqrt{3}$.

(b) The conjugate of $2x - \sqrt{y}$ is $2x + \sqrt{y}$. Then,

$$\frac{x + \sqrt{y}}{2x - \sqrt{y}} = \frac{x + \sqrt{y}}{2x - \sqrt{y}} \cdot \frac{2x + \sqrt{y}}{2x + \sqrt{y}}$$

$$= \frac{(x + \sqrt{y})(2x + \sqrt{y})}{4x^2 - y}$$

$$= \frac{2x^2 + 3x\sqrt{y} + y}{4x^2 - y}.$$

You can use this same technique to rationalize the numerators of certain fractions.

Examples Rationalize the numerator of

(a) $\dfrac{\sqrt{3} + \sqrt{2}}{5}$ (b) $\dfrac{\sqrt{x} - \sqrt{y}}{xy}$

Solutions (a) The conjugate of $\sqrt{3} + \sqrt{2}$ is $\sqrt{3} - \sqrt{2}$. Then,

$$\frac{\sqrt{3} + \sqrt{2}}{5} = \frac{\sqrt{3} + \sqrt{2}}{5} \cdot \frac{\sqrt{3} - \sqrt{2}}{\sqrt{3} - \sqrt{2}}$$

$$= \frac{3 - 2}{5(\sqrt{3} - \sqrt{2})}$$

$$= \frac{1}{5(\sqrt{3} - \sqrt{2})}.$$

(b) The conjugate of $\sqrt{x} - \sqrt{y}$ is $\sqrt{x} + \sqrt{y}$. Then,

$$\frac{\sqrt{x} - \sqrt{y}}{xy} = \frac{\sqrt{x} - \sqrt{y}}{xy} \cdot \frac{\sqrt{x} + \sqrt{y}}{\sqrt{x} + \sqrt{y}}$$

$$= \frac{x - y}{xy(\sqrt{x} + \sqrt{y})}.$$

Radicals with Different Indices

A product or a quotient of two radical expressions that have different indices can be written equivalently as an expression in simplest radical form by application of Definition 3.2-1 (p. 67).

Example 1 Simplify $\sqrt{3}\sqrt[3]{2}$.

Solution By Definition 3.2-1,

$$\begin{aligned}
\sqrt{3}\sqrt[3]{2} &= 3^{1/2} \cdot 2^{1/3} \\
&= 3^{3/6} \cdot 2^{2/6} \\
&= (3^3)^{1/6} \cdot (2^2)^{1/6} \\
&= 27^{1/6} \cdot 4^{1/6} \\
&= 108^{1/6} \\
&= \sqrt[6]{108}.
\end{aligned}$$

Example 2 Write $\dfrac{\sqrt{3}}{\sqrt[3]{2}}$ in simplest radical form.

Solution By Definition 3.2-1,

$$\begin{aligned}
\frac{\sqrt{3}}{\sqrt[3]{2}} &= \frac{3^{1/2}}{2^{1/3}} \\
&= \frac{3^{3/6}}{2^{2/6}} \\
&= \frac{\sqrt[6]{27}}{\sqrt[6]{4}} \\
&= \frac{\sqrt[6]{27}}{\sqrt[6]{4}} \cdot \frac{\sqrt[6]{16}}{\sqrt[6]{16}} \\
&= \frac{\sqrt[6]{432}}{2}.
\end{aligned}$$

EXERCISE 3.6

WRITTEN

In this exercise all variables represent positive real numbers and no denominator is zero.

In Problems 1–16, write each expression as a single radical.

1. $\sqrt{3}\sqrt{9}$
2. $\sqrt{5}\sqrt{15}$
3. $\sqrt{2x}\sqrt{3y}$
4. $\sqrt{5y}\sqrt{8y}$
5. $\sqrt[3]{2}\sqrt[3]{3x}$
6. $\sqrt[3]{3y^2}\sqrt[6]{8y}$
7. $\sqrt[6]{2x^2y}\sqrt[6]{4x^3y^4}$
8. $\sqrt[4]{3p^2q}\sqrt[4]{12p^0q}$

Examples (a) $\dfrac{\sqrt{35}}{\sqrt{5}}$ (b) $\dfrac{\sqrt[5]{27x^3y}}{\sqrt[5]{3x^2y}}$

Solutions (a) $\sqrt{7}$ (b) $\sqrt[5]{9x}$

86 EXPONENTS, ROOTS, AND RADICALS

9. $\dfrac{\sqrt{18}}{\sqrt{3}}$
10. $\dfrac{\sqrt{64}}{\sqrt{2}}$
11. $\dfrac{\sqrt{12x^3}}{\sqrt{6x^2}}$
12. $\dfrac{\sqrt{27x^2y}}{\sqrt{9xy}}$

13. $\dfrac{\sqrt[3]{108x}}{\sqrt[3]{3}}$
14. $\dfrac{\sqrt[3]{27x^3}}{\sqrt[4]{9x^2}}$
15. $\dfrac{\sqrt[10]{108x^4y^3}}{\sqrt[10]{4xy^3}}$
16. $\dfrac{\sqrt[5]{625x^4y^6}}{\sqrt[5]{25x^4y^3}}$

In Problems 17–24, write the products in simplest radical form.

Examples (a) $\sqrt{3x}(\sqrt{2y} - \sqrt{10xy})$ (b) $(\sqrt[3]{3x} + \sqrt[3]{4})(\sqrt[3]{2x} - \sqrt[3]{5})$

Solutions (a) $\sqrt{6xy} - \sqrt{30x^2y} = \sqrt{6xy} - x\sqrt{30y}$

$$\text{(b) } \sqrt[3]{3x}(\sqrt[3]{2x} - \sqrt[3]{5}) + \sqrt[3]{4}(\sqrt[3]{2x} - \sqrt[3]{5})$$
$$= \sqrt[3]{6x^2} - \sqrt[3]{15x} + 2\sqrt[3]{x} - \sqrt[3]{20}$$

17. $\sqrt{5x}(\sqrt{6y} + \sqrt{13z})$
18. $\sqrt{14}(\sqrt{3x} - \sqrt{5y})$
19. $\sqrt[3]{2}(\sqrt[3]{7y} - \sqrt[3]{9z^2})$
20. $\sqrt[5]{15x}(\sqrt[5]{5xy} + \sqrt[5]{4x^2y^2})$
21. $(\sqrt{3} - \sqrt{5})(\sqrt{2} + \sqrt{7})$
22. $(\sqrt{2x} + \sqrt{3y})(\sqrt{5z} - \sqrt{7w})$
23. $(3\sqrt{5} - 4\sqrt{2})(\sqrt{3} + 2\sqrt{7})$
24. $(a + \sqrt{a})(b + \sqrt{b})$

In Problems 25–32, write each expression in simplest radical form.

Examples (a) $\dfrac{5}{\sqrt{2} - \sqrt{3}}$ (b) $\dfrac{\sqrt{5} + \sqrt{3}}{\sqrt{5} - \sqrt{3}}$

Solutions (a) $\dfrac{5}{\sqrt{2} - \sqrt{3}} = \dfrac{5}{\sqrt{2} - \sqrt{3}} \cdot \dfrac{\sqrt{2} + \sqrt{3}}{\sqrt{2} + \sqrt{3}}$

$$= \dfrac{5(\sqrt{2} + \sqrt{3})}{2 - 3}$$
$$= -5(\sqrt{2} + \sqrt{3})$$

(b) $\dfrac{\sqrt{5} + \sqrt{3}}{\sqrt{5} - \sqrt{3}} = \dfrac{\sqrt{5} + \sqrt{3}}{\sqrt{5} - \sqrt{3}} \cdot \dfrac{\sqrt{5} + \sqrt{3}}{\sqrt{5} + \sqrt{3}}$

$$= \dfrac{5 + 2\sqrt{15} + 3}{5 - 3}$$
$$= \dfrac{8 + 2\sqrt{15}}{2}$$
$$= 4 + \sqrt{15}$$

25. $\dfrac{3}{\sqrt{2} - 1}$
26. $\dfrac{-4}{1 + \sqrt{3}}$
27. $\dfrac{1}{2 - \sqrt{2}}$
28. $\dfrac{2}{\sqrt{7} - \sqrt{2}}$

29. $\dfrac{x}{\sqrt{x} - 3}$
30. $\dfrac{\sqrt{y}}{\sqrt{3} - \sqrt{y}}$
31. $\dfrac{\sqrt{x} + \sqrt{y}}{\sqrt{x} - \sqrt{y}}$
32. $\dfrac{\sqrt{x + y}}{1 + \sqrt{x + y}}$

In Problems 33–40, rationalize the numerator of each fractional expression.

Example $\dfrac{\sqrt{3} + \sqrt{x}}{\sqrt{3} - \sqrt{x}}$

Solution A rationalizing factor is $\sqrt{3} - \sqrt{x}$. Then,

$$\dfrac{\sqrt{3} + \sqrt{x}}{\sqrt{3} - \sqrt{x}} = \dfrac{\sqrt{3} + \sqrt{x}}{\sqrt{3} - \sqrt{x}} \cdot \dfrac{\sqrt{3} - \sqrt{x}}{\sqrt{3} - \sqrt{x}}$$

$$= \dfrac{3 - x}{3 - 2\sqrt{3x} + x}.$$

33. $\dfrac{\sqrt{2} + \sqrt{3}}{\sqrt{2} - \sqrt{3}}$ 34. $\dfrac{6 + \sqrt{2}}{2 + \sqrt{3}}$ 35. $\dfrac{\sqrt{x} + y}{x + \sqrt{y}}$ 36. $\dfrac{2x - \sqrt{y}}{x - \sqrt{y}}$

37. $\dfrac{\sqrt{2x} - \sqrt{3y}}{\sqrt{3x} + \sqrt{2y}}$ 38. $\dfrac{\sqrt{xy} - 1}{3 + \sqrt{xy}}$ 39. $\dfrac{x^2 + \sqrt{2xy}}{x - \sqrt{xy}}$ 40. $\dfrac{3 + \sqrt{x + y}}{x + y}$

In Problems 41–56, write each expression equivalently as an expression in simplest radical form.

Examples (a) $\sqrt{2}\sqrt[3]{3}$ (b) $\sqrt{x}\sqrt[4]{y}$

Solutions (a) $\sqrt{2}\sqrt[3]{3} = 2^{1/2}\,3^{1/3}$ (b) $\sqrt{x}\sqrt[4]{y} = x^{1/2}y^{1/4}$
$= 2^{3/6}\,3^{2/6}$ $= x^{2/4}y^{1/4}$
$= 8^{1/6}\,9^{1/6}$ $= (x^2 y)^{1/4}$
$= (8 \cdot 9)^{1/6}$ $= \sqrt[4]{x^2 y}$
$= \sqrt[6]{72}$

41. $\sqrt[3]{2}\sqrt{2}$ 42. $\sqrt[3]{3}\sqrt{2}$ 43. $\sqrt[4]{5}\sqrt{2}$ 44. $\sqrt[4]{5}\sqrt{3}$
45. $\sqrt[3]{x}\sqrt{x}$ 46. $\sqrt[5]{y}\sqrt{y}$ 47. $\sqrt{x}\sqrt[3]{x}\sqrt[4]{x}$ 48. $\sqrt[3]{x}\sqrt[4]{y}\sqrt{z}$

Examples (a) $\dfrac{\sqrt{2}}{\sqrt[3]{3}}$ (b) $\dfrac{\sqrt[3]{x}}{\sqrt{x}}$

Solutions (a) $\dfrac{\sqrt{2}}{\sqrt[3]{3}} = \dfrac{2^{1/2}}{3^{1/3}}$ (b) $\dfrac{\sqrt[3]{x}}{\sqrt{x}} = \dfrac{x^{1/3}}{x^{1/2}}$
$= \dfrac{2^{3/6}}{3^{2/6}}$ $= \dfrac{x^{2/6}}{x^{3/6}}$
$= \dfrac{8^{1/6}}{3^{2/6}} \cdot \dfrac{3^{4/6}}{3^{4/6}}$ $= \dfrac{1}{x^{1/6}}$
$= \dfrac{8^{1/6} \cdot 81^{1/6}}{3}$ $= \dfrac{1}{x^{1/6}} \cdot \dfrac{x^{5/6}}{x^{5/6}}$
$= \dfrac{\sqrt[6]{648}}{3}$ $= \dfrac{\sqrt[6]{x^5}}{x}$

49. $\dfrac{\sqrt{2}}{\sqrt[4]{2}}$ 50. $\dfrac{\sqrt[4]{2}}{\sqrt{2}}$ 51. $\dfrac{\sqrt[4]{2}}{\sqrt[3]{2}}$ 52. $\dfrac{\sqrt{3}}{\sqrt[3]{3}}$

53. $\dfrac{\sqrt{x}}{\sqrt[3]{x}}$ 54. $\dfrac{\sqrt{x}}{\sqrt[3]{y}}$ 55. $\dfrac{\sqrt[4]{x}}{\sqrt{y}}$ 56. $\dfrac{\sqrt[3]{x^2}}{\sqrt{y}}$

3.7 Sums and Differences Involving Radicals

If one or more of the terms in a sum is an irrational number, the sum is sometimes an irrational number. Generally, the terms of a sum involving radicals cannot be combined into a single term. However, if all the terms of a sum contain the same radical as a factor, the distributive law can be applied to write the sum equivalently as an expression which may be in a simpler form.

Examples Write the sums in a simpler form if possible.

(a) $\sqrt[3]{54} + \sqrt[3]{16} - \sqrt[3]{128}$ (b) $\sqrt[3]{2} + \sqrt{3}$

Solutions (a) Theorem 3.5-1 (p. 77) can be used to write

$$\sqrt[3]{54} + \sqrt[3]{16} - \sqrt[3]{128} = \sqrt[3]{27} \cdot \sqrt[3]{2} + \sqrt[3]{8} \cdot \sqrt[3]{2} - \sqrt[3]{64} \cdot \sqrt[3]{2}$$
$$= 3\sqrt[3]{2} + 2\sqrt[3]{2} - 4\sqrt[3]{2}$$
$$= (3 + 2 - 4)\sqrt[3]{2}$$
$$= \sqrt[3]{2}.$$

(b) This sum cannot be simplified, because $\sqrt[3]{2}$ and $\sqrt{3}$ do not contain the same radical as a factor.

EXERCISE 3.7

WRITTEN

In Problems 1–20, write each sum or difference as a single term if possible. All variables $\in R^+$.

Examples (a) $2\sqrt{3} + \sqrt{27}$ (b) $\sqrt{x^3} + \sqrt{xy^2}$

Solutions (a) $2\sqrt{3} + \sqrt{27} = 2\sqrt{3} + \sqrt{9}\sqrt{3}$
$= 2\sqrt{3} + 3\sqrt{3}$
$= 5\sqrt{3}$

(b) $\sqrt{x^3} + \sqrt{xy^2} = \sqrt{x^2}\sqrt{x} + \sqrt{y^2}\sqrt{x}$
$= x\sqrt{x} + y\sqrt{x}$
$= (x + y)\sqrt{x}$

1. $\sqrt{75} - 2\sqrt{27}$
2. $\sqrt{50} + 2\sqrt{32} - \sqrt{2}$
3. $\sqrt{20} + \sqrt{45} - \sqrt{80}$
4. $\sqrt{12} - \sqrt{27} + \sqrt{48}$
5. $\sqrt{4xy} + \sqrt{9xy} - \sqrt{xy}$
6. $\sqrt{4y} - \sqrt{25y} + \sqrt{16y}$
7. $\sqrt{8z^3} - \sqrt{18z^3} + \sqrt{2z^3}$
8. $\sqrt{xy^2} + \sqrt{xz^2} - \sqrt{x^3}$
9. $\sqrt[3]{54} + 2\sqrt[3]{128}$
10. $5\sqrt[3]{2x} + 2\sqrt[3]{16x}$
11. $\sqrt[3]{250y} + 2\sqrt[3]{16y} - \sqrt[3]{2y}$
12. $\sqrt[3]{54x} - \sqrt[3]{128x} + \sqrt[3]{250x}$
13. $\sqrt{20} + \sqrt{180} - \sqrt{40}$
14. $\sqrt{12} + \sqrt{108} - \sqrt{18}$

Examples (a) $\dfrac{2}{3} + \dfrac{\sqrt{2}}{3}$ (b) $\dfrac{x}{\sqrt{2}} + \dfrac{y}{\sqrt{2}}$

Solutions (a) $\dfrac{2+\sqrt{2}}{3}$ (b) $\dfrac{x+y}{\sqrt{2}} = \dfrac{x+y}{\sqrt{2}} \cdot \dfrac{\sqrt{2}}{\sqrt{2}}$
$= \dfrac{(x+y)\sqrt{2}}{2}$

15. $\dfrac{5}{7} - \dfrac{\sqrt{3}}{7}$ 16. $\dfrac{1}{2} + \dfrac{\sqrt{2}}{6}$ 17. $\dfrac{2\sqrt{3}}{3} - \dfrac{\sqrt{2}}{2}$

18. $\dfrac{\sqrt{2}}{5} - \dfrac{\sqrt{3}}{3}$ 19. $\dfrac{2x}{\sqrt{3}} - \dfrac{y}{\sqrt{3}}$ 20. $\dfrac{x^2}{\sqrt{2}} + \dfrac{y^2}{\sqrt{3}}$

Summary

1. If $a \in R$ and $n \in N$, then

$$a^n = \overbrace{a \cdot a \cdot a \cdots a}^{n \text{ factors}}.$$ [3.0]

2. If $a \in R$ and $a \neq 0$, then $a^0 = 1$. [3.1]
3. If $a \in R$, $a \neq 0$, and $n \in N$, then $a^{-n} = \dfrac{1}{a^n}$. [3.1]
4. If $a \in R$ and $n \in N$, then $a^{1/n}$ names a number such that

$$\overbrace{a^{1/n} \cdot a^{1/n} \cdots a^{1/n}}^{n \text{ factors}} = a.$$

$a^{1/n}$ is called the **nth root** of a, and may also be written as $\sqrt[n]{a}$. [3.2]
5. If $a < 0$, then $a^{1/n}$ is a real number if and only if n is a positive odd integer. [3.2]
6. If $a > 0$, then $a^{1/n}$ has two real values (one positive and one negative) if n is a positive even integer, and one real value if n is a positive odd integer. [3.2]
7. If $a \in R$, $a > 0$, $m \in J$, and $n \in N$, then

$$a^{m/n} = (a^{1/n})^m = (a^m)^{1/n}.$$ [3.2]

8. An **algebraic expression** is a collection of symbols that represents a real number for permissible replacements of its variables, provided there are no indicated even roots of negative numbers. [3.3]
9. The parts of an expression formed by products and/or quotients are called **terms**. **Similar terms** have identical variable factors. [3.3]
10. Sums or differences of expressions can be simplified by applying the commutative and associative properties of addition and the distributive law to the collections of similar terms. [3.3]

11. The distributive law is helpful in simplifying some expressions involving products and sums. [3.4]
12. If $a, b \in R$, $a, b > 0$, and $n \in N$, then

$$\sqrt[n]{a} \cdot \sqrt[n]{b} = \sqrt[n]{ab} \quad \text{and} \quad \frac{\sqrt[n]{a}}{\sqrt[n]{b}} = \sqrt[n]{\frac{a}{b}}.$$ [3.5]

13. The theorem stated above is used to change radicals to simplest radical form. [3.5]
14. An expression whose product with another expression containing a radical is rational is called a **rationalizing factor**. [3.5]
15. Theorem 3.5-1 (p. 77) is used to simplify products and/or quotients of radicals that have the same index. [3.6]
16. Theorem 3.2-1 (p. 68) can be used to change radicals having different indices to radicals with the same index so that products and/or quotients of such radicals can be simplified. [3.6]
17. If all terms in a sum (difference) contain the same radical as a factor, it may be possible to simplify the sum (difference) by applying the distributive law. [3.7]

REVIEW EXERCISES

SECTION 3.1

In Problems 1–6, find an equivalent expression in which each variable is written only once and all exponents are positive integers.

1. $x^2 \cdot x^3$
2. $\dfrac{x^5}{x^2}$
3. $\dfrac{y^2}{y^5}$
4. $\dfrac{x^2 y^3}{xy^4}$
5. $(x^2 y^3)^2$
6. $\left(\dfrac{x^2}{y^3}\right)^2$

In Problems 7–12, find an equivalent expression free of fractions in which each variable appears only once.

7. $\dfrac{x^3}{y^{-2}}$
8. $\left(\dfrac{y^2}{y^3}\right)^{-2}$
9. $\left(\dfrac{x^{-1} y^2}{w^0 z^{-2}}\right)^{-1}$
10. $\left(\dfrac{x^{1-n}}{x^{2-n}}\right)^{-2}$
11. $\left(\dfrac{x^n y^{2n-1}}{y^n}\right)^2$
12. $\left(\dfrac{x^{2n} y^{n-1}}{x^{n-1} y^{-1}}\right)^3$

SECTION 3.2

In Problems 13–18, write each expression as a product or quotient of powers in which each variable appears only once and all exponents are positive.

13. $\left(\dfrac{x^{2/3}}{x^{-1/3}}\right)^{1/3}$
14. $(x^{2/3} x^{3/2})^{3/5}$
15. $\left(\dfrac{x^{2/3} y^{1/4}}{z^{1/6}}\right)^{-6}$
16. $(x^{2/3} y^{-4/5})^{-1/2}$
17. $(x^{3/4} x^{-3})^{-1/3}$
18. $\left(\dfrac{x^{1/2} y^{-1/3}}{z^{-1/6}}\right)^{-6}$

SECTION 3.3

Write each sum or difference in Problems 19–22 as a single expression containing no similar terms.

19. $(x^3 + 2x^2 - 3x - 4) + (2x^3 - x^2 + x + 2) + (3 - x + 2x^2 - x^3)$
20. $(x^{-2} + 3x^{-1} - 4) + (6x^{-2} - 5x^{-1} - 1) - (2x^{-2} - 6x^{-1} + 3)$
21. $(6x^{3/2} - 2x^{1/2} + 5) + (3x^{3/2} + 5x^{1/2} + 4)$
22. $(x^{1/2}y^{-2} + 5y^{-3/2} + 6y) - (7y + 3y^{-3/2} - 2x^{1/2}y^{-2})$

SECTION 3.4

In Problems 23–25, write each product as a sum.

23. $x^{1/3}(x^2 - x^{2/3})$
24. $(2x^{1/2} + x^{-1/2})(3x^{1/2} - 2x^{-1/2})$
25. $(x - 3)^{1/2}[(x - 3)^{1/2} + (x - 3)^{-1/2}]$

In Problems 26–28, find the second factor so that each sum is represented as a product.

26. $x + x^0 + x^{-1} = x^{-1}(?)$
27. $x^{3/4} + 2x + x^{1/2} = x^{1/2}(?)$
28. $(2x - 4)^{1/3} - (2x - 4)^{4/3} = (2x - 4)^{1/3}(?)$

SECTION 3.5

Write each expression in Problems 29–34 in simplest radical form.

29. $\sqrt{72x^3y^2}$
30. $\sqrt[3]{-16x^4y^{-3}}$
31. $\sqrt[4]{36y^6}$
32. $\dfrac{2}{\sqrt{5}}$
33. $\dfrac{1}{\sqrt[3]{2x^2}}$
34. $\dfrac{\sqrt[4]{8}}{\sqrt{3}}$

SECTION 3.6

Simplify each product and quotient in Problems 35–38.

35. $(2\sqrt{x} + 3\sqrt{y})(\sqrt{x} - 4\sqrt{y})$
36. $\dfrac{\sqrt{x} + 2}{\sqrt{x} - 2}$
37. $\sqrt{3}\sqrt[3]{4}$
38. $\dfrac{\sqrt{3}}{\sqrt[3]{4}}$

SECTION 3.7

Simplify each sum or difference in Problems 39–41.

39. $\sqrt{54x} - \sqrt{24x} + \sqrt{96x}$
40. $\sqrt[3]{16x^2} + \sqrt[3]{250x^2} - \sqrt[3]{128x^2}$
41. $\dfrac{5}{\sqrt{2}} + 4\sqrt{2} - \dfrac{3\sqrt{2}}{2}$

Polynomials

In Section 3.3 we reviewed the concepts of algebraic expressions and discussed the terms of such expressions. In this chapter we wish to discuss algebraic expressions of a special form called **polynomials**. We first define a polynomial in one variable.

DEFINITION 4.0-1 An expression of the form

$$a_0 x^n + a_1 x^{n-1} + a_2 x^{n-2} + \cdots + a_n$$

where the a's are all real numbers and n is a non-negative integer is called a **real polynomial** in x.

While x can be any kind of an element, in our discussion we are concerned only with real number replacements for x.

If the powers of x are arranged in decreasing order with respect to the exponents, we say that the polynomial is in **standard form**.

4.1 Some Characteristics of Polynomials

It is obvious from Definition 4.0-1 that a polynomial is a sum. Each addend is called a **term**. Some polynomials are given special names depending upon the number of terms. A polynomial of only one term is called a **monomial**, one of two terms a **binomial**, and one of three terms a **trinomial**.

The a factor of each term is called the **coefficient** of the term. The symbol a_i will be used when we wish to refer to the coefficients in a general way. In this situation the i is called a **subscript**, and the replacement set for i is $\{0, 1, 2, 3, \ldots, n\}$. Any, or all, a_i may be zero. If all a_i except a_n are zero, the polynomial is a **constant polynomial**, and if a_n is also zero, we have the **zero polynomial**.

The **degree** of a polynomial in x is determined by the nonzero term containing the highest power of x. For example, $2x^3 + 4x^2 - 7x - 3$ is a third-degree polynomial because the greatest exponent of x is 3. The degree of a constant polynomial is 0; no degree is assigned to the zero polynomial.

It is usually more convenient to arrange the terms of a polynomial in standard form. When the terms are arranged in this order, a_0, the coefficient of the term of highest degree, is called the **leading coefficient**. If the leading coefficient is 1, the polynomial is said to be **monic**.

A symbol such as $P(x)$ [read, "P of x"] is used in a general way to denote a polynomial in x. The symbol $P(a)$ is used to represent the polynomial $P(x)$ where x has been replaced by a. a represents any replacement for x and may be a real number or even another polynomial. If a is a real number, $P(a)$ also refers to the real number named by the polynomial when x has been replaced by a.

Example If $P(x) = 2x^2 - 3x + 7$, find $P(a)$, $P(2)$, and $P(x + 2)$.

Solution
$P(a) = 2a^2 - 3a + 7$

$P(2) = 2(2)^2 - 3(2) + 7$
$= 8 - 6 + 7$
$= 9$

$P(x + 2) = 2(x + 2)^2 - 3(x + 2) + 7$
$= 2(x^2 + 4x + 4) - 3x - 6 + 7$
$= 2x^2 + 8x + 8 - 3x - 6 + 7$
$= 2x^2 + 5x + 9$

In the last example, replacing x with $x + 2$ does not mean that we are letting x equal $x + 2$. Such a replacement indicates that we are concerned with what happens to the polynomial if *any* x is increased by 2. In this example, if we subtract the original polynomial, $P(x)$, from $P(x + 2)$, we see that for any x, $P(x + 2)$ exceeds $P(x)$ by $8x + 2$. Thus,

$P(10) - P(8) = 8(8) + 2 = 66$
$P(8) - P(6) = 8(6) + 2 = 50$
$P(6) - P(4) = 8(4) + 2 = 34$, etc.

You can easily verify these statements by evaluating $P(x) = 2x^2 - 3x + 7$ for $x = 10$, $x = 8$, $x = 6$, and $x = 4$.

Zero of a Polynomial

When x is replaced by some real number c in the polynomial $P(x)$, and $P(c) = 0$, then c is called a **zero of the polynomial**.

Example If $P(x) = x^2 - 5x + 6$, find $P(3)$.

Solution
$$P(3) = 3^2 - 5(3) + 6$$
$$= 9 - 15 + 6$$
$$= 0$$

Hence, 3 is a zero of $x^2 - 5x + 6$. Can you find another zero of this polynomial?

Degree of a Polynomial

So far in our discussion of polynomials, we have considered only those in one variable. While our primary concern is with such polynomials, polynomials containing more than one variable do exist. Let us now consider the degree of a polynomial of this type.

DEFINITION 4.1-1 The degree of a monomial in one or more variables is the sum of the exponents of the separate variables.

Examples The degree of $4x^3$ is 3.
The degree of $3x^2y^3$ is 5 because $2 + 3 = 5$.
The degree of $-4x^2yz^4$ is 7 because $2 + 1 + 4 = 7$.

DEFINITION 4.1-2 The degree of a polynomial in one or more variables is the same as that of the term (monomial) of highest degree.

Example The degree of $x^3y^2 + x^2z^2 + y^4$ is 5.

EXERCISE 4.1

ORAL

Define:
1. Monomial
2. Binomial
3. Trinomial
4. Monic polynomial
5. Constant polynomial
6. Zero polynomial
7. $P(x)$
8. $P(a)$
9. Zero of a polynomial
10. Degree of a polynomial

State the degree of each monomial.
11. $3x^2y$
12. $4x^3y^2$
13. $-2x^2y^3$
14. $-5xy^3$
15. $3xy^2z^3$
16. $2x^3y^2z$
17. $-4x^2y^2z^4$
18. $-8x^3yz^3$

19. Which of these expressions are polynomials and which are not? Explain your answer.
$$x^2 + 1, \quad x + x^{-1}, \quad x^3 - x^2 + 7, \quad x^{2/3} + x^{1/3} + 1$$

20. Which of these expressions are polynomials and which are not? Explain your answer.
$$x - \frac{1}{x}, \quad x^3, \quad 5 - \sqrt{x}, \quad 7$$

WRITTEN

In Problems 1–8, find the value of the polynomial for the given values of x.

Example $P(x) = x^2 - 3x + 1$; $P(2)$

Solution $P(2) = (2)^2 - 3(2) + 1$
$= 4 - 6 + 1$
$= -1$

1. $P(x) = x^2 + 2x + 1$; $P(-2)$, $P(-1)$, $P(0)$, $P(3)$
2. $P(x) = x^2 - 2x + 1$; $P(-3)$, $P(0)$, $P(1)$, $P(2)$
3. $P(x) = x^2 - 2x - 3$; $P(-2)$, $P(-1)$, $P(1)$, $P(3)$
4. $P(x) = 2x^2 - x - 3$; $P(-2)$, $P(-1)$, $P(0)$, $P(\frac{3}{2})$
5. $P(x) = 6x^2 - x - 1$; $P(-\frac{1}{3})$, $P(0)$, $P(\frac{1}{2})$
6. $P(x) = 8x^2 + 2x - 1$; $P(-\frac{1}{2})$, $P(-2)$, $P(\frac{1}{4})$
7. $P(x) = x^3 - 3x^2 - 4x + 12$; $P(-2)$, $P(-1)$, $P(3)$
8. $P(x) = x^3 - x^2 - 5x + 5$; $P(\sqrt{5})$, $P(0)$, $P(1)$

In Problems 9–12, find the indicated polynomial.

9. $P(x) = x^2 + x + 1$; $P(a + 1)$
10. $P(x) = x^2 + 2x + 1$; $P(a - 1)$
11. $P(x) = x^2 - 3x + 2$; $P(x + h)$
12. $P(x) = x^2 - 3x + 2$; $P(x - h)$

13. Use the result of Problem 11 to find $\dfrac{P(x + h) - P(x)}{h}$.

14. Use the result of Problem 12 to find $\dfrac{P(x - h) - P(x)}{h}$.

4.2 Binary Operations with Real Polynomials

A real polynomial names a real number for every real number replacement of its variable. Therefore, we can apply any of the properties of real numbers to operations with real polynomials.

Sums and Differences

The techniques discussed in Section 3.3 for simplifying sums and differences of algebraic expressions are applicable to polynomials, so no further discussion should be necessary here. However, Exercise 4.2 does include some problems of this type which you should work to increase your manipulative skills.

Products

Recall from Section 3.3 that repeated application of the distributive law is necessary if the product of two expressions is to be written as a sum of terms. Then if c_i and d_i represent the corresponding terms of two polynomials, we have

$$(c_1 + c_2 + c_3 + \cdots + c_n)(d_1 + d_2 + d_3 + \cdots + d_m)$$
$$= c_1 d_1 + c_1 d_2 + \cdots + c_1 d_m + c_2 d_1 + c_2 d_2 + \cdots + c_2 d_m$$
$$+ c_3 d_1 + c_3 d_2 + \cdots + c_3 d_m + \cdots + c_n d_1 + c_n d_2 + \cdots + c_n d_m,$$

which denotes the sum of the products of each term of the first polynomial with every term of the second polynomial. This sum can be simplified by combining any two or more similar terms (see p. 70). We specify the simplest form of a polynomial to be an equivalent polynomial with no similar terms.

Certain products involving polynomials occur with sufficient frequency that special mention should be made of them. If you are not already familiar with any of these, you should verify them by performing the necessary multiplications.

$(a + b)^2 = a^2 + 2ab + b^2$
$(a - b)^2 = a^2 - 2ab + b^2$ (Square of a binomial)

$(a + b)(a - b) = a^2 - b^2$ (Product of sum and difference of two numbers)

$(a + b + c)^2 = a^2 + b^2 + c^2 + 2ab + 2ac + 2bc$ (Square of a trinomial)

$(a + b)^3 = a^3 + 3a^2b + 3ab^2 + b^3$
$(a - b)^3 = a^3 - 3a^2b + 3ab^2 - b^3$ (Cube of a binomial)

EXERCISE 4.2

WRITTEN

Write each sum, difference, or product as a polynomial in simplest form. Do as much of the work mentally as you can.

Examples (a) $(3x^2 - 2x + 4) + (5x^2 + 7x - 13)$
(b) $(7x^3 + 5x - 7) - (2x^2 - 2x + 3)$

Solutions (a) $(3x^2 - 2x + 4) + (5x^2 + 7x - 13)$
$= 3x^2 - 2x + 4 + 5x^2 + 7x - 13$
$= 3x^2 + 5x^2 - 2x + 7x + 4 - 13$
$= 8x^2 + 5x - 9$

(b) $(7x^3 + 5x - 7) - (2x^2 - 2x + 3)$
$= (7x^3 + 5x - 7) + (-2x^2 + 2x - 3)$
$= 7x^3 - 2x^2 + 5x + 2x - 7 - 3$
$= 7x^3 - 2x^2 + 7x - 10$

1. $(5x^2 + 3x + 2) + (8x^2 - 7x + 7)$
2. $(4x^2 - 6x + 1) + (3x^2 + 2x - 8)$
3. $(6x^2 + 2x - 7) + (7 - 2x - 6x^2)$
4. $(4x^2 + 3x - 5) + (4x^2 - 3x + 5)$
5. $(2x^2 + 3x + 8) - (x^2 + 4x - 2)$
6. $(6x^3 + 5x - 1) - (2x^3 - 3x + 4)$
7. $(4x^5 - 3x^3 + x) + (x^4 - 2x^2 + 5)$
8. $(3x^5 + 4x^3 - 7x) - (2x^4 - 3x^2 + 8)$
9. $(7x^2 + 3xy - 5y^2) + (2x^2 - 4xy + 3y^2)$
10. $(8x^3 - 6x^2y + 4xy^2 - 3y^3) + (x^3 + 7x^2y - 3xy^2 + 5y^3)$
11. $(4x^2 + 2x^2y - 3xy^2 + xy) + (2x^2y - 3xy + 2x + 5y)$
12. $(2x^2 + 2y^2 - 3xy + 7x) - (x^2y + xy^2 + 2x - 3y)$

Example

$(2x^2 - 3x + 1)(3x^2 + 4x - 3)$
$= 2x^2(3x^2) + 2x^2(4x) + 2x^2(-3) - 3x(3x^2) - 3x(4x) - 3x(-3)$
$\qquad\qquad\qquad\qquad\qquad\qquad\qquad + 1(3x^2) + 1(4x) + 1(-3)$
$= 6x^4 + 8x^3 - 6x^2 - 9x^3 - 12x^2 + 9x + 3x^2 + 4x - 3$
$= 6x^4 + 8x^3 - 9x^3 - 6x^2 - 12x^2 + 3x^2 + 9x + 4x - 3$
$= 6x^4 - x^3 - 15x^2 + 13x - 3$

13. $(x^2 + x + 1)(x^2 - x - 1)$
14. $(2x^2 + 3x - 2)(x^2 - 2x + 3)$
15. $(x^3 - 2x^2)(3x^2 + 5x - 4)$
16. $(2x^4 + 3x)(5x^3 + 4x + 1)$
17. $(3x^3 + 6x^2 - 5x + 4)(2x^3 - 3x^2 + 2x + 1)$
18. $(4x^3 - 2x^2 - 4x + 3)(3x^3 + 4x^2 - 3x + 2)$
19. $(2x^4 - 3x^2 + 6)(3x^3 + 4x + 1)$
20. $(x^5 + 2x^3 + 3x)(x^4 - x^2 + 2)$
21. $(x + y + z)(x + y - z)$
22. $(2x - 3y + 4z)(2x - 3y - 4z)$

Examples (a) $(2x + 3y - 2z)^2$ (b) $(3x - 2y)^3$

Solutions (a) $(2x + 3y - 2z)^2$
$= (2x)^2 + (3y)^2 + (-2z)^2 + 2(2x)(3y) + 2(2x)(-2z) + 2(3y)(-2z)$
$= 4x^2 + 9y^2 + 4z^2 + 12xy - 8xz - 12yz$

(b) $(3x - 2y)^3$
$= (3x)^3 - 3(3x)^2(2y) + 3(3x)(2y)^2 - (2y)^3$
$= 27x^3 - 54x^2y + 36xy^2 - 8y^3$

23. $(2x - 4)^2$
24. $(3x + 2y)^2$
25. $(2x^2 + 5y)(2x^2 - 5y)$
26. $(x^3 - 9y^2)(x^3 + 9y^2)$
27. $(2x - 4y + 3z)^2$
28. $(3x^2 - 2x + 5)^2$
29. $(x^2 - 2xy + y^2)^2$
30. $(4x^2 + 4xy + y^2)^2$
31. $(x + 3y)^3$
32. $(2x + 5y)^3$
33. $(x^2 + y^2)^3$
34. $(x^2 + xy)^3$

4.3 Quotients—Synthetic Division

If $P(x)$ and $Q(x)$ are two polynomials arranged in decreasing powers of x, the quotient $P(x)/Q(x)$ can be expressed in an equivalent form by applying the following theorem.

THEOREM 4.3-1 If $P(x)$ and $Q(x)$ are polynomials, $Q(x) \neq 0$, and the degree of $Q(x)$ is less than or equal to the degree of $P(x)$, then

$$\frac{P(x)}{Q(x)} = S(x) + \frac{T(x)}{Q(x)}$$

where $S(x)$ and $T(x)$ are unique polynomials such that $S(x) \cdot Q(x)$ is a polynomial of the same degree as $P(x)$, and $T(x)$ is either 0 or a polynomial whose degree is less than that of $Q(x)$.

The proof of this theorem is beyond the scope of this book.

Division Algorithm

Division is accomplished by selecting a suitable monomial $S_1(x)$ such that the product of $S_1(x)$ and the term of highest degree of the divisor is equal to the term of highest degree of the dividend. This gives a remainder

$$T_1(x) = P(x) - [Q(x) \cdot S_1(x)]$$

that is of lesser degree than $P(x)$. $T_1(x)$ is then used as a new dividend and a second monomial $S_2(x)$ is selected as above. The new remainder

$$T_2(x) = T_1(x) - [Q(x) \cdot S_2(x)]$$

is of lesser degree than $T_1(x)$. This process is continued until a remainder $T_n(x)$ is obtained such that $T_n(x)$ is of lesser degree than $Q(x)$, or $T_n(x) = 0$.

The complete technique, called the **division algorithm**, is illustrated in the following example.

Example $\quad (2x^3 + 2x + 3) \div (2x^2 - 3x + 1)$

Solution $\quad S_1(x) = x$ since $(2x^2)(x) = 2x^3$.

$$\begin{array}{r} x \\ 2x^2 - 3x + 1 \overline{\smash{)}2x^3 + 2x + 3} \\ Q(x) \cdot S_1(x) = 2x^3 - 3x^2 + x \\ T_1(x) = 3x^2 + x + 3 \end{array}$$

With $T_1(x)$ as a new dividend, $S_2(x) = \frac{3}{2}$ since $(2x^2)(\frac{3}{2}) = 3x^2$

$$\begin{array}{r} x + \tfrac{3}{2} \\ 2x^2 - 3x + 1 \overline{\smash{)}2x^3 + 2x + 3} \\ 2x^3 - 3x^2 + x \\ \hline 3x^2 + x + 3 \\ Q(x) \cdot S_2(x) = 3x^2 - \tfrac{9}{2}x + \tfrac{3}{2} \\ \hline T_2(x) = \tfrac{11}{2}x + \tfrac{3}{2} \end{array}$$

Then $\dfrac{P(x)}{Q(x)} = x + \tfrac{3}{2} + \dfrac{\tfrac{11}{2}x + \tfrac{3}{2}}{2x^2 - 3x + 1}$.

Synthetic Division

The process of dividing one polynomial by another is rather tedious. If the divisor is a monic binomial of degree 1—that is, if it is of the form $x - a$, $a \in \mathbf{R}$, $a \neq 0$—then the labor involved can be lessened considerably by a process known as **synthetic division**. If $2x^3 - x^2 - 8x + 15$ is divided by $x - 2$, the ordinary algorithm appears as follows:

$$
\begin{array}{r}
2x^2 + 3x - 2 \\
x - 2 \overline{\smash{\big)}\, 2x^3 - x^2 - 8x + 15} \\
\underline{2x^3 - 4x^2 } \\
3x^2 - 8x \\
\underline{3x^2 - 6x } \\
-2x + 15 \\
\underline{-2x + 4} \\
11
\end{array}
$$

In this calculation, many of the symbols had to be written repeatedly. The numbers 2, 3, and -2 in the quotient polynomial each appear three times, and the last two terms of the dividend each appear twice. Also, the powers of x serve no useful purpose other than to separate the terms. If we omit all repetitions, powers of x, and the coefficient of x in the divisor, the arrangement becomes

$$
\begin{array}{r}
-2 \overline{\smash{\big)}\, 2 - 1 - 8 + 15} \\
-4 \\
\hline
3 \\
-6 \\
\hline
-2 \\
+ 4 \\
\hline
+ 11
\end{array}
$$

If the scattered terms are now aligned in three rows, and the leading coefficient is repeated in the third row, we have

$$
\begin{array}{r}
-2 \overline{\smash{\big)}\, 2 - 1 - 8 + 15} \\
-4 - 6 + 4 \\
\hline
2 + 3 - 2 + 11
\end{array}
\quad
\begin{array}{l}
\text{(first row)} \\
\text{(second row)} \\
\text{(third row)}
\end{array}
$$

Observe that each term in the second row is the product of the divisor -2 and the preceding term in the third row. Also note that each term of the third row, except the first, is the difference of the two corresponding terms in the first and second rows.

The same third row is obtained if we replace the divisor -2 with 2 and *add* the terms in the second row (determined in the same way as above) to the corresponding terms in the first row.

$$
\begin{array}{r}
2 \overline{\smash{\big)}\, 2 - 1 - 8 + 15} \\
4 + 6 - 4 \\
\hline
2 + 3 - 2 + 11
\end{array}
$$

If we compare the third row with the quotient polynomial $S(x)$ in the ordinary division example, we see that the number 11 on the extreme right is the remainder, and that the other numbers are the coefficients of $S(x)$. Thus, we can interpret these three lines as

$$(2x^3 - x^2 - 8x + 15) \div (x - 2) = 2x^2 + 3x - 2 + \frac{11}{x-2}.$$

In both methods of division of polynomials, it is important that the divisor and the dividend be arranged in order of decreasing powers of the variable or, if the polynomials contain more than one variable, in descending order with respect to powers of the same variable. If the dividend does not contain all non-negative powers of degree less than n, it is helpful to write all missing terms as powers with 0 as a coefficient. This is essential in the case of synthetic division.

The technique of synthetic division is summarized as follows. In the division of a polynomial $P(x)$ by $x - r$:

(1) The **synthetic divisor** is r.
(2) The first row is composed of the coefficients of the powers of x in $P(x)$, with 0 used as the coefficient of any missing power.
(3) The first term in the third row is the leading coefficient of $P(x)$.
(4) The first term in the second row (placed under the second term of the first row) is the product of r and the first term in the third row.
(5) The second term in the third row is the sum of the second term in the first row and the first term in the second row.
(6) The remaining terms in the second and third rows are found in the same way.

Example Use synthetic division to find

$$(3x^4 + 2x^2 + 3x - 5) \div (x + 1)$$

Solution The divisor must be of the form $x - r$, hence, $r = -1$.

$$\begin{array}{r|rrrrr}
-1 & 3 + 0 + 2 + 3 - 5 \\
 & - 3 + 3 - 5 + 2 \\
\hline
 & 3 - 3 + 5 - 2 - 3
\end{array}$$

Thus,

$$(3x^4 + 2x^2 + 3x - 5) \div (x + 1) = 3x^3 - 3x^2 + 5x - 2 - \frac{3}{x+1}$$

EXERCISE 4.3

WRITTEN

In Problems 1–20, express each quotient in the form $S(x) + T(x)/Q(x)$. Assume that no divisor equals zero.

Example $(3x^3 + 5x^2 - x - 1) \div (3x^2 + 2x - 3)$

Solution

$$\begin{array}{r} x+1 \\ 3x^2+2x-3 \overline{\smash{\big)}\, 3x^3+5x^2-x-1} \\ \underline{3x^3+2x^2-3x} \\ 3x^2+2x-1 \\ \underline{3x^2+2x-3} \\ +2 \end{array}$$

Thus, $(3x^3+5x^2-x-1) \div (3x^2+2x-3) = x+1+\dfrac{2}{3x^2+2x-3}$.

1. $(x^2+5x+6) \div (x+2)$
2. $(x^2-5x-6) \div (x+1)$
3. $(x^2+4x+7) \div (x+3)$
4. $(x^2-6x+9) \div (x-5)$
5. $(2x^2+7x+8) \div (x+2)$
6. $(2x^2-13x+15) \div (x-5)$
7. $(x^4+4x^2+12) \div (x^2-3)$
8. $(x^4+3x^2-8) \div (x^2-1)$
9. $(4x^3-8x^2+13x-5) \div (2x-1)$
10. $(4x^3+8x^2-9x-7) \div (2x+3)$
11. $(6x^3-17x^2+17x+22) \div (3x-1)$
12. $(2x^3+4x^2+2x-6) \div (2x-4)$
13. $(4x^4-4x^3+3x^2-4x+12) \div (x^2-2x-3)$
14. $(3x^4+5x^3+x^2-x-7) \div (3x^2+2x-4)$
15. $(x^3-8) \div (x-2)$
16. $(8x^3+27) \div (2x+3)$
17. $(6x^2-13xy-5y^2) \div (2x-5y)$ (HINT: Treat as polynomial in x.)
18. $(8x^2+2xy-3y^2) \div (4x+3y)$
19. $(6x^4+17x^2y-14y^2) \div (3x^2-2y)$
20. $(x^3-3x^2y+3xy^2-y^3) \div (x^2-2xy+y^2)$

Use synthetic division to express each quotient in Problems 21–36 in the form

$$S(x) + T(x)/Q(x).$$

Example $(5x^3-11x^2-14x-10) \div (x-3)$

Solution
$$\begin{array}{r} 3 \rfloor \; 5 \; -11 \; -14 \; -10 \\ +15 \; +12 \; -6 \\ \hline 5 \; +4 \; -2 \; -16 \end{array}$$

The first three numbers in line 3 are the coefficients of $S(x)$, and $T(x)$ is -16. Hence, the quotient expressed in the desired form is

$$5x^2+4x-2-\dfrac{16}{x-3}.$$

21. $(4x^2-2x+3) \div (x-3)$
22. $(2x^2+3x-7) \div (x+4)$
23. $(-2x^2-2x+7) \div (x-2)$
24. $(-3x^2+4x-5) \div (x+2)$
25. $(3x^3-x^2+2x+7) \div (x+2)$
26. $(2x^3+4x^2-3x+5) \div (x-3)$
27. $(-2x^3-5x^2+7) \div (x+3)$
28. $(3x^3-2x-75) \div (x-3)$
29. $(2x^3+5x^2-4x-5) \div (x+1)$
30. $(2x^3-3x^2-x-5) \div (x-1)$
31. $(3x^4-2x^3+x^2-x+7) \div (x-2)$
32. $(-2x^4+5x^3-2x^2-7x+5) \div (x+3)$
33. $(x^3+3x^2-x+3) \div (x-3)$

34. $(x^3 - 4x + 8) \div (x + 2)$
35. $(x^4 + 2x^2 + 2) \div (x - 2)$
36. $(x^4 - x^2 - 6) \div (x - 3)$

4.4 Completing the Square

In Section 4.2 we reviewed the facts that

$$\left.\begin{array}{l}(a + b)^2 = (a + b)(a + b) = a^2 + 2ab + b^2 \\ (a - b)^2 = (a - b)(a - b) = a^2 - 2ab + b^2\end{array}\right\}. \tag{1}$$

The square of a binomial is a polynomial that is called a **perfect square trinomial**. You will find that the ability to recognize that a given trinomial is a perfect square, and thus the square of a binomial, is an important asset in working with expressions involving polynomials.

Recognizing Perfect Square Trinomials

In order to help you develop this ability, let us examine the trinomials in Equations (1), above. In either case you should note that the first and third terms, a^2 and b^2, are perfect squares. Thus, in order for a trinomial to be the square of a binomial, it must contain two terms that are perfect squares. [Remember that the square of any nonzero real number is always positive.] Now examine the second terms of the above trinomials. The following statement should be evident:

> *A trinomial is the square of a binomial if two of its terms are perfect squares and the other term is exactly ± 2 times the product of the square roots of the perfect squares.*

Examples Why is it evident that each of the following is a perfect square trinomial?

(a) $x^2 + 6x + 9$
(b) $x^2 - 10x + 25$
(c) $4x^2 + 12xy + 9y^2$

Solutions (a) The square root of x^2 is x, the square root of 9 is 3, and $6x = 2 \cdot x \cdot 3$.
(b) The square root of x^2 is x, the square root of 25 is 5, and $-10x = -2 \cdot x \cdot 5$.
(c) The square root of $4x^2$ is $2x$, the square root of $9y^2$ is $3y$, and $12xy = 2 \cdot 2x \cdot 3y$.

Forming Perfect Square Trinomials

The perfect square trinomial has many applications in mathematics. Some of these are: to make certain polynomials easier to factor; to find the solution sets of quadratic equations; and to simplify some expressions that occur in the calculus. In view of its importance, we consider it worthwhile to spend some time on the techniques of forming perfect square trinomials.

Once you have learned to recognize whether or not a trinomial is the square of a binomial, you can use this knowledge to transform some binomials into perfect square trinomials by the addition of a new term.

Let us return to Equations (1) at the beginning of this section. We see that a perfect square trinomial is the square of a binomial that is either the sum or the difference of the square roots of the two perfect square terms of the trinomial. In fact, the sign of the middle term of the trinomial is determined by the form of the binomial being squared. This leads us directly to our method for forming a perfect square trinomial, given the sum of two squares. All that is necessary to do is to find the square roots of the two terms and multiply the product of these square roots by 2 or -2. Either of these products is a term that can be added to the given terms to form a perfect square trinomial.

Examples Find the term that can be added to the two given terms to form a perfect square trinomial.

(a) $x^2 + 4$ (b) $9x^2 + 1$

Solutions (a) The square root of x^2 is x, and the square root of 4 is 2. Hence, we can add $2 \cdot x \cdot 2 = 4x$, or $-2 \cdot x \cdot 2 = -4x$.

(b) The square root of $9x^2$ is $3x$, and the square root of 1 is 1. Hence we can add either $2 \cdot 3x \cdot 1 = 6x$ or $-2 \cdot 3x \cdot 1 = -6x$.

Completing the Square

More frequently, we find that a polynomial from which a perfect square trinomial is to be formed contains a term of second degree and a term of first degree with respect to the variable. In such cases, it is the second square term that must be found.

It is not particularly difficult to find this term if the given term of second degree is a perfect square. Simply take the square root of this term, multiply it by 2, and then divide the first-degree term by this product. The result is the *square root* of the term to be added. In terms of the expression $a^2 + 2ab$, the square root of a^2 is a, and $2ab$ divided by $2a$ is b, which is the square root of b^2, the other square term of $a^2 + 2ab + b^2$.

Examples What term must be added to the binomial to form a perfect square trinomial? This trinomial is the square of what binomial?

(a) $x^2 + 8x$ (b) $4x^2 - 12x$ (c) $x^2 + 5x$ (d) $9x^2 + 12xy$

Solutions (a) The square root of x^2 is x. Hence, we divide $8x$ by $2x$ to obtain 4, which is the square root of the term to be added. $4^2 = 16$, and the perfect square trinomial is $x^2 + 8x + 16$. This trinomial is the square of $x + 4$.

(b) The square root of $4x^2$ is $2x$. Then, $-12x \div 4x = -3$, so that $(-3)^2$, or 9, is the term to add. The trinomial is $4x^2 - 12x + 9$, which is the square of $2x - 3$.

(c) $5x \div 2x = \frac{5}{2}$. Thus, $(\frac{5}{2})^2$, or $\frac{25}{4}$, is the term to add. The trinomial is $x^2 + 5x + \frac{25}{4}$, which is the square of $x + \frac{5}{2}$.

(d) $12xy \div 6x = 2y$. Thus, $(2y)^2$ or $4y^2$ is the term to add. The trinomial is $9x^2 + 12xy + 4y^2$, which is the square of $3x + 2y$.

Completing the square is most frequently applied to polynomials of the form $x^2 + px$. In such a case, we simply square one-half the coefficient of x (first-degree term) to find the required term. Thus, to complete the square for $x^2 + px$, we square $p/2$ to obtain $p^2/4$. When this term is added, we have

$$x^2 + px + \frac{p^2}{4},$$

which is a trinomial square.

When you use the technique of completing the square in problem solving, the new expression must have the same value as the original. For example, to convert $x^2 + 2x$ to a perfect square trinomial we must add 1. However, $x^2 + 2x + 1$ is not equal to $x^2 + 2x$. How, then, can a perfect square trinomial be formed without changing the value of the original expression? Answer: By using the properties of the additive inverse and zero. Recall that for any real number a, $a + (-a) = 0$, and $a + 0 = a$. Hence, if we add both 1 and -1 to $x^2 + 2x$, we form a new polynomial that has the same value as the original but contains a perfect square trinomial. That is,

$$x^2 + 2x = (x^2 + 2x + 1) - 1.$$

When it is necessary to add a term to a polynomial, the negative of that term must also be added if the polynomial is to retain its original value.

Sometimes the second-degree term of a polynomial is not a perfect square. In some such cases we can form a perfect square trinomial without using irrational numbers if we first factor out the leading coefficient. This often leads to fractions as coefficients, but we can still apply the technique in the same manner as before.

Example Change the form of $2x^2 + 5x + 3$ so that part of the polynomial is a perfect square trinomial.

Solution We first factor out 2, the leading coefficient.

$$2x^2 + 5x + 3 = 2(x^2 + \tfrac{5}{2}x + \tfrac{3}{2}).$$

Next, we devote our attention to $x^2 + \tfrac{5}{2}x$. One-half of $\tfrac{5}{2}$ is $\tfrac{5}{4}$, and the square of $\tfrac{5}{4}$ is $\tfrac{25}{16}$. Thus, $x^2 + \tfrac{5}{2}x + \tfrac{25}{16}$ is a perfect square trinomial. Then, to avoid changing the value of the original polynomial, we add $-\tfrac{25}{16}$ inside the parentheses. This gives

$$2x^2 + 5x + 3 = 2(x^2 + \tfrac{5}{2}x + \tfrac{25}{16} - \tfrac{25}{16} + \tfrac{3}{2})$$
$$= 2[(x^2 + \tfrac{5}{2}x + \tfrac{25}{16}) - \tfrac{1}{16}],$$

which is the required form.

4.4 Completing the Square

EXERCISE 4.4

ORAL

State the trinomial form of each expression.

1. $(a + b)^2$
2. $(a - b)^2$
3. $(x + y)^2$
4. $(x - y)^2$
5. $(x + 1)^2$
6. $(x + 2)^2$
7. $(y - 1)^2$
8. $(y - 2)^2$
9. $(x + 3)^2$
10. $(y + 4)^2$
11. $(x - 4)^2$
12. $(y - 3)^2$
13. $(2x + 1)^2$
14. $(3y - 1)^2$
15. $(2x - 1)^2$
16. $(3y + 1)^2$
17. $(2x + 3)^2$
18. $(3x + 2)^2$
19. $(3x - 2)^2$
20. $(2x - 3)^2$

Which of the following trinomials are perfect squares?

21. $x^2 + 2x + 1$
22. $x^2 - 2x + 1$
23. $x^2 + 2x - 1$
24. $x^2 + x + 2$
25. $x^2 + 6x + 9$
26. $x^2 - 10x + 25$
27. $x^2 + 4x - 4$
28. $x^2 - 6x - 9$
29. $4x^2 + 4x + 1$
30. $9y^2 - 6y + 1$
31. $4x^2 + 12x + 9$
32. $9x^2 - 12x + 4$

WRITTEN

In Problems 1–20, what term must be added to the binomial so that the result is a perfect square trinomial?

Examples (a) $x^2 + 9$ (b) $x^2 + 4x$

Solutions (a) Both x^2 and 9 are perfect squares. Hence, we need to add the first-degree term which is ± 2 times the products of the square roots of the given terms. The square root of x^2 is x, and the square root of 9 is 3. Hence, we can add either $6x$ or $-6x$.

(b) In this case the second square term must be added. 2 times the square root of x^2 is $2x$, and $4x \div 2x = 2$, which is the square root of the term to be added. Thus, $(2)^2$ or 4 is the required term. Alternatively, since the leading coefficient is 1, we can square one-half the coefficient of x to obtain the same number.

1. $x^2 + 16$
2. $x^2 + 25$
3. $x^2 + 49$
4. $x^2 + 36$
5. $x^2 + 81$
6. $x^2 + 121$
7. $x^2 + 2x$
8. $x^2 - 2x$
9. $x^2 - 6x$
10. $x^2 + 6x$
11. $x^2 + 8x$
12. $x^2 + 10x$
13. $x^2 - 10x$
14. $x^2 - 8x$
15. $x^2 + 2xy$
16. $x^2 - 4xy$
17. $4x^2 + 4x$
18. $9x^2 - 6x$
19. $x^2 + 3x$
20. $x^2 - 7x$

In Problems 21–32, change the form of each expression so that some of its terms form a perfect square trinomial.

Example $3x^2 + 6x - 5$

Solution   Factor 3, the leading coefficient, from the polynomial to obtain

$3(x^2 + 2x - \tfrac{5}{3})$.

$x^2 + 2x$ can be made into a perfect square trinomial by adding 1. However, adding 1 changes the value of the expression, so we add $1 - 1$ inside the parentheses, which gives

$$3(x^2 + 2x + 1 - 1 - \tfrac{5}{3}) = 3[(x^2 + 2x + 1) - \tfrac{8}{3}].$$

21. $x^2 + 2x - 3$
22. $x^2 + 4x - 5$
23. $x^2 - 6x - 16$
24. $x^2 - 10x - 11$
25. $2x^2 + 4x - 7$
26. $2x^2 - 12x + 4$
27. $5x^2 + 30x - 4$
28. $3x^2 - 12x - 10$
29. $2x^2 + 3x - \tfrac{7}{4}$
30. $2x^2 - 5x + \tfrac{9}{4}$
31. $3x^2 + 2x - \tfrac{2}{3}$
32. $3x^2 - 4x - \tfrac{5}{3}$

4.5 Factoring Quadratic Trinomials

A polynomial of the form

$$ax^2 + bx + c \qquad (a \neq 0)$$

is called a **quadratic trinomial in standard form**. In elementary algebra you learned how to factor certain expressions of this type over the set of integers. (A discussion of one such technique appears in Appendix 1.) The technique that you studied involved considerable guesswork, or trial and error. While we agree that factoring by inspection is the best method if you can recognize the factors, the technique of completing the squre can be used to eliminate the guesswork as well as permit you to factor many quadratic trinomials over the set of irrational numbers.

Factoring the Difference of Two Squares

Before we begin this discussion we ask you to recall from Section 4.2 that

$$(a + b)(a - b) = a^2 - b^2,$$

which by the symmetric property of equality is

$$a^2 - b^2 = (a + b)(a - b).$$

Thus, the difference of the square of two numbers equals the product of the sum and the difference of the square roots of the two squares.

Examples (a) $x^2 - 4 = (x + 2)(x - 2)$ (b) $9x^2 - 4y^2 = (3x + 2y)(3x - 2y)$
(c) $(x + 2)^2 - 9 = [(x + 2) + 3][(x + 2) - 3] = (x + 5)(x - 1)$
(d) $4x^2 + 12x + 9 - 36 = (2x + 3)^2 - 36$
$\qquad = [(2x + 3) + 6][(2x + 3) - 6]$
$\qquad = (2x + 9)(2x - 3)$

Factoring Trinomials as a Difference of Squares

Examples (c) and (d), above, suggest a procedure by which you can use the technique of completing the square to first write the polynomial as the difference of two squares and then in factored form.

Examples Use the technique of completing the square to factor the quadratic trinomial.
(a) $x^2 + 5x + 6$ (b) $x^2 - 2x - 8$ (c) $2x^2 - 3xy - 2y^2$

Solutions (a) Since the coefficient of x^2 is 1, we add the square of one-half the coefficient of $x(\frac{25}{4})$ and the negative of this number to obtain

$$x^2 + 5x + \tfrac{25}{4} + 6 - \tfrac{25}{4} = (x^2 + 5x + \tfrac{25}{4}) - \tfrac{1}{4}$$
$$= (x^2 + \tfrac{5}{2})^2 - (\tfrac{1}{2})^2$$
$$= (x + \tfrac{5}{2} + \tfrac{1}{2})(x + \tfrac{5}{2} - \tfrac{1}{2})$$
$$= (x + 3)(x + 2).$$

(b) As in Example (a), we add 1 and -1 to obtain

$$x^2 - 2x + 1 - 1 - 8 = (x^2 - 2x + 1) - 9$$
$$= (x - 1)^2 - (3)^2$$
$$= (x - 1 + 3)(x - 1 - 3)$$
$$= (x + 2)(x - 4).$$

(c) Since the coefficient of x^2 is not 1, we first factor out 2 to obtain $2(x^2 - \tfrac{3}{2}xy - y^2)$. We can now form a perfect square trinomial of the first two terms inside the parentheses by adding $\tfrac{9}{16}y^2$ and $-\tfrac{9}{16}y^2$ to obtain

$$2(x^2 - \tfrac{3}{2}xy + \tfrac{9}{16}y^2 - \tfrac{9}{16}y^2 - y^2)$$
$$= 2[(x^2 - \tfrac{3}{2}xy + \tfrac{9}{16}y^2) - (\tfrac{9}{16}y^2 + y^2)]$$
$$= 2[(x - \tfrac{3}{4}y)^2 - (\tfrac{5}{4}y)^2]$$
$$= 2[(x - \tfrac{3}{4}y + \tfrac{5}{4}y)(x - \tfrac{3}{4}y - \tfrac{5}{4}y)]$$
$$= 2(x + \tfrac{1}{2}y)(x - 2y)$$
$$= (2x + y)(x - 2y).$$

Not all quadratic trinomials are factorable by this technique. For example, consider

$$x^2 + 4x + 8.$$

If we complete the square in x, we have

$$x^2 + 4x + 4 + 4$$
$$(x + 2)^2 + (2)^2,$$

which is the *sum* of two squares, not the difference. This kind of an expression cannot be factored in the set of real numbers. The quadratic formula, discussed in Section 5.2, can be used to determine the factorability of quadratic trinomials.

Factoring in the Set of Irrational Numbers*

We agree that it might have been easier to factor the trinomials in the examples above by other techniques. However, not all trinomials are factorable over the set of integers. Sometimes the factors may contain rational numbers or even irrational numbers.

---

*Optional.

Examples Factor: (a) $2x^2 - \frac{1}{6}x - \frac{1}{6}$; (b) $x^2 + 2x - 4$.

Solutions (a) We first factor out 2 to obtain

$$2(x^2 - \tfrac{1}{12}x - \tfrac{1}{12}).$$

We next change the form of the expression within the parentheses by adding both the square of one-half the coefficient of x and its negative. This gives

$$2(x^2 - \tfrac{1}{12}x + \tfrac{1}{576} - \tfrac{1}{576} - \tfrac{1}{12}) = 2[(x^2 - \tfrac{1}{12}x + \tfrac{1}{576}) - (\tfrac{49}{576})]$$
$$= 2[(x - \tfrac{1}{24})^2 - (\tfrac{7}{24})^2]$$
$$= 2[(x - \tfrac{1}{24} + \tfrac{7}{24})(x - \tfrac{1}{24} - \tfrac{7}{24})]$$
$$= 2[(x + \tfrac{1}{4})(x - \tfrac{1}{3})].$$

(b) We begin by completing the square involving the first two terms by adding $1 - 1$. We then have

$$x^2 + 2x + 1 - 1 - 4 = (x^2 + 2x + 1) - 5.$$

5 is not the square of a rational number, but we can think of $5 = (\sqrt{5})^2$. Hence,

$$(x^2 + 2x + 1) - 5 = (x + 1)^2 - (\sqrt{5})^2$$
$$= (x + 1 + \sqrt{5})(x + 1 - \sqrt{5}).$$

EXERCISE 4.5

WRITTEN

In Problems 1–20, write the given expression in factored form.

Examples (a) $4x^2 - 9 = (2x + 3)(2x - 3)$
(b) $(x + 2)^2 - 25 = (x + 2 + 5)(x + 2 - 5) = (x + 7)(x - 3)$
(c) $x^2 - 5 = (x + \sqrt{5})(x - \sqrt{5})$
(d) $(2x + 1)^2 - 13 = (2x + 1 + \sqrt{13})(2x + 1 - \sqrt{13})$
(e) $(x + 3)^2 - (y - 2)^2 = [(x + 3) + (y - 2)][(x + 3) - (y - 2)]$
$= (x + y + 1)(x - y + 5)$

1. $x^2 - 9$ 2. $x^2 - 16$ 3. $9x^2 - 4$ 4. $25x^2 - 36$
5. $x^2 - 2$ 6. $x^2 - 7$ 7. $4x^2 - 11$ 8. $16x^2 - 17$

9. $(x + 3)^2 - 16$ 10. $(x - 4)^2 - 25$
11. $(3x + 2)^2 - 9$ 12. $(4x - 7)^2 - 16$
13. $(2x - 3)^2 - 25$ 14. $(3x + 1)^2 - 81$
15. $(x + 1)^2 - (y + 2)^2$ 16. $(x - 3)^2 - (y - 3)^2$
17. $(2x + 1)^2 - (3y + 2)^2$ 18. $(3x - 4)^2 - (2y - 3)^2$
19. $(4x + 3)^2 - (2y + 1)^2$ 20. $(3x - 2)^2 - (2y - 3)^2$

In Problems 21–32, use the technique of completing the square to write the polynomial in the form of the difference of two squares and then in factored form.

Examples (a) $x^2 - 3x + 2 = x^2 - 3x + \frac{9}{4} - \frac{9}{4} + 2$
$= (x - \frac{3}{2})^2 - (\frac{1}{2})^2$
$= (x - \frac{3}{2} + \frac{1}{2})(x - \frac{3}{2} - \frac{1}{2})$
$= (x - 1)(x - 2)$

(b) $3x^2 - x - 2 = 3(x^2 - \frac{1}{3}x - \frac{2}{3}) = 3[x^2 - \frac{1}{3}x + \frac{1}{36} - \frac{1}{36} - \frac{2}{3}]$
$= 3[(x - \frac{1}{6})^2 - (\frac{5}{6})^2]$
$= 3(x - \frac{1}{6} + \frac{5}{6})(x - \frac{1}{6} - \frac{5}{6})$
$= 3(x + \frac{2}{3})(x - 1)$
$= (3x + 2)(x - 1)$

21. $x^2 + 7x + 12$
22. $x^2 - 5x + 6$
23. $x^2 + 4x + 3$
24. $x^2 + 3x + 2$
25. $x^2 - 6x + 5$
26. $x^2 - x - 2$
27. $3x^2 - 5x - 12$
28. $6x^2 + x - 15$
29. $6x^2 + 11x + 4$
30. $x^2 + 4x - 12$
31. $x^2 - 5x + 4$
32. $x^2 + 4x - 32$
*33. $x^2 - 7x + 2$
*34. $3x^2 - 9x + 4$
*35. $2x^2 + 8x - 5$
*36. $4x^2 + 16x - 15$
*37. $2x^2 + 3x - 1$
*38. $2x^2 - 5x - 1$

Example $x^2 + 4xy - 5y^2$

Solution Complete the square of the terms involving x by adding $4y^2$ and $-4y^2$ to obtain
$x^2 + 4xy + 4y^2 - 4y^2 - 5y^2 = (x + 2y)^2 - 9y^2$
$= (x + 2y + 3y)(x + 2y - 3y)$
$= (x + 5y)(x - y)$.

39. $x^2 - 2xy - 8y^2$
40. $x^2 + 2xy - 15y^2$
41. $x^2 + 6xy - 7y^2$
42. $x^2 - 10xy + 9y^2$
43. $2x^2 - xy - 3y^2$
44. $3x^2 + 2xy - 5y^2$

4.6 Other Methods of Factoring

Sum or Difference of Two Cubes

A binomial that is the sum or difference of two cubes can be factored by a technique similar to that used in factoring quadratic trinomials (Section 4.5). This involves completing the cube instead of completing the square.

Recall from Section 4.2 that
$$(a + b)^3 = a^3 + 3a^2b + 3ab^2 + b^3.$$
The polynomial $a^3 + b^3$ can now be written equivalently by adding $(3a^2b + 3ab^2)$ and $(-3a^2b - 3ab^2)$, as
$$a^3 + 3a^2b + 3ab^2 + b^3 - 3a^2b - 3ab^2,$$

and then as

$$(a + b)^3 - 3ab(a + b) = (a + b)[(a + b)^2 - 3ab]$$
$$= (a + b)(a^2 + 2ab + b^2 - 3ab)$$
$$= (a + b)(a^2 - ab + b^2).$$

The difference of two cubes can be treated similarly to produce

$$a^3 - b^3 = (a - b)(a^2 + ab + b^2).$$

Examples Factor: (a) $x^3 + 8$; (b) $64x^3 - z^3$.

Solutions (a) Use the equation $a^3 + b^3 = (a + b)(a^2 - ab + b^2)$ where $a = x$ and $b = 2$ to obtain

$$x^3 + 8 = x^3 + (2)^3 = (x + 2)(x^2 - 2x + 4).$$

(b) Use the equation $a^3 - b^3 = (a - b)(a^2 + ab + b^2)$, where $a = 4x$ and $b = z$, to obtain

$$64x^3 - z^3 = (4x)^3 - z^3$$
$$= (4x - z)(16x^2 + 4xz + z^2).$$

Fourth-Degree Trinomials as a Difference of Squares

Fourth-degree trinomials of the form

$$a^2x^4 + bx^2y^2 + c^2y^4$$

can sometimes be factored by completing the square and then factoring as the difference of two squares.

Example $x^4 + x^2y^2 + y^4$

Solution
$$x^4 + x^2y^2 + y^4 = x^4 + 2x^2y^2 + y^4 - x^2y^2$$
$$= (x^2 + y^2)^2 - (xy)^2$$
$$= (x^2 + y^2 + xy)(x^2 + y^2 - xy)$$
$$= (x^2 + xy + y^2)(x^2 - xy + y^2)$$

Factoring by Grouping

The method of factoring in the example above involves a technique known as factoring by grouping. You will find this technique to be useful in factoring certain polynomials which contain four or more terms.

Example Express $xy + 2y - 3x - 6 + yz - 3z$ as the product of two polynomials.

Solution Since the last two terms both contain z but do not contain x, we first group the terms as

$$(xy + 2y - 3x - 6) + (yz - 3z).$$

We can now see that in the first of these groupings, the first two terms contain y as a factor, and the next two terms contain -3 as a factor. Factoring these groups of terms separately, we obtain

$$[y(x + 2) - 3(x + 2)] + z(y - 3).$$

Next we observe that $x + 2$ is a factor of both terms enclosed in the brackets. Thus, we obtain

$$(x + 2)(y - 3) + z(y - 3).$$

Finally, we see that $y - 3$ is a common factor of the two expressions, which leads to

$$(y - 3)(x + 2 + z) \quad \text{or} \quad (y - 3)(x + z + 2).$$

Example Factor $x^2 - y^2 + 4x + 6y - 5$.

Solution Regroup the terms to obtain

$$(x^2 + 4x) - (y^2 - 6y) - 5.$$

Next, complete the squares in both parentheses, making certain that you do not change the value of the expression. This gives

$$(x^2 + 4x + 4) - (y^2 - 6y + 9) - 5 - 4 + 9$$
$$= (x + 2)^2 - (y - 3)^2$$
$$= [(x + 2) + (y - 3)][(x + 2) - (y - 3)]$$
$$= (x + y - 1)(x - y + 5).$$

When an expression is factored, the polynomial obtained by multiplying the factors together should be the same as the original. If we multiply the factors in the preceding example, we have by repeated applications of the distributive law

$$(x + y - 1)(x - y + 5) = x(x - y + 5) + y(x - y + 5) - 1(x - y + 5)$$
$$= x^2 - xy + 5x + xy - y^2 + 5y - x + y - 5$$
$$= x^2 - y^2 + 4x + 6y - 5,$$

the original polynomial. You should observe that it is necessary to multiply by each of the terms in the first factor, including the -1.

EXERCISE 4.6

WRITTEN

Factor each polynomial.

1. $x^3 - y^3$
2. $x^3 + y^3$
3. $x^3 + 8$
4. $y^3 - 27$
5. $125 + x^3$
6. $1000 - y^3$
7. $1 - x^3$
8. $1 + x^3 y^3$
9. $8x^3 - 27$
10. $216x^3 + 8y^3$
11. $64x^3 - 27y^3$
12. $125x^3 - 8y^3$

Example $x^4 + 3x^2 + 4$

Solution
$$x^4 + 3x^2 + 4 = x^4 + 4x^2 + 4 - x^2$$
$$= (x^2 + 2)^2 - (x)^2$$
$$= (x^2 + 2 + x)(x^2 + 2 - x)$$
$$= (x^2 + x + 2)(x^2 - x + 2)$$

13. $x^4 + 5x^2 + 9$
14. $x^4 + x^2 + 1$
15. $9x^4 + 2x^2 + 1$
16. $y^4 - 3y^2 + 1$
17. $x^4 - 8x^2 + 4$
18. $y^4 + 4$
19. $4x^4 + 1$
20. $9y^4 - 10y^2 + 1$
21. $9x^4 - 7x^2 + 1$
22. $4y^2 + 8y^2 + 9$
23. $4x^4 - 16x^2 + 9$
24. $4 + 11y^2 + 9y^4$

Examples (a) $3xy + xz + 3yw + zw$ (b) $9x^2 - y^2 + 2yz - z^2$

Solutions (a) $3xy + xz + 3yw + zw = x(3y + z) + w(3y + z)$
$$= (x + w)(3y + z)$$

(b) $9x^2 - y^2 + 2yz - z^2 = 9x^2 - (y^2 - 2yz + z^2)$
$$= (3x)^2 - (y - z)^2$$
$$= (3x + y - z)(3x - y + z)$$

25. $2ax + 2ay + bx + by$
26. $cx - cy + 3dx - 3dy$
27. $4hx - 4hy + 5cx - 5cy$
28. $ax^2 + bx^2 + ay^2 + by^2$
29. $x^3 - 3x^2 + x - 3$
30. $1 - 4x + 2x^2 - 8x^3$
31. $3x^3 - 2x^2 + 6x - 4$
32. $4 - 5x - 8x^2 + 10x^3$
33. $x^3 - 3x^2 + 3 - x$
34. $2 + 4x - 10x^4 - 5x^3$
35. $x^2 + 2x + 1 - 9y^2$
36. $4w^2 + 20w + 25 - 81z^2$
37. $y^2 + 2yz + z^2 - 4x^2$
38. $9x^2 - 4y^2 - 4yz - z^2$
39. $4x^2 - 9y^2 - 6y - 1$
40. $16x^2 - y^2 + 2yz - z^2$

41. $x^2 + 6x + 9 - y^2 - 2y - 1$ [HINT: See example (b) above.]
42. $4x^2 + 4x + 1 - 9y^2 - 12y - 4$

Summary

1. Our discussion of polynomials was limited to expressions of the form
$$a_0 x^n + a_1 x^{n-1} + \cdots + a_{n-1} x + a_n$$
where each a is a real number and the replacement set for x is $\mathbf{R}$. [4.0]

2. A polynomial is a sum. Each addend is called a **term**. The words **monomial**, **binomial**, and **trinomial** refer to polynomials of one, two, and three terms, respectively. Each a is called a **coefficient**, and the nonzero coefficient of the term containing the highest power of x is called the **leading coefficient**. The exponent of x in this term denotes the **degree** of the polynomial. [4.1]

3. The symbol $P(x)$ refers to a polynomial in x. The symbol $P(a)$ represents

the polynomial when x has been replaced with a. a can be a real number or even another polynomial. [4.1]
4. Sums of polynomials are simplified in the same manner as sums of expressions were in Section 3.3. [4.2]
5. Repeated applications of the distributive law may be necessary to simplify the product of two polynomials. [4.2]
6. To divide one polynomial by another of the same or lesser degree, select a suitable monomial $S_1(x)$ such that the product of $S_1(x)$ and the term of highest degree of the divisor is equal to the term of highest degree of the dividend. Subtracting the product of $S_1(x)$ and the divisor from the dividend gives a remainder $T_1(x)$ that is of lesser degree than $P(x)$. The process is repeated, using each remainder as a new dividend until the remainder is of lesser degree than the divisor. [4.3]
7. **Synthetic division** is an abbreviated method of dividing a polynomial by a *monic binomial*. [4.3]
8. A trinomial is the square of a binomial if two of its terms are perfect squares and the other term is exactly ± 2 times the product of the square roots of the squares. [4.4]
9. **Completing the square** refers to adding a term to a binomial so that the resulting trinomial is the square of some binomial. The technique is ordinarily applied to a binomial of the form $x^2 + px$, in which case the term to be added is the square of one-half the coefficient of the first-degree term $(p/2)^2$. [4.4]
10. The difference of two squares can be written as the product of the sum and the difference of the square roots of the respective squares. Symbolically,
$$a^2 - b^2 = (a+b)(a-b).$$ [4.5]
11. Some real polynomials of the form
$$ax^2 + bx + c,$$
called **quadratic trinomials**, can be expressed as the product of two binomials by completing the square and factoring the resulting polynomial as the difference of two squares. [4.5]
12. Other methods of factoring include the sum or difference of two cubes, fourth-degree trinomials, and factoring by grouping. [4.6]

REVIEW EXERCISES

SECTION 4.1

1. $P(x) = x^2 - 3x - 4$; find $P(-1)$, $P(0)$, $P(2)$, $P(4)$.
2. $P(x) = 2x^3 - 3x^2 - 11x + 6$; find $P(-2)$, $P(\frac{1}{2})$, $P(2)$, $P(3)$.
3. $P(x) = x^2 + 5x + 4$; find $\dfrac{P(x+h) - P(x)}{h}$.

SECTION 4.2

Simplify the sums and products in Problems 4–12.

4. $(2x^2 - x - 3) + (6x^2 + 8x - 1)$
5. $(8x^2 + 2x - 1) - (2x^2 - x - 3)$
6. $(3x^2 - 2xy + 5y^2) + (x^2 + 4xy - 3y^2) - (2x^2 - 2xy + 2y^2)$
7. $(2x + 3)(3x - 2)$
8. $(3x^2 - 2x + 4)(5x^2 + 7x + 2)$
9. $(4x - 5)^2$
10. $(2x + 3y)(2x - 3y)$
11. $(x^2 + x + 1)^2$
12. $(2x - 3y)^3$

SECTION 4.3

In Problems 13–16, express each quotient in the form $P(x) \div Q(x) = S(x) + \dfrac{T(x)}{Q(x)}$.

13. $(x^2 - 5x + 6) \div (x - 3)$
14. $(2x^2 - 12x + 15) \div (2x + 4)$
15. $(2x^3 + 11x^2 + 22x + 7) \div (2x + 3)$
16. $(4x^4 - x^2 + x - 7) \div (2x^2 + x - 3)$

Use synthetic division to express each quotient in Problems 17–20 in the same form as above.

17. $(x^2 - 7x + 12) \div (x - 3)$
18. $(4x^3 - 2x^2 - 9x + 10) \div (x - 1)$
19. $(x^4 + 3x^2 - 8) \div (x + 2)$
20. $(x^5 - 1) \div (x - 1)$

SECTION 4.4

In Problems 21–24, find the term to add to the binomial to form a perfect square trinomial.

21. $x^2 + 4$
22. $x^2 + 25$
23. $x^2 - 12x$
24. $16x^2 + 8x$

For each trinomial in Problems 15–28, find an equivalent expression that is a sum or difference containing the square of a binomial and a constant.

25. $x^2 + 8x - 12$
26. $x^2 + 6x + 12$
27. $3x^2 + 6x - 13$
28. $2x^2 + 6x + 3$

SECTION 4.5

In Problems 29–34, factor each polynomial if possible.

29. $4x^2 - 9$
30. $9x^2 - 16$
31. $(2x + 3)^2 - 16$
32. $x^2 + 4x - 5$
33. $x^2 - 7x + 12$
34. $3x^2 + 5x - 2$

SECTION 4.6

Factor each polynomial in Problems 35–40.

35. $27x^3 - 8y^3$
36. $125x^3 + 64y^3$
37. $x^4 - 10x^2 + 9$
38. $16x^2 - 4y^2 + 4yz - z^2$
39. $2xy - 8x + 3y - 12$
40. $x^2 - 4x + 4 - 9y^2 + 12y - 4$

Quadratic Equations and Inequalities in One Variable

An open sentence in which both members are polynomials such that one of them is of second degree and the other is of second or lower degree is called a second-degree open sentence. As with open sentences of first degree, these may be either equations or inequalities. Second-degree open sentences are called **quadratic equations** or **quadratic inequalities**. In this chapter, we are concerned with solving such open sentences in one variable.

Standard Form

Any of the techniques used in Chapter 2 for generating equivalent open sentences can be applied to second-degree open sentences. Thus, by adding the same term, or terms, to both members, an equivalent open sentence can be formed such that one member is zero (the zero polynomial). This leads to what we designate as the **standard form**. If a, b, and c are any real numbers, $a \neq 0$, then

$$ax^2 + bx + c = 0$$

is the standard form of a quadratic equation, and

$$ax^2 + bx + c < 0$$

is the standard form of a quadratic inequality. Note that in either case, the left-hand member is a quadratic trinomial in standard form. The "less than" symbol in the inequality is used in a general way to denote any order symbol, $<, \leq, >, \geq$.

Examples   Write the given open sentence in standard form.

    (a) $3x^2 + 4x = x^2 - 8$ (b) $4x + 7 < 3x - x^2$

Solutions   (a) We add $-x^2 + 8$ to both members to obtain

$$3x^2 - x^2 + 4x + 8 = x^2 - x^2 - 8 + 8,$$

115

which, upon combining the similar terms, becomes

$$2x^2 + 4x + 8 = 0.$$

(b) We add $x^2 - 3x$ to both members to obtain

$$x^2 - 3x + 4x + 7 < 3x - 3x - x^2 + x^2,$$

which, upon combining the similar terms, becomes

$$x^2 + x + 7 < 0.$$

5.1 Solution of Quadratic Equations by Factoring

If the left-hand member of a quadratic equation in standard form is factorable over the set of real numbers into two binomial factors, you can find the solution set of the equation by setting each factor containing the variable equal to zero and then finding the solutions of the resulting first-degree equations. The basis for this technique is established by the proof of the following theorem.

THEOREM 5.1-1 For every $a, b \in \mathbf{R}$, if $ab = 0$, then either $a = 0$, $b = 0$, or both $a = 0$ and $b = 0$.

PROOF:

If $a = 0$, then $ab = 0$ by Theorem 1.4-3 (If $a \in \mathbf{R}$, $a \cdot 0 = 0$), and no further proof is required.

If $a \neq 0$, then

STATEMENTS	JUSTIFICATIONS
1. $a, b \in \mathbf{R}, \ ab = 0, \ a \neq 0$	1. Hypothesis
2. $\dfrac{1}{a} \in \mathbf{R}$	2. Multiplicative inverse axiom
3. $\dfrac{1}{a} \cdot ab = \dfrac{1}{a} \cdot 0$	3. Theorem 1.4-2 (Multiplication law)
4. $\left(\dfrac{1}{a} \cdot a\right) \cdot b = \dfrac{1}{a} \cdot 0$	4. Associative property of multiplication
5. $b = \dfrac{1}{a} \cdot 0$	5. Multiplicative inverse and identity axioms
6. $b = 0$	6. Theorem 1.4-3 ($a \cdot 0 = 0$)

In finding the solution set of a quadratic equation, you should factor the left-hand member by inspection if possible. Otherwise, you can use the technique discussed in Section 4.5.

Examples Find the solution set of each equation.

(a) $x^2 - 5x + 6 = 0$ (b) $x^2 + 5x - 4 = 0$

Solutions (a) Factor the left-hand member to obtain

$$(x - 2)(x - 3) = 0.$$

If $x - 2 = 0$, then $x = 2$.
If $x - 3 = 0$, then $x = 3$.
Hence, the solution set is $\{2, 3\}$.

(b) Complete the square to obtain

$$x^2 + 5x + \tfrac{25}{4} - \tfrac{25}{4} - 4 = 0.$$

$$(x + \tfrac{5}{2})^2 - \tfrac{41}{4} = 0$$

$$\left[\left(x + \frac{5}{2}\right) + \frac{\sqrt{41}}{2}\right]\left[\left(x + \frac{5}{2}\right) - \frac{\sqrt{41}}{2}\right] = 0$$

$$\left(x + \frac{5 + \sqrt{41}}{2}\right)\left(x + \frac{5 - \sqrt{41}}{2}\right) = 0$$

If $x + \dfrac{5 + \sqrt{41}}{2} = 0$, then $x = \dfrac{-5 - \sqrt{41}}{2}$.

If $x + \dfrac{5 - \sqrt{41}}{2} = 0$, then $x = \dfrac{-5 + \sqrt{41}}{2}$.

Hence, the solution set is $\left\{\dfrac{-5 - \sqrt{41}}{2}, \dfrac{-5 + \sqrt{41}}{2}\right\}$.

If the decimal form of solution is desired, you can use the table of square roots (Appendix 2) to obtain $\{-5.702, 0.702\}$.

If an equation is not in standard form, you should first form an equivalent equation that is in standard form.

EXERCISE 5.1

WRITTEN

Find the solution set in **R** of each quadratic equation by factoring. Complete the square if necessary. If the solutions are irrational numbers, leave in simplest radical form or put in decimal form, as directed by your instructor.

Example $2x = 3 - x^2$

Solution Write the equation in standard form,

$$x^2 + 2x - 3 = 0.$$

Factor the left-hand member.

$$(x + 3)(x - 1) = 0$$

The solution set is seen by inspection to be $\{-3, 1\}$.

118 QUADRATIC EQUATIONS AND INEQUALITIES IN ONE VARIABLE

1. $x^2 - 3x - 10 = 0$
2. $y^2 - 5y - 14 = 0$
3. $x^2 + x = 12$
4. $y^2 + 3y = 28$
5. $21x^2 - 14x = 0$
6. $3y^2 - 7y = 0$

(HINT: In Problems 5–8 apply the distributive law to the left-hand member.)

7. $9x^2 - 144 = 0$
8. $5x^2 - 125 = 0$

(HINT: $b = 0$)

9. $x^2 + 8 = 6x$
10. $8y^2 + 3 = 10y$
11. $x^2 + 3x - 5 = 0$
12. $y^2 - 2y = 5$
13. $x^2 + 6x = -9$
14. $4y^2 - 4y = -1$
15. $3x^2 + 2 = -7x$
16. $6y^2 + 15 = 19y$
17. $2 - 2x - x^2 = 0$
18. $2 + 5y - y^2 = 0$
19. $9x^2 + 6x + 1 = 0$
20. $2y^2 + 5y - 7 = 0$
21. $x^2 + 1 = 3x$
22. $y^2 - 15 = 3y$
23. $8x^2 - 2x = 45$
24. $12y^2 + 19y = 18$

Example $2(x + 1)^2 - 3(x + 1) + 1 = 0$

Solution Factor the left-hand member.

$[2(x + 1) - 1][(x + 1) - 1] = 0$
$(2x + 2 - 1)(x + 1 - 1) = 0$
$(2x + 1)(x) = 0,$

for which the solution set is $\{-\tfrac{1}{2}, 0\}$.

25. $(x + 2)^2 + 3(x + 2) - 10 = 0$
26. $(y - 3)^2 - 5(y - 3) - 14 = 0$
27. $(x - 1)^2 + (x - 1) - 12 = 0$
28. $(y + 4)^2 + 3(y + 4) - 28 = 0$
29. $8(x - 1)^2 - 2(x - 1) - 1 = 0$
30. $4(y - 2)^2 - 4(y - 2) + 1 = 0$

5.2 The Quadratic Formula

In the previous section you solved quadratic equations in one variable by factoring, but in the case of an expression not factorable over the set of integers you found this technique to be unwieldy. This difficulty can be avoided by using the **quadratic formula**, which we derive from the general quadratic equation

$$ax^2 + bx + c = 0, \quad a \neq 0.$$

Since $a \neq 0$, we divide both members of this equation by a to obtain

$$x^2 + \frac{b}{a}x + \frac{c}{a} = 0.$$

We then form a trinomial square in x by adding

$$\frac{b^2}{4a^2} - \frac{b^2}{4a^2},$$

which gives us

$$x^2 + \frac{b}{a}x + \frac{b^2}{4a^2} - \frac{b^2}{4a^2} + \frac{c}{a} = 0$$

5.2 The Quadratic Formula

$$\left(x + \frac{b}{2a}\right)^2 - \left(\frac{b^2 - 4ac}{4a^2}\right) = 0$$

$$\left(x + \frac{b}{2a}\right)^2 - \left(\frac{\sqrt{b^2 - 4ac}}{2a}\right)^2 = 0.$$

Factoring the left-hand member, we have

$$\left(x + \frac{b}{2a} - \frac{\sqrt{b^2 - 4ac}}{2a}\right)\left(x + \frac{b}{2a} + \frac{\sqrt{b^2 - 4ac}}{2a}\right) = 0.$$

We then equate each of these factors to 0 and solve the resulting equations for x, to obtain

$$x = \frac{-b + \sqrt{b^2 - 4ac}}{2a} \quad \text{or} \quad x = \frac{-b - \sqrt{b^2 - 4ac}}{2a}.$$

These two equations are ordinarily combined into one equation by using the sign $\pm$ (read "plus or minus").

$$x = \frac{-b \pm \sqrt{b^2 - 4ac}}{2a} \qquad \text{(The quadratic formula)}$$

Using the Quadratic Formula

To use the quadratic formula, we replace a, b, and c with the proper constants and carry out the arithmetic.

Examples Find the solution set by the quadratic formula.

(a) $x^2 + 7x - 8 = 0$ (b) $2x^2 - 3x - 4 = 0$

Solutions (a) $a = 1$, $b = 7$, and $c = -8$. Substituting these values into the formula gives

$$x = \frac{-7 \pm \sqrt{49 + 32}}{2}$$

$$= \frac{-7 \pm \sqrt{81}}{2}$$

$$= \frac{-7 \pm 9}{2},$$

from which we obtain $\{1, -8\}$ as the solution set.

(b) $a = 2$, $b = -3$, and $c = -4$. Substituting these values into the formula gives

$$x = \frac{3 \pm \sqrt{9 + 32}}{4}$$

$$= \frac{3 \pm \sqrt{41}}{4}$$

for which the solution set is

$$\left\{\frac{3+\sqrt{41}}{4}, \frac{3-\sqrt{41}}{4}\right\}.$$

The decimal form is $\{2.351, 0.851\}$.

The Discriminant

The quadratic formula involves the square root of a number, $b^2 - 4ac$, which by the first axiom of order can be negative, zero, or positive. Because it determines the nature of the solutions of the equation, $b^2 - 4ac$ is called the **discriminant** of the quadratic equation.

(1) If $b^2 - 4ac < 0$, the equation has no solutions in the set of real numbers, because the square root of a negative number is undefined in the real numbers.

(2) If $b^2 - 4ac = 0$, the formula reduces to $x = \dfrac{-b}{2a}$ and thus the solution set contains only one member.

(3) If $b^2 - 4ac > 0$, the solution set contains two members, which may be rational or irrational. (See Section 1.8.)

Examples Discuss the nature of the solution set of the quadratic equation without solving.

(a) $x^2 + 2x + 5 = 0$
(b) $x^2 - 6x + 9 = 0$
(c) $2x^2 + 3x - 2 = 0$

Solutions (a) $a = 1$, $b = 2$, and $c = 5$. Hence, $b^2 - 4ac = 4 - 20 = -16$. Thus, the equation has no real number solutions.

(b) $a = 1$, $b = -6$, and $c = 9$. Hence, $b^2 - 4ac = 36 - 36 = 0$. Thus, the solution set contains one member.

(c) $a = 2$, $b = 3$, and $c = -2$. Hence, $b^2 - 4ac = 9 + 16 = 25$. Thus, the solution set contains two members. Also, since $\sqrt{25}$ is a rational number, the two solutions are rational numbers.

Example Find a value for k so that the solution set of $x^2 + 8x - 3 + k = 0$ will contain only one member.

Solution For the solution set to contain only one member, $b^2 - 4ac = 0$. Then, since $a = 1$, $b = 8$, and $c = -3 + k$,

$$b^2 - 4ac = 64 - 4(-3 + k) = 0$$
$$64 + 12 - 4k = 0$$
$$-4k = -76$$
$$k = 19.$$

EXERCISE 5.2

WRITTEN

Without solving, state whether or not the solutions of each equation in Problems 1-6 are real numbers.

Examples (a) $x^2 - 6x + 8 = 0$ (b) $4x^2 + 2x + 3 = 0$

Solutions (a) $a = 1$, $b = -6$, $c = 8$
$$b^2 - 4ac = (-6)^2 - 4(1)(8)$$
$$= 36 - 32$$
$$= 4$$

Thus, $b^2 - 4ac > 0$, and the solutions are real numbers.

(b) $a = 4$, $b = 2$, $c = 3$
$$b^2 - 4ac = 2^2 - 4(4)(3)$$
$$= 4 - 48$$
$$= -44$$

Thus, $b^2 - 4ac < 0$, and there are no real number solutions.

1. $x^2 - 2x + 5 = 0$
2. $9y^2 - 13y + 5 = 0$
3. $6 + 5x - 6x^2 = 0$
4. $9y^2 + 5y + 4 = 0$
5. $16x^2 + 24x - 9 = 0$
6. $y^2 + 3y - 28 = 0$

In Problems 7-24, find the solution set of each quadratic equation by using the quadratic formula. Write any irrational solutions in simplest radical form or in decimal form, as directed by your instructor.

Example $2x^2 - x - 10 = 0$

Solution $a = 2$, $b = -1$, $c = -10$. Substitution of these values in the formula gives
$$x = \frac{-(-1) \pm \sqrt{(-1)^2 - 4(2)(-10)}}{2(2)}$$
$$= \frac{1 \pm \sqrt{1 + 80}}{4}$$
$$= \frac{1 \pm \sqrt{81}}{4}$$
$$= \frac{1 \pm 9}{4}.$$

The solution set is $\{-2, \frac{5}{2}\}$.

7. $6x^2 - x - 12 = 0$
8. $3y^2 + y - 4 = 0$
9. $9x^2 + 4 = 12x$
10. $4y^2 + 1 = 4y$
11. $9x^2 - 6x = 2$
12. $4y^2 + 8y = 5$
13. $10x^2 - 11x + 3 = 0$
14. $8y^2 + 8y - 7 = 0$

15. $49x^2 - 16 = 0$
17. $5x^2 + 2x = 0$
19. $x^2 = 2x + 24$
21. $x^2 = 4x - 2$
23. $9 - 10x + 2x^2 = 0$

16. $9y^2 - 49 = 0$
18. $25y^2 + 20y = 0$
20. $y^2 = 8 - 2y$
22. $2y^2 - 9 = -6y$
24. $1 + 2x - 5x^2 = 0$

In Problems 25–30 find a value for k such that the solution set of the quadratic equation will contain one and only one member.

Examples   (a) $x^2 + 5x + k = 0$ (b) $2x^2 + kx + 5 = 0$

Solutions   (a) $a = 1$, $b = 5$, and $c = k$. Hence, $b^2 - 4ac = 25 - 4k$. If the solution set is to contain one and only one member, then

$$25 - 4k = 0,$$

from which

$$k = \tfrac{25}{4}.$$

(b) $a = 2$, $b = k$, and $c = 5$. Hence, $b^2 - 4ac = k^2 - 40 = 0$. Hence, $k = 2\sqrt{10}$ or $-2\sqrt{10}$.

25. $x^2 + 4x + k = 0$
27. $4x^2 + 12x + k = 0$
29. $x^2 + kx + 3 = 0$

26. $x^2 - 9x + k = 0$
28. $9x^2 - 30x + k = 0$
30. $x^2 + kx + 8 = 0$

In Problems 31–36, find a set of values for k such that the solution set of the quadratic equation will contain two real numbers.

Example $x^2 + 6x + k = 0$

Solution $a = 1$, $b = 6$, and $c = k$. Hence, $b^2 - 4ac = 36 - 4k$. If the solution set of the equation is to contain two real numbers, then

$$36 - 4k > 0$$
$$-4k > -36$$
$$4k < 36$$
$$k < 9.$$

Therefore, the required set of values for k is $\{k \mid k < 9\}$.

31. $x^2 - 8x + k = 0$
33. $x^2 - kx + 9 = 0$
35. $x^2 + 2x + k + 3 = 0$

32. $2x^2 + 4x + k = 0$
34. $2x^2 + kx + 8 = 0$
36. $x^2 - 2x + k - 2 = 0$

5.3   Equations Involving Radicals

The principal nth root ($n \in N$) of a, $\sqrt[n]{a}$, has been defined for odd indices (n an odd integer) as a positive number if $a > 0$ and as a negative number if $a < 0$. However, an even nth root of a is a real number if and only if $a > 0$. Since we are considering only those expressions that represent real numbers, the nth root of an expression containing one variable can be defined if the replacement set is restricted so that the expression represents a non-negative number when n is even.

Examples State the restrictions on the replacement set for x so that each indicated root will represent a real number.

(a) $\sqrt{x-7}$ (b) $\sqrt[4]{x^2-2}$ (c) $\sqrt[6]{x^2+2}$

Solutions (a) $x \geq 7$
(b) $|x| \geq \sqrt{2}$
(c) No restriction, because $x^2 + 2$ is always positive.

Now turn your attention to the solution of equations containing radicals. The following theorem can be used as a basis for finding the solution sets of such equations.

THEOREM 5.3-1 If $a, b \in \mathbf{R}$, $n \in \mathbf{N}$, and $a = b$, then $a^n = b^n$.

PROOF:

STATEMENTS	JUSTIFICATIONS
1. $a = a$	1. Reflexive property of equality
2. $\underbrace{a \cdot a \cdot a \cdots a}_{n \text{ factors}} = \underbrace{a \cdot a \cdot a \cdots a}_{n \text{ factors}}$	2. Repeated applications of Theorem 1.4-2 (If $a = b$, $c \neq 0$, then $a \cdot c = b \cdot c$)
3. $a = b$	3. Hypothesis
4. $\underbrace{a \cdot a \cdot a \cdots a}_{n \text{ factors}} = \underbrace{b \cdot b \cdot b \cdots b}_{n \text{ factors}}$	4. Substitution principle
5. $a^n = b^n$	5. Definition 3.0-1: Definition of an exponent

Clearing Radicals from Equations

Theorem 5.3-1 permits us to raise both members of an equation to the same power. This will enable us to eliminate radical expressions from some equations. For example, to eliminate a square root we can square both members, to eliminate a cube root we can cube both members, etc. Our discussion is limited to simplifying equations containing square roots.

Before we simplify such an equation, we must consider what happens to the solution set of an equation when both of its members are raised to some power. An equation formed in this way will have all the solutions of the original equation. However, the new equation may also have one or more solutions that are not solutions of the original. This is due to the fact that an even-numbered power of any nonzero number is always positive.

124 QUADRATIC EQUATIONS AND INEQUALITIES IN ONE VARIABLE

Example   Are $x + 4 = 1$ and $(x + 4)^2 = (1)^2$ equivalent equations?

Solution   We can see by inspection that the only solution of $x + 4 = 1$ is -3. For the second equation, we have

$$(x + 4)^2 = (1)^2$$
$$x^2 + 8x + 16 = 1$$
$$x^2 + 8x + 15 = 0$$
$$(x + 5)(x + 3) = 0$$

for which the solution set is $\{-5, -3\}$. However, -5 is not a solution of $x + 4 = 1$, because if we replace x with -5 we obtain

$$-1 = 1,$$

which is false. Therefore, the equations are not equivalent.

The following example illustrates the technique of first simplifying and then solving an equation that contains one square root.

Example   Find the solution set of $\sqrt{2x + 3} = x + 2$, $x \geq -\frac{3}{2}$.

Solution   If we square both members of the equation we obtain

$$2x + 3 = x^2 + 4x + 4$$
$$x^2 + 2x + 1 = 0,$$

whose solution set is $\{-1\}$.

However, squaring both members of an equation may not yield an equivalent equation, so the members of the solution set obtained may not be solutions of the original equation. Therefore, we must check each possible solution. The solution set in this example is checked by answering the question, "Does $\sqrt{-2 + 3} = -1 + 2$?" Since the answer is yes, the solution set of $\sqrt{2x + 3} = x + 2$ is $\{-1\}$.

If an equation contains two terms involving radicals, you will find it more practical to generate an equivalent equation with one radical in each member. However, it still may be necessary for you to raise both members of the equation to a power more than one time to obtain an equation free of radicals.

Example   Find the solution set of $\sqrt{x} + 2 - \sqrt{x + 2} = 0$.

Solution   Squaring both members of this equation would result in a term containing the product of the two radicals. To avoid such a term, we form the equivalent equation

$$\sqrt{x} + 2 = \sqrt{x + 2}.$$

Now we can square both members and equate the results to obtain

$$x + 4\sqrt{x} + 4 = x + 2,$$

from which

$$4\sqrt{x} = -2.$$

This equation obviously has no solution, since $4\sqrt{x}$ must be a non-negative number. Hence, the solution set is $\emptyset$.

Extraneous Solutions

Any solution of an equation that is not a solution of the equation from which it was derived is sometimes called **extraneous**. The first example in the previous section illustrates how an extraneous solution can be introduced by squaring both members of an equation. Whenever the squaring process is used, all possible solutions must be checked by substitution in the original equation.

EXERCISE 5.3

WRITTEN

Find the solution set in **R** for each equation.

Examples (a) $\sqrt{x+1} = 4$ (b) $\sqrt{x}\sqrt{x+1} = \sqrt{6}$

Solutions (a) $\sqrt{x+1} = 4$
$$x + 1 = 16$$
$$x = 15$$
Since $\sqrt{15 + 1} = 4$, the solution set is $\{15\}$.

(b) $\sqrt{x}\sqrt{x+1} = \sqrt{6}$
$$\sqrt{x^2 + x} = \sqrt{6}$$
$$x^2 + x = 6$$
$$x^2 + x - 6 = 0$$
$$(x+3)(x-2) = 0$$
$$x = -3, \quad x = 2$$

-3 is not a solution, because $\sqrt{-3}$ and $\sqrt{-3+1}$ are not real numbers. 2 is a solution since $\sqrt{2}\sqrt{3} = \sqrt{6}$.
The solution set is $\{2\}$.

1. $\sqrt{x} = 4$
2. $\sqrt{z-1} = 5$
3. $\sqrt{x+3} = 5$
4. $\sqrt{y-5} = 2$
5. $\sqrt{7x-3} = 2x - 3$
6. $\sqrt{3x+10} = x + 4$
7. $(2 + 3x)^{1/2} = -4$
8. $(3y - 1)^{1/2} = -2$
9. $\sqrt{x}\sqrt{2x-3} = 3$
10. $\sqrt{x-3}\sqrt{x+3} = 4$
11. $\sqrt{x+3}\sqrt{x-9} = 8$
12. $\sqrt{x-4}\sqrt{x+4} = 3$

Example $\sqrt{3x+1} = \sqrt{x} - 1$

Solution Since the left-hand member contains a single radical, we square both members and equate the results to obtain

$$3x + 1 = x - 2\sqrt{x} + 1$$
$$2x = -2\sqrt{x}$$
$$4x^2 = 4x$$
$$4x^2 - 4x = 0$$
$$4x(x - 1) = 0$$
$$x = 0, \quad x = 1.$$

If $x = 0$, $\sqrt{0+1} \neq \sqrt{0} - 1$. If $x = 1$, $\sqrt{4} \neq \sqrt{1} - 1$. The solution set is $\emptyset$.

13. $\sqrt{x+5} - 1 = \sqrt{x}$
14. $\sqrt{4x+1} = \sqrt{2x} + 1$
15. $2\sqrt{x+1} = 1 - 2\sqrt{x}$
16. $\sqrt{2x-2} = 2 + \sqrt{4x+3}$
17. $\sqrt{7-4x} = \sqrt{3-2x} + 1$
18. $3\sqrt{2+x} = \sqrt{4x+1} + 3$
19. $\sqrt{2-8x} + 2\sqrt{1-6x} = 2$
20. $4\sqrt{y} + \sqrt{1+16y} = 5$
21. $\sqrt{4x+17} + \sqrt{x+1} = 4$
22. $\sqrt{y+7} + \sqrt{y+4} = 3$
23. $(x-3)^{1/2} + (x+5)^{1/2} = 4$
24. $(x-2)^{1/2} + (2x+5)^{1/2} = 3$

Example $\sqrt{3x+2} = 3\sqrt{x} - \sqrt{2}$

Solution Square both members.

$$3x + 2 = 9x - 6\sqrt{2x} + 2$$
$$6\sqrt{2x} = 6x$$
$$\sqrt{2x} = x$$

Square both members again:

$$2x = x^2$$
$$x^2 - 2x = 0$$
$$x(x - 2) = 0$$
$$x = 0, \quad x = 2$$

If $x = 0$, $\sqrt{2} \neq -\sqrt{2}$. If $x = 2$, $\sqrt{8} = 3\sqrt{2} - \sqrt{2}$, or $2\sqrt{2} = 2\sqrt{2}$. Thus, the solution set is $\{2\}$.

25. $\sqrt{x} + \sqrt{2} = \sqrt{x+2}$
26. $\sqrt{3+2x} - \sqrt{3-2x} = \sqrt{2x}$
27. $\sqrt{2z-1} + \sqrt{4z+3} = \sqrt{5}$
28. $\sqrt{2x} + \sqrt{6x+4} = 2\sqrt{4x+1}$
29. $\sqrt{x} + \sqrt{x-4} = 2$
30. $\sqrt{2\sqrt{2x}+5} - \sqrt{2x} = 2$

5.4 Equations Quadratic in Form

An equation is said to be **quadratic in form** if it can be written equivalently as a quadratic equation in terms of the same expression involving a variable. Some examples are

(a) $x^4 + 3x^2 - 5 = 0$, which can be written equivalently as $(x^2)^2 + 3(x^2) - 5 = 0$.

(b) $x + 2x^{1/2} - 3 = 0$, which can be written equivalently as $(x^{1/2})^2 + 2(x^{1/2}) - 3 = 0$.

(c) $(x - 2)^2 - 5(x - 2) + 6 = 0$.

The first example is a quadratic equation in x^2, the second is a quadratic equation in $x^{1/2}$, and the third is a quadratic equation in $(x - 2)$.

Any of the techniques for solving quadratic equations can be used for finding the solution sets of equations quadratic in form. However, to avoid some of the difficulties which may arise in completing the square or in using the quadratic formula, this discussion is limited to factoring over the set of integers.

Examples Find the solution set in **R**.

(a) $x - 5x^{1/2} + 6 = 0$
(b) $9x^4 - 13x^2 + 4 = 0$
(c) $(x - 2)^2 - 3(x - 2) - 4 = 0$

Solutions (a) The equation $x - 5x^{1/2} + 6 = 0$ can be written equivalently as

$(x^{1/2})^2 - 5(x^{1/2}) + 6 = 0$.

The left-hand member of this equation can be factored to obtain

$(x^{1/2} - 3)(x^{1/2} - 2) = 0$.

$x^{1/2} - 3 = 0$ or $x^{1/2} - 2 = 0$

$x^{1/2} = 3$ or $x^{1/2} = 2$

Then, by squaring both members of each equation (Theorem 5.3-1), we find

$x = 9$ or $x = 4$.

Since the squaring process has been used, it is necessary to check each value by substituting in the original equation.

Is $9 - 5(9^{1/2}) + 6 = 0$? $[9 - 5(3) + 6 = 9 - 15 + 6 = 0]$. Yes.

Is $4 - 5(4^{1/2}) + 6 = 0$? $[4 - 5(2) + 6 = 4 - 10 + 6 = 0]$. Yes.

Since both values make the original equation a true statement, the solution set is $\{4, 9\}$.

(b) The equation $9x^4 - 13x^2 + 4 = 0$ can be written equivalently as

$9(x^2)^2 - 13(x^2) + 4 = 0$.

$(9x^2 - 4)(x^2 - 1) = 0$

$9x^2 - 4 = 0$ or $x^2 - 1 = 0$.

$x^2 = \frac{4}{9}$ or $x^2 = 1$

Then, equating the square roots of both members,

$|x| = \frac{2}{3}$ or $|x| = 1$,

from which the solution set $\{-1, -\frac{2}{3}, \frac{2}{3}, 1\}$ is obtained. You can show by substituting in the original equation that each of these is a solution.

(c) You may find it helpful in some cases to replace an expression with another variable. If you let $y = x - 2$, the equation $(x - 2)^2 - 3(x - 2) - 4 = 0$ becomes

$$y^2 - 3y - 4 = 0.$$
$$(y + 1)(y - 4) = 0$$
$$y = -1 \quad \text{or} \quad y = 4$$

Then,

$x - 2 = -1$ or $x - 2 = 4$.

The solution set is $\{1, 6\}$, which you can prove by substituting these values for x in the original equation.

EXERCISE 5.4

WRITTEN

Follow the appropriate preceding example to find the solution set in R for each equation.

1. $2x - 7x^{1/2} + 3 = 0$
2. $3y + 2y^{1/2} - 1 = 0$
3. $x^4 - 5x^2 + 4 = 0$
4. $2y^4 + 17y^2 - 9 = 0$
5. $x^4 - 2x^2 - 24 = 0$
6. $8x^4 + 7x^2 - 1 = 0$
7. $x^{2/3} - 2x^{1/3} - 8 = 0$
8. $y^{2/3} - 2y^{1/3} - 15 = 0$
9. $x^{-2} - x^{-1} - 6 = 0$
10. $y^{-2} + 9y^{-1} - 10 = 0$
11. $x^{-4} - 10x^{-2} + 9 = 0$
12. $y^{-4} - 2y^{-2} - 3 = 0$
13. $3x^{2/3} + 7x^{1/3} - 20 = 0$
14. $4x^{2/3} - 12x^{1/3} + 9 + 0$
15. $4y^{-1} + 3y^{-1/2} - 1 = 0$
16. $3x^{-1} - 5x^{-1/2} + 2 = 0$
17. $(x - 1)^2 - 2(x - 1) - 15 = 0$
18. $(z + 1)^2 - 5(z + 1) + 6 = 0$
19. $(y^2 + 5y)^2 - 8(y^2 + 5y) - 84 = 0$
20. $(x^2 - 1)^2 + 3(x^2 - 1) + 2 = 0$
21. $(x^2 - 5) - 5(x^2 - 5)^{1/2} + 6 = 0$
22. $(x^2 - 1) - 2(x^2 - 1)^{1/2} - 8 = 0$

5.5 Quadratic Inequalities

In Section 2.1 you learned how to generate equivalent first-degree inequalities from a given inequality. You can use these same techniques to generate equivalent inequalities from an inequality of second degree, which is also called a **quadratic inequality** That is, if $a, b, c \in R$, then

$a < b$ is equivalent to $a + c < b + c$,
$a < b$ is equivalent to $ac < bc$ if $c > 0$,
$a < b$ is equivalent to $ac > bc$ if $c < 0$.

Remember that these theorems can be applied to inequalities of either sense. The standard form of a quadratic inequality is either

$$ax^2 + bx + c < 0 \quad \text{or} \quad ax^2 + bx + c > 0, \qquad a \neq 0.$$

In either case equality may also be included ($\leq, \geq$).

As with linear inequalities, the solution set must be specified in such a way that it is easily determined that any number in the set is a solution and any number not in the set is not a solution.

Any of the techniques for solving quadratic equations can be applied to a quadratic inequality in standard form, but we will limit this discussion to those where the left-hand member ($ax^2 + bx + c$) is factorable over the set of integers. The following example illustrates this procedure.

Example   Specify the solution set of $x^2 + 4x > 12$.

Solution   We first form the equivalent inequality in standard form, which is

$$x^2 + 4x - 12 > 0.$$

Next, we factor the left-hand member over the integers to obtain

$$(x + 6)(x - 2) > 0.$$

This last statement asserts that the product of two numbers is a positive number. Recall that there are two possibilities for such a product: either both factors are positive or both factors are negative.

If we consider both factors to be positive, then

$$x + 6 > 0 \quad \text{and} \quad x - 2 > 0,$$

from which

$$x > -6 \quad \text{and} \quad x > 2.$$

This means that any member of

$$\{x \mid x > -6\} \cap \{x \mid x > 2\} = \{x \mid x > 2\}$$

is a solution of $x^2 + 4x > 12$. Figure 5.5-1, called a **sign graph**, illustrates that if x is replaced with any number greater than 2, both factors are positive.

If we consider both factors to be negative, then

$$x + 6 < 0 \quad \text{and} \quad x - 2 < 0,$$

from which

$$x < -6 \quad \text{and} \quad x < 2.$$

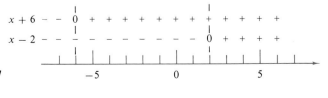

Figure 5.5-1

Therefore, any member of

$$\{x \mid x < -6\} \cap \{x \mid x < 2\} = \{x \mid x < -6\}$$

is a solution of $x^2 + 4x > 12$. The sign graph shows that if x is replaced with any number less than -6, both factors are negative.

Finally, since the two factors can both be positive *or* they can both be negative, we specify the solution set to be

$$\{x \mid x > 2\} \cup \{x \mid x < -6\}.$$

The graph of the solution set is shown in Figure 5.5-2.

Figure 5.5-2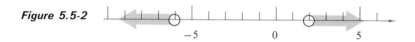

The example above shows two techniques for specifying the solution set, an analytic method and a graphic method. Both methods depend upon factoring. In the following example, only the graphic technique is used.

Example Specify the solution set of $x^2 + x - 6 < 0$.

Solution The inequality is in standard form, so we can proceed by factoring the left-hand member to obtain

$$(x + 3)(x - 2) < 0.$$

This statement asserts that the product of two numbers is a negative number. To obtain such a product, one factor must be positive and the other must be negative.

If we prepare a sign graph as in the preceding example, it will look like Figure 5.5-3. By inspecting the sign graph we can observe that only

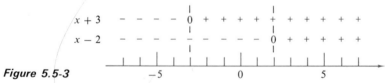

Figure 5.5-3

numbers that are greater than -3 and less than 2 meet the requirements to obtain a negative product. Hence, we specify the solution set to be

$$\{x \mid x > -3\} \cap \{x \mid x < 2\} = \{x \mid -3 < x < 2\}.$$

The graph of the solution set is shown in Figure 5.5-4.

Figure 5.5-4

EXERCISE 5.5

Specify the solution set of each inequality and then represent the solution set on a line graph.

In Problems 1–10 use the sign graph technique.

Example $(x - 1)(x + 2) > 0$

Solution The sign graph for $x - 1$ and $x + 2$ is given in Figure 5.5-5. Both factors must be positive or they must both be negative for the product to be positive. Hence, the solution set is

$$\{x \mid x < -2\} \cup \{x \mid x > 1\}.$$

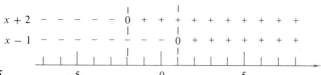

Figure 5.5-5

The graphic representation of the solution set is shown in Figure 5.5-6.

Figure 5.5-6

1. $x(x - 1) > 0$
2. $(x + 1)(x - 2) > 0$
3. $(x + 2)(x - 3) > 0$
4. $(x - 3)(x - 6) > 0$
5. $(2x + 3)(x - 2) > 0$
6. $(3x + 2)(2x + 5) > 0$
7. $x^2 - 7x + 10 > 0$
8. $x^2 + 5x + 6 > 0$
9. $2x^2 + 5x > -3$
10. $x^2 - 6x > -8$

In Problems 11–20 use the analytic method.

Example $3x^2 - 5x \leq 0$

Solution Factor the left-hand member to obtain

$$x(3x - 5) \leq 0.$$

Since the product is negative, one factor must be positive and the other must be negative.

If $x \geq 0$ and $3x - 5 \leq 0$, we have

$$\{x \mid x \geq 0\} \cap \{x \mid 3x - 5 \leq 0\} = \{x \mid x \geq 0\} \cap \{x \mid x \leq \tfrac{5}{3}\}$$
$$= \{x \mid 0 \leq x \leq \tfrac{5}{3}\}.$$

If $x \leq 0$ and $3x - 5 \geq 0$, we have

$$\{x \mid x \leq 0\} \cap \{x \mid x \geq \tfrac{5}{3}\}.$$

Since there is no number that satisfies both conditions, this is the empty set.

Hence, we specify the solution set to be

$\{x \mid 0 \leq x \leq \frac{5}{3}\}$.

The graph of this set is shown in Figure 5.5-7.

Figure 5.5-7

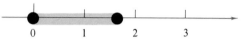

11. $x(2x - 7) \leq 0$
12. $x(x + 3) \leq 0$
13. $(x + 2)(x + 5) < 0$
14. $(x + 1)(x - 2) < 0$
15. $(x - 2)(x - 3) < 0$
16. $(x + 5)(x - 5) < 0$
17. $16 - x^2 < 0$
18. $(1 + x - 2x^2) \leq 0$
19. $3x^2 - 2x - 5 \leq 0$
20. $3x^2 - 5x + 2 \leq 0$

Use either method to specify the solution sets of the inequalities in Problems 21–26.

21. $(x + 2)(x - 2)(x - 5) < 0$
22. $x(x - 2((x + 3) > 0$
23. $(x - 1)(x + 3)(x + 4) > 0$
24. $x(x + 1)(x - 3) < 0$
25. $(x + 1)(x - 3)(x - 4) < 0$
26. $(x - 4)(x + 2)(x + 5) < 0$

27. If $x > 1$, show without graphing that $x < x^2$.
28. If $0 < x < 1$, show that $x > x^2$, without graphing.

5.6 Word Problems

In Section 2.4 we discussed a technique for the analysis and solution (if a solution exists) of word problems whose mathematical models could be expressed as equations of the first degree in one variable. A similar technique can be applied to word problems whose mathematical models are equations of the second degree in one variable.

The procedure consists of the following four steps:

(1) Represent each quantity to be determined in terms of the same variable.
(2) Translate the word statement into an algebraic statement (equation or inequality).
(3) Find the solution set of the mathematical statement.
(4) Interpret the solution set in terms of the original statement of the problem.

This technique is illustrated in the following example.

Example   The sum of the squares of two consecutive positive integers is 41. Find the integers.

Solution   (1) If we let x represent an integer, then $x + 1$ represents the next consecutive integer.
(2) $x^2 + (x + 1)^2 = 41$
(3) $x^2 + x^2 + 2x + 1 = 41$
$2x^2 + 2x - 40 = 0$

$$2(x^2 + x - 20) = 0$$
$$(x + 5)(x - 4) = 0$$

The solution set is $\{-5, 4\}$.

(4) -5 cannot be a solution to the problem because the statement calls for positive integers. The required integers are 4 and 5, since $16 + 25 = 41$.

EXERCISE 5.6

For each problem:

(1) Write a statement or statements to express the required quantity or quantities in terms of one variable.
(2) Write a suitable equation or inequality.
(3) Find the solution set of the mathematical statement.
(4) Interpret this solution set in terms of the stated problem.

Example   Find two numbers whose sum is 27 and whose product is 92.

Solution (1) If we let x represent one number, then $27 - x$ represents the other number.
(2) The product of the two numbers is

$$x(27 - x) = 92.$$

(3) This may be transformed into an equivalent quadratic equation:

$$27x - x^2 = 92$$
$$x^2 - 27x + 92 = 0$$
$$(x - 4)(x - 23) = 0$$

The solution set is $\{4, 23\}$.
(4) The two numbers are 4 and 23, since $4 + 23 = 27$ and $4 \cdot 23 = 92$.

1. Find two consecutive positive even integers whose product is 168.
2. Find two numbers whose sum is 27 and whose product is 180.
3. Find two numbers whose difference is 7 and whose product is 198.
4. The sum of two numbers is 12, and the sum of their squares is 80. What are the numbers?
5. The product of two consecutive even integers is 288. Find the integers.
6. The product of two consecutive odd integers is 143. Find the integers.
7. Separate 42 into two parts such that the first is the square of the second.
8. Separate the number a into two parts such that the first is the square of the second.
9. Find three positive integers such that their sum is 44, the second is twice the first, and the third is twice the square of the first.
10. The sum of three positive integers is 21. Find the three numbers if the second is three times the first and the third is the square of the first.

Example   The base of a triangle is 4 feet shorter than its altitude, and the area of the triangle is 126 square feet. Find the measures of the base and altitude.

Solution (1) Let x represent the measure of the base in feet. Then $x + 4$ represents the measure of the altitude in feet.

(2) Since the area of a triangle is one-half the product of the base and the altitude, we have

$\frac{1}{2}(x)(x + 4) = 126.$

(3) $\qquad x^2 + 4x = 252$

$x^2 + 4x - 252 = 0$

$(x + 18)(x - 14) = 0$

The solution set is $\{-18, 14\}$.

(4) Since the measure of the base must be positive, the measure of the base is 14 feet, and the measure of the altitude is $14 + 4 = 18$ feet.

11. The base of a triangle is 5 feet longer than the altitude. Find the measures of the base and the altitude in feet if the area of the triangle is 42 square feet.
12. Find the dimensions of a rectangle whose perimeter is 64 feet and whose area is 252 square feet.
13. 42 square yards of material were required to lay wall-to-wall carpeting in a rectangular room that is 3 feet longer than it is wide. What are the dimensions of the room?
14. A square piece of floor covering lacked 6 feet of being long enough to completely cover the floor of a room whose area is 135 square feet. What are the dimensions of a piece of the same material added so that the floor could be completely covered?
15. If the side of a square is increased by 2 inches, the area is increased by 24 square inches. Find the side of the original square.
16. If the radius of a circle is increased by 3.5 inches, the area is increased by 192.5 square inches. Find the radius of the original circle. (Use the value $\pi = \frac{22}{7}$, which is a rational approximation of the irrational number 3.1416 . . . .)
17. An open box is to be made by cutting a 2-inch square from each corner of a square piece of cardboard and then turning up the sides. If the volume of the resulting box is to be 128 cubic inches, what is the length of the side of the original square?
18. The area of a rectangle whose dimensions are 12 and 8 inches is doubled by adding two strips of equal width, one along one end and the other along one side to form another rectangle. Find the width of the strips.
19. A rectangular piece of cardboard twice as long as it is wide is to be made into an open-topped box by cutting a 2-inch square from each corner and then turning up the sides. Find the dimensions of the cardboard if the box measures 480 cubic inches.
20. A rectangular building twice as long as it is wide is divided into two rooms by a partition 30 feet from and parallel to one of the shorter walls. If the area of the larger room is 3500 square feet, find the dimensions of the building.

Example It took a certain number of men twice as many days to pave a street as there were men. If 12 more men had been employed, the paving would have been done in 5 days more than half the time. How many men were employed?

Solution (1) Let x represent the number of men. Then $2x$ represents the number of days required to pave the street.
(2) Here we introduce the concept of work called *man-days*. The job requires a number of man-days equal to the product of the number of men and the number of days. Thus,

$x(2x) = (x + 12)(x + 5).$

(3) $\quad\quad\quad 2x^2 = x^2 + 17x + 60$

$\quad\quad x^2 - 17x - 60 = 0$

$\quad\quad (x - 20)(x + 3) = 0$

The solution set is $\{-3, 20\}$.

(4) Since a number of men cannot be a negative number, the number of men employed was 20.

21. A man made a trip by automobile that required a number of hours equal to one-fifth his average speed. If his average speed had been 10 miles per hour faster, he would have made the trip in $6\frac{2}{5}$ hours. Find his average speed.

22. The number of hours required for a man to make a trip by automobile was equal to one-twelfth his average speed. It would have taken him 1 hour longer if his speed had been 10 miles per hour less. Find his average speed on the trip.

23. The measure of one dimension of a rectangle is three times the measure of the other. If the shorter dimension is increased by 4 inches and the longer dimension is decreased by 2 inches, the area is doubled. Find the dimensions of the original rectangle.

24. A man bought a number of packing crates and found that the cost of each in cents was three times the number of crates. If the price had been one-third less per crate, he could have purchased 10 more for the same amount of money. How many crates did he buy?

Example   The product of two consecutive odd integers is less than 35. Between what two integers must the smaller lie?

Solution   (1) Let x represent the smaller integer. Then $x + 2$ represents the larger integer.

(2) Since the product is less than 35, we have

$\quad\quad x(x + 2) < 35.$

(3) $\quad\quad\quad x^2 + 2x < 35$

$\quad\quad x^2 + 2x - 35 < 0$

$\quad\quad (x + 7)(x - 5) < 0$

Since $x + 7 > x - 5$ for any real number replacement of x, $x + 7 > 0$ and $x - 5 < 0$. Hence, the solution set can be designated as

$\{x \mid -7 < x < 5\}.$

(4) The smaller integer must be between -7 and 5.

25. The product of two consecutive positive integers is less than 56. Between what two non-negative integers must the smaller lie?

26. The product of two consecutive even integers is less than 48. Between what two integers must the smaller lie?

27. One real number is 2 less than another, and their product is less than 63. Specify the greatest integer less than, and the least integer greater than, the smaller number.

28. The sum of two positive real numbers is 15, and their product is less than 50. Specify the greatest integer less than, and the least integer greater than, the lesser number.

29. Between what two integers must the measure of the side of the square in Problem 17 lie so that the volume of the box will be between 108 and 128 cubic inches?

30. The base of a triangle is to be 4 inches shorter than its altitude. Between what two integers must the measure of the base be if the area is between 6 and 16 square inches?

Summary

1. An open sentence of the form

$$ax^2 + bx + c = 0$$

   (the $=$ may be replaced by either $>$, $<$, $\geq$, or $\leq$) is a **quadratic equation (inequality)** in standard form. [5.0]

2. If the left-hand member of a quadratic equation in standard form is factorable over the real numbers into two first-degree factors, the solution set can be determined by setting each factor containing the variable equal to zero and then finding the solutions of the resulting linear equations. [5.1]

3. If the left-hand member of the equation $ax^2 + bx + c = 0$ is factored by completing the square, and each factor is equated to zero, the resulting equations can be combined to form

$$x = \frac{-b \pm \sqrt{b^2 - 4ac}}{2a}.$$

   This equation is called the **quadratic formula**. [5.2]

4. Since square roots of negative numbers are not real numbers, the quantity $b^2 - 4ac$, called the **discriminant**, determines the nature of the solutions of a quadratic equation as follows:

$$b^2 - 4ac < 0; \text{ no real solutions}$$
$$b^2 - 4ac = 0; \text{ one solution}$$
$$b^2 - 4ac > 0; \text{ two distinct solutions}$$
   [5.2]

5. An equation free of radicals can sometimes be formed by squaring both members of an equation containing square roots of expressions. However, such an equation *may* not be equivalent to the original equation. [5.3]

6. An equation that can be written equivalently as a quadratic equation in terms of a single expression is said to be **quadratic in form**. [5.4]

7. If the left-hand member of a quadratic inequality in standard form can be factored over the real numbers, the properties of the product of two signed numbers can be helpful in specifying the solution set. That is, the two factors of a positive product are either both positive or both negative, and the two factors of a negative product have different signs. [5.5]

8. A **sign graph** is a device that shows the intervals of values in which an expression is negative and in which it is positive. [5.5]

9. Some word problems can be solved by solving an appropriate quadratic equation or quadratic inequality. [5.6]

REVIEW EXERCISES

SECTION 5.1

In Problems 1–6, find the solution set of each equation by factoring. Rewrite the equation in standard form if necessary.

1. $x^2 - 5x + 6 = 0$ 2. $8x^2 + 6x - 9 = 0$
3. $3x^2 - 7 = 4x$ 4. $8x^2 - 10x = 3$

5. $2(x - 1)^2 - 3(x - 1) - 5 = 0$
6. $3(x + 1)^2 + 7(x + 1) + 2 = 0$

SECTION 5.2

Without solving, state whether or not the solutions of the equations in Problems 7–10 are real numbers.

7. $3x^2 - x - 5 = 0$ 8. $x^2 + 4x + 6 = 0$
9. $x^2 + x - 1 = 0$ 10. $x^2 - 8x - 13 = 0$

In Problems 11–16, solve each equation by using the quadratic formula. Give irrational solutions in simplest radical form.

11. $x^2 + 2x - 15 = 0$ 12. $x^2 + 4x - 12 = 0$ 13. $2x^2 - 5 = 0$
14. $6x^2 + 5x - 6 = 0$ 15. $3x^2 + 7x - 5 = 0$ 16. $2x^2 + 3x - 8 = 0$

17. Find a value for k such that $x^2 - 8x + k = 0$ has only one solution.
18. Find a set of values for k such that $kx^2 + 6x + k = 0$ has two distinct solutions.

SECTION 5.3

Find the solution set of each equation in Problems 19–24.

19. $\sqrt{x + 8} = 1$ 20. $\sqrt{x + 1} - 3 = 2$
21. $\sqrt{3x + 10} = x + 4$ 22. $\sqrt{x + 5} + \sqrt{x} = 5$
23. $\sqrt{x + 4} + \sqrt{x - 4} = 4$ 24. $\sqrt{x + 4} + \sqrt{x + 7} = 3$

SECTION 5.4

Find the solution set of each equation in Problems 25–30.

25. $2x - 5x^{1/2} + 3 = 0$ 26. $x^{2/3} - 5x^{1/3} + 6 = 0$
27. $3x^{-2} - 7x^{-1} + 2 = 0$ 28. $x^4 - 13x^2 + 36 = 0$
29. $(x^2 - 1)^2 - 11(x^2 - 1) + 24 = 0$ 30. $(x^2 + 2) - 5\sqrt{x^2 + 2} + 6 = 0$

SECTION 5.5

Specify the solution set of each inequality in Problems 31–35.

31. $(x + 3)(x - 2) > 0$ 32. $(2x + 3)(3x - 2) < 0$
33. $x^2 - 7x + 10 < 0$ 34. $2x^2 - x - 3 > 0$
35. $(x + 2)(x - 1)(x - 5) > 0$

SECTION 5.6

Use the four-part method of analysis to solve Problems 36 and 37.

36. A man bought a number of trees for $100. At a later date, when the trees were on sale for $5 less per tree, the man was able to buy 10 more than the original number for the same amount ($100). How many trees did he buy at each time?
37. The product of two positive consecutive even integers is less than 48. What is the greatest value of the smaller integer?

Fractions

In the general discussion of polynomials, we have used the symbols $P(x)$, $Q(x)$, $S(x)$, and $T(x)$ to represent unspecified polynomials, and an expression of the form $P(x)/Q(x)$ to represent the quotient of two polynomials. In Section 4.3, we discussed such quotients, and, by the division algorithm, we represented them in the form

$$\frac{P(x)}{Q(x)} = S(x) + \frac{T(x)}{Q(x)}, \qquad Q(x) \neq 0.$$

In this chapter, we again discuss quotients of polynomials, but we consider $P(x)/Q(x)$ as a fraction.

Our discussion has been limited to real polynomials, polynomials that represent a real number for every real number replacement of the variable. As a result, all the theorems that have been stated about real numbers of the form a/b, $a, b \in \mathbf{R}$, $b \neq 0$, are true for $P(x)/Q(x)$ where $Q(x) \neq 0$. Thus, we will not need to prove additional theorems. However, we will restate each theorem as the need arises and give the proof of one of them.

6.1 Equivalent Fractions

Whenever the form of a fraction is changed or when two or more fractions are combined into one fraction by an operation, we must be certain that the result is equivalent to the original. Theorem 1.4-9 (p. 14) is the basis for determining whether or not two fractions are equal. To apply this theorem, we find the products of the numerator of one fraction and the denominator of the other. If these two products are equal, the fractions are equal. Of course, we must restrict the replacements for x so that no denominator is equal to zero.

THEOREM 6.1-1 If $P(x)$, $Q(x)$, $S(x)$, and $T(x)$ are real polynomials, then for every real number replacement of x such that $Q(x)$ and $T(x)$ are not zero,

$$\frac{P(x)}{Q(x)} = \frac{S(x)}{T(x)}$$

if and only if $P(x) \cdot T(x) = Q(x) \cdot S(x)$.

PROOF ("If" part):

STATEMENTS	JUSTIFICATIONS
1. For any real number replacement of x, let $P(x) = a$, $Q(x) = b$, $S(x) = c$, and $T(x) = d$, where a, b, c, $d \in \mathbf{R}$.	1. Definition of a real polynomial
2. If $b, d \neq 0$ and $ad = bc$, then $a/b = c/d$.	2. Theorem 1.4-9
3. Therefore, if $Q(x)$, $T(x) \neq 0$ and $P(x) \cdot T(x) = Q(x) \cdot S(x)$, then $\dfrac{P(x)}{Q(x)} = \dfrac{S(x)}{T(x)}$.	3. Substitution

The "only if" part is proved in a similar manner.

Examples Are the following statements true?

(a) $\dfrac{x^2 - 5x + 6}{x^2 - 4x + 4} = \dfrac{x - 3}{x - 2}$ (b) $\dfrac{x + 2}{x - 1} = \dfrac{x + 9}{13 - x}$

Solutions (a) If $\dfrac{x^2 - 5x + 6}{x^2 - 4x + 4} = \dfrac{x - 3}{x - 2}$, then

$(x^2 - 5x + 6)(x - 2) = (x^2 - 4x + 4)(x - 3).$

Applying the distributive law to both members of the latter equation gives

$(x^2 - 5x + 6)x + (x^2 - 5x + 6)(-2)$
$\qquad = (x^2 - 4x + 4)x + (x^2 - 4x + 4)(-3)$
$x^3 - 5x^2 + 6x - 2x^2 + 10x - 12$
$\qquad = x^3 - 4x^2 + 4x - 3x^2 + 12x - 12.$

Then, collecting the similar terms,

$x^3 - 7x^2 + 16x - 12 = x^3 - 7x^2 + 16x - 12,$

which is an identity. Therefore, the original statement is true.

(b) If $\dfrac{x+2}{x-1} = \dfrac{x+9}{13-x}$, then

$(x+2)(13-x) = (x-1)(x+9)$.

Multiplying the factors in the members of the latter equation gives

$-x^2 + 11x + 26 = x^2 + 8x - 9$,

which is not an identity. If this equation is transformed into a quadratic equation in standard form, application of the quadratic formula shows the equation to have real number solutions. Thus, the original statement is a conditional equation. However, since it is not true for every permissible replacement for x, we say that the two fractions are not equivalent.

Simplest Form of a Fraction

The word "simplify" and the phrase "write in simplest form" occur rather frequently in algebra. It is usually necessary to explain what is meant by the simplest form of an expression. In our present discussion, the simplest form of a fraction is an equivalent fraction in which the numerator and denominator have no common polynomial factors. Two polynomials that do not have a common factor are called **relatively prime polynomials**.

If a fraction is not in simplest, or lowest, terms, it can be reduced by dividing both the numerator and denominator by all of their common factors. The replacement set for the variable must be restricted so that no divisor is equal to zero.

THEOREM 6.1-2 If $P(x)$, $Q(x)$, and $S(x)$ are real polynomials, then for every real number replacement of x for which $Q(x)$, $S(x) \neq 0$,

$$\frac{P(x) \cdot S(x)}{Q(x) \cdot S(x)} = \frac{P(x)}{Q(x)} \quad \text{and} \quad \frac{P(x)}{Q(x)} = \frac{P(x) \cdot S(x)}{Q(x) \cdot S(x)}.$$

This theorem is proved in a manner similar to that of Theorem 6.1-1 by using Theorem 1.4-10.

Example Reduce $\dfrac{x^2 - 5x - 14}{x^2 - 9x + 14}$ to lowest terms.

Solution We begin by factoring both the numerator and denominator.

$$\frac{x^2 - 5x - 14}{x^2 - 9x + 14} = \frac{(x+2)(x-7)}{(x-2)(x-7)}$$

Then, if $x \neq 7$, we divide both the numerator and denominator by $x - 7$ to obtain $\dfrac{x+2}{x-2}$, which is in lowest terms because $x + 2$ and $x - 2$ have no common factor.

Building Fractions to Higher Terms

Theorem 6.1-2 can also be applied to build a fraction to higher terms, by multiplying both the numerator and denominator by the same nonzero polynomial. One technique for building fractions that you can use in Section 6.2 is illustrated in the following example.

Example Find a fraction equivalent to $\dfrac{x-1}{x+2}$, $x \neq -2$, such that the denominator is $x^2 - 3x - 10$.

Solution Evidently, the denominator $x + 2$ must be multiplied by some polynomial to produce the new denominator $x^2 - 3x - 10$. One way in which the multiplier can be determined is to factor the new denominator.

$$x^2 - 3x - 10 = (x + 2)(x - 5).$$

Thus, we see that $x + 2$ must be multiplied by $x - 5$. Then, if $x \neq 5$, we multiply both the numerator and denominator of the original fraction by $x - 5$ to obtain

$$\frac{x-1}{x+2} = \frac{(x-1)(x-5)}{(x+2)(x-5)} = \frac{x^2 - 6x + 5}{x^2 - 3x - 10},$$

which is the required fraction.

Changing Signs

Since every nonzero real number is either positive or negative, and the quotient of two such numbers is also positive or negative, there are three signs associated with a fraction: the sign of the quotient (the fraction); the sign of the numerator; and the sign of the denominator. Any two of these three signs can be changed without affecting the value of the fraction. However, since the numerator and the denominator may be polynomials of more than one term, to change the sign of one of these we must change the sign of each of its terms.

THEOREM 6.1-3 If $P(x)$ and $Q(x)$ are real polynomials, then for every real number replacement of x for which $Q(x) \neq 0$,

$$\frac{P(x)}{Q(x)} = -\frac{-P(x)}{Q(x)} = -\frac{P(x)}{-Q(x)} = \frac{-P(x)}{-Q(x)}$$

and

$$-\frac{P(x)}{Q(x)} = \frac{-P(x)}{Q(x)} = \frac{P(x)}{-Q(x)} = -\frac{-P(x)}{-Q(x)}.$$

This theorem can be proved by using Theorems 1.4-15 and 1.4-16.

Examples (a) $\dfrac{x-1}{x-2} = -\dfrac{-(x-1)}{x-2} = -\dfrac{1-x}{x-2}$

$= -\dfrac{x-1}{-(x-2)} = -\dfrac{x-1}{2-x}$

$= \dfrac{-(x-1)}{-(x-2)} = \dfrac{1-x}{2-x}$

(b) $-\dfrac{x^2 - 2x - 1}{x+2} = \dfrac{-(x^2 - 2x - 1)}{x+2}$

$= \dfrac{-x^2 + 2x + 1}{x+2}$

Many operations with fractions are more easily accomplished if the polynomials are in factored form. How can we change the sign of a polynomial if it is in factored form? Recall that in multiplying real numbers, the product is positive if an even number of the factors are negative, and negative if an odd number of the factors are negative. Since the polynomials we are using represent real numbers, we can change the sign of a polynomial by changing the signs of an odd number of its factors. On the other hand, if we change the signs of an even number of factors, we do not change the sign of the polynomial.

Examples (a) $(x+1)(1-x) = -(x+1)[-(1-x)]$
$= -(x+1)(-1+x)$
$= -(x+1)(x-1)$

(b) $(1-x)(2-x) = [-(1-x)][-(2-x)]$
$= (-1+x)(-2+x)$
$= (x-1)(x-2)$

In Example (a), since we changed the sign of only one factor, we changed the sign of the polynomial as denoted by the minus sign in front of the first parenthesis. In Example (b), we changed the signs of two factors, and thus did not change the sign of the polynomial.

Some problems can be more easily solved if the signs of one or more factors are changed. These factors may be in either the numerator or the denominator of a fraction. In such a case, we must be certain that the new fraction is equivalent to the original fraction. Hence, we must change the signs of an even number of factors, or if this is not convenient, change the signs of an odd number of the factors *and* the sign in front of the fraction.

Examples (a) $\dfrac{(x-1)(x-2)}{(x-3)(x-4)} = \dfrac{(1-x)(2-x)}{(x-3)(x-4)} = \dfrac{(x-1)(x-2)}{(3-x)(4-x)}$

$= \dfrac{(x-1)(2-x)}{(x-3)(4-x)} = \dfrac{(1-x)(2-x)}{(3-x)(4-x)}$

(b) $\dfrac{(x-1)(x-2)}{(x-3)(x-4)} = -\dfrac{(x-1)(2-x)}{(x-3)(x-4)}$

$= -\dfrac{(x-1)(x-2)}{(x-3)(4-x)}$

$= -\dfrac{(x-1)(2-x)}{(3-x)(4-x)}$

In the first three forms in Example (a) the signs of two factors were changed, and in the fourth form the signs of four factors were changed. In the first two forms of Example (b) the sign of one factor was changed, and in the third form the signs of three factors were changed. It was necessary to change the sign in front of each of these forms to avoid changing the value of the fraction. There are other ways in which signs can be changed in either example. Can you find others?

EXERCISE 6.1

WRITTEN

Assume that all denominators are nonzero in this exercise.

In Problems 1–10, use Theorem 6.1-1 to show whether or not the fractions in each pair are equivalent.

Examples (a) $\dfrac{2x^2 + 5x + 3}{2x^2 + 7x + 6}$; $\dfrac{x+1}{x+2}$ (b) $\dfrac{x-1}{x-2}$; $\dfrac{x+1}{x+2}$

Solutions (a) The fractions are equivalent if

$(x+2)(2x^2 + 5x + 3) = (x+1)(2x^2 + 7x + 6)$.

By performing the multiplications in the members of this equation, we obtain

$2x^3 + 9x^2 + 13x + 6 = 2x^3 + 9x^2 + 13x + 6$,

which is an identity. Therefore, the two fractions are equivalent.

(b) The fractions are equivalent if

$(x-1)(x+2) = (x-2)(x+1)$.

Performing the multiplications in the members of the equation gives

$x^2 + x - 2 = x^2 - x - 2$,

which is not an identity. However, if we solve the equation, we find the solution set to be $\{0\}$. Thus, the two fractions do have the same value if $x = 0$, but for any other permissible replacement of x they do not have the same value. Therefore, we say that the two fractions are not equivalent.

1. $\dfrac{2x+2}{4x-2}$; $\dfrac{x+1}{2x+1}$ 2. $\dfrac{3x-6}{6x+3}$; $\dfrac{x-2}{2x+1}$

3. $\dfrac{x^2 + 2x + 1}{x^2 - 2x + 1}$; $\dfrac{x + 1}{x - 1}$

4. $\dfrac{x^2 - 5x + 6}{x^2 - 4x + 4}$; $\dfrac{x - 3}{x - 2}$

5. $\dfrac{x^2 - 1}{x^2 - 2x + 1}$; $\dfrac{x + 1}{x - 1}$

6. $\dfrac{x^2 - 1}{x^2 + 1}$; $\dfrac{-1}{1}$

7. $\dfrac{(x + 1)^2}{(x - 1)^2}$; $\dfrac{x^2 + 1}{x^2 - 1}$

8. $\dfrac{x^2 + 2x + 1}{2x}$; $\dfrac{x^2 + 1}{1}$

9. $\dfrac{x^3 + 2x - x - 2}{x^3 + x^2 - 2x - 2}$; $\dfrac{x + 2}{x + 1}$

10. $\dfrac{x^3 - 6x^2 + 11x - 6}{x^3 - 7x - 6}$; $\dfrac{x^2 - 3x + 2}{x^2 + 3x + 2}$

Use Theorems 6.1-2 and 6.1-3 to write each fraction in Problems 11–18 in simplest form.

Examples (a) $\dfrac{x^2 - 5x + 6}{x^2 - 6x + 9}$ (b) $\dfrac{x^2 - 4}{6 - x - x^2}$

Solutions (a) Factoring both the numerator and the denominator gives

$$\dfrac{(x - 2)(x - 3)}{(x - 3)(x - 3)}.$$

Since no denominators are 0, $x \neq 3$, we divide both the numerator and the denominator by $x - 3$ to obtain the equivalent fraction

$$\dfrac{x - 2}{x - 3}.$$

(b) Factoring both the numerator and the denominator gives

$$\dfrac{(x + 2)(x - 2)}{(3 + x)(2 - x)}.$$

Note that, while the numerator and the denominator do not contain a common factor, $x - 2$ and $2 - x$ are negatives of each other. If we change the sign of an odd number of factors, in this case one, the sign of the product is changed. Thus,

$$\dfrac{(x + 2)(x - 2)}{(3 + x)(2 - x)} = \dfrac{-(x + 2)[-(x - 2)]}{(3 + x)(2 - x)}$$

$$= \dfrac{-(x + 2)(2 - x)}{(3 + x)(2 - x)} = \dfrac{-x - 2}{x + 3}.$$

11. $\dfrac{2x + 2}{6x + 4}$

12. $\dfrac{3x - 9}{9x + 6}$

13. $\dfrac{2x - 2}{x - x^2}$

14. $\dfrac{2x^2 - 4x}{6 - 3x}$

15. $\dfrac{x^2 - 4}{x^2 - 4x + 4}$

16. $\dfrac{x^2 + 2x + 1}{x^2 - 1}$

17. $\dfrac{x^2 - 7x + 10}{10 - 2x}$

18. $\dfrac{x^2 - 7x + 12}{9 - 3x}$

Find the missing numerator or denominator so that the fractions of each pair in Problems 19–28 are equivalent.

Examples (a) $\dfrac{x + 2}{x - 3}$; $\dfrac{?}{x^2 + x - 12}$ (b) $\dfrac{2x - 3}{x + 2}$; $\dfrac{9 - 4x^2}{?}$

Solutions (a) Determine the multiplier by factoring the new denominator.

$$x^2 + x - 12 = (x - 3)(x + 4)$$

This indicates that $x - 3$, the denominator of the first fraction, must be multiplied by $x + 4$. If we multiply the denominator of a fraction by a nonzero expression, we must multiply the numerator by the same expression. Hence,

$$\frac{x+2}{x-3} = \frac{(x+2)(x+4)}{(x-3)(x+4)} = \frac{x^2+6x+8}{x^2+x-12}.$$

(b) Factor the numerator of the second fraction.

$$9 - 4x^2 = (3 + 2x)(3 - 2x)$$

While neither of these factors is the numerator of the first fraction, $3 - 2x$ is the negative of $2x - 3$. Hence

$$9 - 4x^2 = -(3 + 2x)[-(3 - 2x)]$$
$$= -(3 + 2x)(2x - 3)$$

and the necessary multiplier is $-(3 + 2x)$. Multiplying both the numerator and the denominator of the given fraction by $-(3 + 2x)$, we have

$$\frac{2x-3}{x+2} = \frac{(2x-3)[-(3+2x)]}{(x+2)[-(3+2x)]} = \frac{9-4x^2}{-2x^2-7x-6}.$$

19. $\dfrac{x+1}{x-1};\ \dfrac{?}{3-3x}$

20. $\dfrac{x-4}{2x+3};\ \dfrac{8-2x}{?}$

21. $\dfrac{2x+1}{x-2};\ \dfrac{2x^2+5x+2}{?}$

22. $\dfrac{x+3}{3x-2};\ \dfrac{?}{3x^2-23x+14}$

23. $\dfrac{2x+5}{x+4};\ \dfrac{?}{2x^2+2x-24}$

24. $\dfrac{x-3}{2x-1};\ \dfrac{18+3x-3x^2}{?}$

25. $\dfrac{(x-1)(2+x)}{(x-2)(x-3)(x-4)};\ \dfrac{?}{(2-x)(3-x)(4-x)}$

26. $\dfrac{(2x+3)(3x-2)}{(2x+1)(x+2)(2x-3)};\ \dfrac{(3+2x)(2-3x)}{?}$

27. $\dfrac{x+1}{x-1};\ \dfrac{?}{x^3-1}$

28. $\dfrac{2x+3}{3x-2};\ \dfrac{8x^3+27}{?}$

6.2 Sums and Differences

You can find the sum or difference of two fractions in a manner similar to that in which you found sums and differences of rational numbers. If the fractions have the same denominator, simply write the sum or difference of the numerators over the common denominator and reduce the result to lowest terms.

THEOREM 6.2-1 If $P(x)$, $Q(x)$, and $S(x)$ are real polynomials, then for every real number replacement of x for which $Q(x) \neq 0$,

$$\frac{P(x)}{Q(x)} + \frac{S(x)}{Q(x)} = \frac{P(x) + S(x)}{Q(x)}$$

and

$$\frac{P(x)}{Q(x)} - \frac{S(x)}{Q(x)} = \frac{P(x) - S(x)}{Q(x)}.$$

Examples Write each sum or difference as a fraction in lowest terms.

(a) $\dfrac{x+2}{x} + \dfrac{4}{x}$

(b) $\dfrac{2x-3}{x^2+x+1} + \dfrac{x+4}{x^2+x+1}$

(c) $\dfrac{2x-5}{3x} - \dfrac{3}{3x}$

(d) $\dfrac{3x+4}{x^2+5} - \dfrac{2x-3}{x^2+5}$

Solutions We note that in each case the two fractions have the same denominator.

(a) $\dfrac{x+2}{x} + \dfrac{4}{x} = \dfrac{x+2+4}{x} = \dfrac{x+6}{x}$

(b) $\dfrac{2x-3}{x^2+x+1} + \dfrac{x+4}{x^2+x+1} = \dfrac{2x-3+(x+4)}{x^2+x+1}$

$$= \dfrac{3x+1}{x^2+x+1}$$

(c) $\dfrac{2x-5}{3x} - \dfrac{3}{3x} = \dfrac{2x-5-3}{3x} = \dfrac{2x-8}{3x}$

(d) $\dfrac{3x+4}{x^2+5} - \dfrac{2x-3}{x^2+5} = \dfrac{3x+4-(2x-3)}{x^2+5}$

$$= \dfrac{3x+4-2x+3}{x^2+5}$$

$$= \dfrac{x+7}{x^2+5}$$

In every case, the numerator and denominator of the result have no common factors. Hence, all the answers are in lowest terms.

The technique illustrated in the preceding examples can be used only if the fractions have the same denominator. However, since a fraction that is equivalent to a given fraction can be formed by multiplying both the numerator and denominator of the given fraction by the same nonzero number, a sum or difference of two fractions having different denominators can be simplified by replacing one or both fractions with equivalent fractions that do have the same denominator.

Least Common Multiple

Some algebra textbooks state a theorem concerning the sum of two fractions that have different denominators. The gist of the theorem is that if a/b and c/d are two fractions, then

$$\frac{a}{b} + \frac{c}{d} = \frac{ad + bc}{bd},$$

indicating that all we have to do is to change both fractions to equivalent fractions whose denominator is the product of the denominators of the given fractions. While this can be done in all cases, quite often a sum or a difference can be simplified more easily if the **least common denominator** is used. The least common denominator (LCD) is the **least common multiple** (LCM) of the denominators.

DEFINITION 6.2-1 The *least common multiple* (LCM) of a set of polynomials is the polynomial of lowest degree that is exactly divisible by every polynomial in the set.

The first step in finding the LCM of a set of polynomials is to factor each polynomial. Next, write the product of all the different factors, each factor being used in the product the greatest number of times that it appears as a factor in any one of the polynomials. If the polynomials have no common factor, then the LCM is the product of the polynomials. The LCM is usually left in factored form.

Examples (a) $\{x^2y, xy^2, y^3\}$
(b) $\{x^2 - 2x + 1, x^2 - 1, x^2 + 2x + 1\}$
(c) $\{x^2 - x - 2, x^3 + 1, x^2 - x + 1\}$
(d) $\{2x + 3, 3x - 2\}$

Solutions (a) We note that x occurs as a factor twice in one of the polynomials and y occurs as a factor three times. Hence, the LCM is x^2y^3.

(b) $x^2 - 2x + 1 = (x - 1)(x - 1)$
$x^2 - 1 = (x + 1)(x - 1)$
$x^2 + 2x + 1 = (x + 1)(x + 1)$

We note that $x - 1$ and $x + 1$ are the only factors, each occurring twice in one of the polynomials, but no more than that in any polynomial. Thus, the LCM is

$(x - 1)(x - 1)(x + 1)(x + 1)$.

(c) $x^2 - x - 2 = (x + 1)(x - 2)$
$x^3 + 1 = (x + 1)(x^2 - x + 1)$,

but the third polynomial, $x^2 - x + 1$, is not factorable over the set of integers. We note that the different factors are $x + 1$, $x - 2$, and $x^2 - x + 1$. Then, since no factor occurs more than once in any of the polynomials, the LCM is

$(x + 1)(x - 2)(x^2 - x + 1)$.

(d) Neither of these polynomials is factorable over the set of integers. Thus, they have no common factors, and the LCM is their product,

$(2x + 3)(3x - 2)$.

Sums of Fractions with Different Denominators

After you have learned how to find the LCM of a set of two or more polynomials, you are ready to simplify sums of fractions that have different denominators. A technique for such problems is outlined below.

(1) Find the least common denominator (LCD).
(2) Replace each fraction with an equivalent fraction whose denominator is the LCD.
(3) Add the numerators and write the result over the LCD.
(4) Reduce to lowest terms.

Examples (a) $\dfrac{x+3}{x} + \dfrac{y-2}{y}$ (b) $\dfrac{x+2}{x^2 - x - 2} + \dfrac{x-1}{x^2 + x - 6}$

Solutions (a) The LCD is xy because the two denominators have no common factors.

$$\frac{x+3}{x} = \frac{(x+3)y}{xy} = \frac{xy + 3y}{xy}$$

$$\frac{y-2}{y} = \frac{(y-2)x}{xy} = \frac{xy - 2x}{xy}.$$

Then,

$$\frac{xy + 3y}{xy} + \frac{xy - 2x}{xy} = \frac{xy + 3y + xy - 2x}{xy}$$

$$= \frac{2xy - 2x + 3y}{xy},$$

which is in lowest terms.

(b) The factored forms of the denominators are

$$x^2 - x - 2 = (x - 2)(x + 1)$$

and

$$x^2 + x - 6 = (x - 2)(x + 3).$$

The LCD is $(x - 2)(x + 1)(x + 3)$, because these factors are the only ones that occur and no factor appears more than once in either polynomial. We next replace each fraction with an equivalent fraction having the LCD as its denominator.

$$\frac{x+2}{(x-2)(x+1)} + \frac{x-1}{(x-2)(x+3)}$$

$$= \frac{(x+2)(x+3)}{(x-2)(x+1)(x+3)} + \frac{(x-1)(x+1)}{(x-2)(x+1)(x+3)}$$

$$= \frac{(x^2 + 5x + 6) + (x^2 - 1)}{(x - 2)(x + 1)(x + 3)}$$

$$= \frac{2x^2 + 5x + 5}{(x - 2)(x + 1)(x + 3)}$$

Since the numerator is not divisible by any of the factors of the denominator, the fraction is in lowest terms.

Differences of Fractions

The difference of two fractions is the difference of the numerators over the common denominator. You will avoid considerable difficulty and make fewer errors if you will recall the definition of a difference $[a - b = a + (-b)]$. You must also recall that the negative of a polynomial is the polynomial consisting of the negative of each term in the original.

Example The negative of $3x^2 - 2x - 7$ is $-3x^2 + 2x + 7$.

With the exception of changing the numerator of the subtrahend to its negative, finding the difference of two fractions is essentially the same as finding the sum.

Example Simplify $\dfrac{x + 4}{x^2 - 5x + 6} - \dfrac{x - 5}{x^2 + x - 6}$.

Solution We first find the LCD.

$$x^2 - 5x + 6 = (x - 2)(x - 3)$$
$$x^2 + x - 6 = (x - 2)(x + 3)$$

Since no factor occurs more than once in either polynomial, the LCD is

$$(x - 2)(x - 3)(x + 3).$$

Then,

$$\frac{x + 4}{x^2 - 5x + 6} - \frac{x - 5}{x^2 + x - 6}$$

$$= \frac{(x + 4)(x + 3)}{(x - 2)(x - 3)(x + 3)} - \frac{(x - 5)(x - 3)}{(x - 2)(x - 3)(x + 3)}$$

$$= \frac{(x^2 + 7x + 12) - (x^2 - 8x + 15)}{(x - 2)(x - 3)(x + 3)}$$

$$= \frac{x^2 + 7x + 12 - x^2 + 8x - 15}{(x - 2)(x - 3)(x + 3)}$$

$$= \frac{15x - 3}{(x - 2)(x - 3)(x + 3)},$$

which is in lowest terms.

EXERCISE 6.2

WRITTEN

Find each sum or difference in Problems 1–20 as a fraction in lowest terms. State the necessary restriction(s) on x so that each denominator is a nonzero polynomial.

Examples (a) $\dfrac{x^2 + 3x + 5}{x + 3} + \dfrac{x^2 - x - 17}{x + 3}$ (b) $\dfrac{3x^2 - 4x + 1}{x^2 - 6x + 8} - \dfrac{x^2 + 4x + 1}{x^2 - 6x + 8}$

Solutions (a) Since the only denominator is $x + 3$, $x \neq -3$. Thus, we have the sum of the numerators over the common denominator.

$$\dfrac{(x^2 + 3x + 5) + (x^2 - x - 17)}{x + 3} = \dfrac{2x^2 + 2x - 12}{x + 3}$$
$$= \dfrac{2(x^2 + x - 6)}{x + 3}$$
$$= \dfrac{2(x - 2)(x + 3)}{x + 3}$$
$$= 2(x - 2)$$
$$= 2x - 4$$

(b) $x^2 - 6x + 8 = (x - 2)(x - 4)$. Thus, $x \neq 2$ or 4. Since the denominators are the same, we write the difference of the numerators over the common denominator.

$$\dfrac{(3x^2 - 4x + 1) - (x^2 + 4x + 1)}{(x - 2)(x - 4)} = \dfrac{3x^2 - 4x + 1 - x^2 - 4x - 1}{(x - 2)(x - 4)}$$
$$= \dfrac{2x^2 - 8x}{(x - 2)(x - 4)}$$
$$= \dfrac{2x(x - 4)}{(x - 2)(x - 4)}$$
$$= \dfrac{2x}{x - 2}$$

1. $\dfrac{x + 3}{y} + \dfrac{2x - 1}{y}$

2. $\dfrac{4y - 5}{x} + \dfrac{3y + 2}{x}$

3. $\dfrac{4x - 3y}{xy} + \dfrac{x + 7y}{xy}$

4. $\dfrac{x^2 + 2}{2x} + \dfrac{x^2 - 5}{2x}$

5. $\dfrac{2x^2 + 4x - 3}{y^2} + \dfrac{5 - 3x - x^2}{y^2}$

6. $\dfrac{8x + 3xy - 4y}{4ab} + \dfrac{3y - 4xy - 5x}{4ab}$

7. $\dfrac{2x + 5}{y} - \dfrac{4}{y}$

8. $\dfrac{3y - 2}{y} - \dfrac{7y}{y}$

9. $\dfrac{5x + 2y}{z} - \dfrac{2x + y}{z}$

10. $\dfrac{4x - 5y}{z} - \dfrac{2x - 6y}{z}$

11. $\dfrac{2x + 3}{x + 1} + \dfrac{x + 2}{x + 1}$

12. $\dfrac{3x - 4}{x - 3} + \dfrac{2x + 7}{x - 3}$

13. $\dfrac{x + 4}{x + 5} - \dfrac{3 - 2x}{x + 5}$

14. $\dfrac{4x - 7}{x + 4} - \dfrac{3x - 1}{x + 4}$

15. $\dfrac{x^2 - 7x + 2}{x + 2} + \dfrac{14x + 8}{x + 2}$

16. $\dfrac{x^2 - 3x - 20}{x + 5} + \dfrac{x^2 + 7x - 10}{x + 5}$

17. $\dfrac{2x^2 + x - 3}{x - 1} - \dfrac{2 - 2x - x^2}{x - 1}$

18. $\dfrac{5x^2 - 8x + 12}{x - 4} - \dfrac{x^2 + 8x - 4}{x - 4}$

19. $\dfrac{x^2 + 2x - 17}{x^2 + 5x + 6} + \dfrac{x^2 - 6x - 13}{x^2 + 5x + 6}$

20. $\dfrac{2x^2 - x + 5}{x^2 - 3x - 4} - \dfrac{x^2 + x - 13}{x^2 - 3x - 4}$

In Problems 21-32, find the least common multiple (LCM) of each set of polynomials.

Examples (a) $\{2xy, 3x^2, 6y^2\}$
(b) $\{x^2 - 2x - 3, x^2 - x - 2, 6 - 5x + x^2\}$

Solutions (a) We note that 2 and 3 occur no more than once as factors in any polynomial and that x and y occur no more than twice. Hence, the LCM is $6x^2y^2$.

(b) The factors of the polynomials are

$$x^2 - 2x - 3 = (x - 3)(x + 1)$$
$$x^2 - x - 2 = (x - 2)(x + 1)$$
$$6 - 5x + x^2 = (2 - x)(3 - x).$$

Since $2 - x = -(x - 2)$ and $3 - x = -(x - 3)$,

$$(2 - x)(3 - x) = (x - 2)(x - 3).$$

Thus, the different factors are $x - 3$, $x - 2$, and $x + 1$. Since no factor occurs more than once in any of the polynomials, the LCM is

$$(x - 3)(x - 2)(x + 1).$$

21. $\{2xy^2, 4x^2y, 6x^2y^3\}$
22. $\{5x^3y^2, 4xy^3, 10x^2y\}$
23. $\{6x, 9y^2, 18x^2y^3\}$
24. $\{7xy^2, 5x, 35x^2y^2\}$
25. $\{x + 1, x^2 - 2x + 1, x^2 - 1\}$
26. $\{x - 2, x^2 - 4, x^2 + 4x + 4\}$
27. $\{x^2 - x - 6, x^2 + 5x + 6, x^2 - 9\}$
28. $\{x^2 + 6x + 5, x^2 - 2x - 3, x^2 - 1\}$
29. $\{x^2 - 1, x^3 - 1, x^3 + 1\}$
30. $\{x^2 - 4, x^3 - 8, x^3 + 8\}$
31. $\{x^4 + x^2 + 1, x^3 + 1, x^3 - 1\}$
32. $\{x^4 + 4x^2 + 16, x^3 + 8, x^3 - 8\}$

Simplify each sum or difference in Problems 33-60. Assume that all denominators are nonzero.

Examples (a) $\dfrac{4}{xy^2} + \dfrac{3}{x^3y}$ (b) $\dfrac{x + 1}{x^2 + 2x - 3} + \dfrac{x - 2}{x^2 + 5x + 6}$

(c) $\dfrac{x}{2x + 3} - \dfrac{x - 2}{3x + 2}$

Solutions (a) The LCD is x^3y^2. Then,

$$\dfrac{4}{xy^2} + \dfrac{3}{x^3y} = \dfrac{4(x^2) + 3(y)}{x^3y^2} = \dfrac{4x^2 + 3y}{x^3y^2}$$

(b) $x^2 + 2x - 3 = (x - 1)(x + 3)$
$x^2 + 5x + 6 = (x + 2)(x + 3)$

The LCD is $(x - 1)(x + 2)(x + 3)$.

Replace each fraction with an equivalent fraction with the LCD as its denominator.

$$\frac{(x+1)(x+2)}{(x-1)(x+2)(x+3)} + \frac{(x-2)(x-1)}{(x-1)(x+2)(x+3)}$$

$$= \frac{x^2 + 3x + 2 + x^2 - 3x + 2}{(x-1)(x+2)(x+3)}$$

$$= \frac{2x^2 + 4}{(x-1)(x+2)(x+3)}$$

(c) Since $2x + 3$ and $3x + 2$ have no common factors, the LCD is $(2x + 3)(3x + 2)$. Replace each fraction with an equivalent fraction having the LCD as its denominator.

$$\frac{x(3x+2)}{(2x+3)(3x+2)} - \frac{(x-2)(2x+3)}{(2x+3)(3x+2)} = \frac{(3x^2+2x) - (2x^2-x-6)}{(2x+3)(3x+2)}$$

$$= \frac{3x^2 + 2x - 2x^2 + x + 6}{(2x+3)(3x+2)}$$

$$= \frac{x^2 + 3x + 6}{(2x+3)(3x+2)}$$

33. $\dfrac{2x}{3y} + \dfrac{4}{xy^2}$

34. $\dfrac{5}{2xy^2} + \dfrac{8y}{3x}$

35. $\dfrac{7}{3x^2y} - \dfrac{4x}{5y}$

36. $\dfrac{5x^2}{7y} - \dfrac{2y}{3x^2}$

37. $\dfrac{6}{3x} + \dfrac{4}{5y} - \dfrac{1}{xy^2}$

38. $\dfrac{2x}{7y} - \dfrac{3y}{2x} + \dfrac{4}{x^2y^2}$

39. $\dfrac{2}{3x-2} + \dfrac{3}{3x+2}$

40. $\dfrac{3}{7x+1} + \dfrac{2}{x-5}$

41. $\dfrac{2x+1}{2x+3} - \dfrac{x+2}{3x-2}$

42. $\dfrac{x}{4x^2-1} - \dfrac{4}{6x-3}$

43. $\dfrac{5x}{x+4} - \dfrac{4x^2+2x-1}{x^2+x-12}$

44. $\dfrac{2x+1}{x^2+4x-60} - \dfrac{2}{x-6}$

45. $\dfrac{3x-5x^2}{4x^2+12x+9} + \dfrac{3-x}{4x+6}$

46. $\dfrac{2y-3}{2y^2-18} + \dfrac{4}{-3y^2+11y-6}$

47. $\dfrac{x-2}{x^2-16} - \dfrac{x+2}{x^2+8x+16}$

48. $\dfrac{2y-5}{2y^2-50} + \dfrac{4}{3y^2-13y-10}$

49. $\dfrac{3}{x^2-xy-6y^2} - \dfrac{x-2y}{2x^2+6xy+4y^2}$

50. $\dfrac{x-2y}{x^2-16y^2} - \dfrac{x-2y}{x^2+8xy+16y^2}$

51. $\dfrac{x+5x^2}{x^3-1} - \dfrac{3}{2x-2}$

52. $\dfrac{y^2}{y^3+8} - \dfrac{2y}{y^2-2y+4}$

53. $\dfrac{2x}{8x^3-27} + \dfrac{5}{4x^2-12x+9}$

54. $\dfrac{x^2+x+1}{x^3-64} + \dfrac{x}{x^2-5x+4}$

55. $\dfrac{x-1}{x^4+x^2+1} + \dfrac{x+1}{x^2+x+1}$

56. $\dfrac{1}{x^4+4} - \dfrac{x-2}{x^3+8}$

57. $\dfrac{3x-2}{8x^2-18} + \dfrac{3+x}{2} - \dfrac{4-x}{2x^2-9x+9}$

58. $\dfrac{3x-2}{2x^2-x-3} - \dfrac{2x+5}{3x^2+6x+3} + \dfrac{2}{x+1}$

59. $\dfrac{1}{2xy - 6x + yz - 3z} - \dfrac{xy + yz}{(z^2 - 4x^2)(y^2 - 6y + 9)}$

60. $\dfrac{3}{x^2 - xy - 6y^2} + \dfrac{x + 3y}{x^3 + 8y^3} - \dfrac{x - 2y}{2x^2 - 4xy + 8y^2}$

6.3 Products and Quotients

The product of two fractions is the product of the numerators divided by the product of the denominators. The quotient of two fractions is the product of the dividend and the reciprocal of the divisor.

THEOREM 6.3-1 If $P(x)$, $Q(x)$, $S(x)$, and $T(x)$ are real polynomials, then for every replacement of x for which $Q(x)$ and $T(x)$ are not zero,

$$\frac{P(x)}{Q(x)} \cdot \frac{S(x)}{T(x)} = \frac{P(x) \cdot S(x)}{Q(x) \cdot T(x)},$$

and if the replacements of x are further restricted so that $S(x) \neq 0$, then

$$\frac{P(x)}{Q(x)} \div \frac{S(x)}{T(x)} = \frac{P(x)}{Q(x)} \cdot \frac{T(x)}{S(x)} = \frac{P(x) \cdot T(x)}{Q(x) \cdot S(x)}.$$

Simplifying Products of Fractions

The product of two fractions can be simplified by first factoring all numerators and denominators and then dividing the numerator and the denominator of the product by any common factors.

Examples Simplify.

(a) $\dfrac{3y}{2x} \cdot \dfrac{8x^2}{5y^2}$ 　　(b) $\dfrac{x^2 - x - 6}{x^2 + 2x + 1} \cdot \dfrac{x^2 - 1}{x^2 - 6x + 9}$

Solutions (a) Ordinarily, monomials are not written in factored form.

$$\frac{3y}{2x} \cdot \frac{8x^2}{5y^2} = \frac{3y \cdot 8x^2}{2x \cdot 5y^2} = \frac{24x^2 y}{10xy^2} \qquad (x, y \neq 0)$$

We now note that both the numerator and denominator can be divided by $2xy$. Thus, the simplest form is

$$\frac{12x}{5y}.$$

(b) After factoring each polynomial, the product becomes

$$\frac{x^2 - x - 6}{x^2 + 2x + 1} \cdot \frac{x^2 - 1}{x^2 - 6x + 9} = \frac{(x - 3)(x + 2)(x + 1)(x - 1)}{(x + 1)(x + 1)(x - 3)(x - 3)},$$

$$(x \neq -1, 3).$$

We note that $x - 3$ and $x + 1$ are factors common to both the numerator and the denominator. Hence, we divide by these factors to obtain

$$\frac{(x + 2)(x - 1)}{(x + 1)(x - 3)} = \frac{x^2 + x - 2}{(x + 1)(x - 3)}.$$

If desired, the denominator can also be written in polynomial form, in which case the answer is

$$\frac{x^2 + x - 2}{x^2 - 2x - 3}.$$

Simplifying Quotients

The reciprocal of a fraction is found by inverting the fraction:

$$\frac{1}{a/b} = \frac{b}{a}$$

Thus, to simplify the quotient of two fractions, we invert the divisor and then proceed as in the simplification of a product.

Examples (a) $\dfrac{10xy^2}{3z} \div \dfrac{5xz}{2y}$ (b) $\dfrac{x^2 - 5x + 6}{x^2 + 4x + 4} \div \dfrac{x^2 - x - 6}{x^2 + 6x + 8}$

Solutions (a) We invert the divisor to obtain

$$\frac{10xy^2}{3z} \div \frac{5xz}{2y} = \frac{10xy^2}{3z} \cdot \frac{2y}{5xz} = \frac{20xy^3}{15xz^2}.$$

Both the numerator and denominator can be divided by $5x$:

$$\frac{20xy^3 \div 5x}{15xz^2 \div 5x} = \frac{4y^3}{3z^2}.$$

(b) We invert the divisor and factor all numerators and denominators.

$$\frac{x^2 - 5x + 6}{x^2 + 4x + 4} \div \frac{x^2 - x - 6}{x^2 + 6x + 8} = \frac{x^2 - 5x + 6}{x^2 + 4x + 4} \cdot \frac{x^2 + 6x + 8}{x^2 - x - 6}$$

$$= \frac{(x - 2)(x - 3)(x + 2)(x + 4)}{(x + 2)(x + 2)(x - 3)(x + 2)}$$

Note that $x - 3$ and $x + 2$ are factors of both the numerator and denominator. Hence, we divide the numerator and denominator by these factors to obtain

$$\frac{(x - 2)(x + 4)}{(x + 2)(x + 2)},$$

which can be written as

$$\frac{x^2 + 2x - 8}{(x + 2)(x + 2)} \quad \text{or as} \quad \frac{x^2 + 2x - 8}{x^2 + 4x + 4}.$$

EXERCISE 6.3

In this exercise, assume that no denominator is equal to zero. Simplify each product and quotient.

1. $\dfrac{6x}{5y} \cdot \dfrac{10y^2}{3x^2}$
2. $\dfrac{8xy}{7z} \cdot \dfrac{21z^2}{16y^2}$
3. $\dfrac{5x^2y}{2vw^3} \cdot \dfrac{12v^2w}{35xy^2}$
4. $\dfrac{9x^2y^3}{4v^2w} \cdot \dfrac{16vw^2}{27xy^4}$
5. $\dfrac{10x}{7y} \div \dfrac{5x}{14y^2}$
6. $\dfrac{25x^2y}{12xz} \div \dfrac{15xy^2}{24z}$
7. $\dfrac{3x^2z}{22y^2w} \div \dfrac{18xw}{11yz}$
8. $\dfrac{27x^3y^3}{35wz^4} \div \dfrac{18x^2w}{25z^3}$
9. $\dfrac{x+2}{x+6} \cdot \dfrac{x^2-36}{x^2-4}$
10. $\dfrac{x}{x^2-4} \cdot \dfrac{2x^2+3x-14}{4x+14}$
11. $\dfrac{4x^2-25}{3x+1} \cdot \dfrac{9x^2-1}{4x^2+10x}$
12. $\dfrac{3x^2}{x^2+2x+1} \cdot \dfrac{4x^2-4}{x^3-x^2}$
13. $\dfrac{3x-x^2}{x^2+1} \div \dfrac{9-x^2}{4x^3+4x}$
14. $\dfrac{2x^2-9x-35}{6x+4} \div \dfrac{x^2-49}{9x^2+12x+4}$
15. $\dfrac{x^3+8}{2x-6} \cdot \dfrac{4x^2-36}{2x+4}$
16. $\dfrac{27x^3-1}{4x^2-2x+1} \cdot \dfrac{8x^3+1}{9x^2+3x+1}$
17. $\dfrac{x^2-4x+16}{x^2+2x} \div \dfrac{x-2}{x^2+4x}$
18. $\dfrac{3x^3-81}{4x^2-16} \div \dfrac{6x-18}{8x+16}$

Example

$\dfrac{x^2-x-6}{x^2-x-2} \cdot \dfrac{2-3x-2x^2}{3+2x-x^2}$

Solution

Factor the polynomials to obtain

$$\dfrac{(x-3)(x+2)(2+x)(1-2x)}{(x-2)(x+1)(3-x)(1+x)}.$$

Before we attempt to reduce this fraction, we should examine the factors carefully. Note that $x-3$ in the numerator and $3-x$ in the denominator are negatives of each other. Thus, we can change the sign of another factor, say $1-2x$ in the numerator, and change $3-x$ to $x-3$ in the denominator. We now have

$$\dfrac{(x-3)(x+2)(x+2)(2x-1)}{(x-2)(x+1)(x-3)(x+1)},$$

which, when reduced to lowest terms, is

$$\dfrac{(x+2)(x+2)(2x-1)}{(x-2)(x+1)(x+1)}.$$

The fraction can be left in this form, or the numerator can be written in polynomial form as

$$\dfrac{2x^3+7x^2+4x-4}{(x-2)(x-1)(x+1)}.$$

The denominator is usually left in factored form.

19. $\dfrac{2x-6}{5x+5} \cdot \dfrac{10-5x}{3-x}$

20. $\dfrac{x^2-4x}{3-x} \cdot \dfrac{x-3}{4-x}$

21. $\dfrac{x^2-4}{x^2-9} \div \dfrac{2-x}{3+x}$

22. $\dfrac{x^2-16}{x-3} \div \dfrac{4-x}{9-x^2}$

23. $\dfrac{x^2-x-6}{2-x} \cdot \dfrac{x^2-x-2}{3x-x^2}$

24. $\dfrac{x^2-6x+8}{4-x} \cdot \dfrac{x-3}{6-5x+x^2}$

25. $\dfrac{x^2-6x+5}{x^2-8x+7} \div \dfrac{10+3x-x^2}{2-x-x^2}$

26. $\dfrac{x^2-8x+15}{x^2-9x+14} \div \dfrac{21-4x-x^2}{16-6x-x^2}$

27. $\dfrac{x^3-1}{1-x^2} \cdot \dfrac{x-2}{x^2+x+1}$

28. $\dfrac{8-x^3}{x^2-4x+4} \div \dfrac{x^2+2x+2}{x^2+2x+4}$

29. $\dfrac{x^4+4}{y^4+y^2+1} \cdot \dfrac{y^2-y+1}{x^2+2x+2}$

30. $\dfrac{x^4-26x^2+25}{125-x^3} \div \dfrac{x^2+4x-5}{x-5}$

Find a fraction in simplest form that is equivalent to the expression involving products and quotients in each of Problems 31–38.

Example $\dfrac{3x+2}{5x^2-x} \cdot \dfrac{2x^2-x}{2x^2-x-1} \div \dfrac{6x^2+x-2}{10x^2+3x-1}$

Solution Factor each polynomial and invert the fraction following the division sign to obtain

$$\dfrac{3x+2}{x(5x-1)} \cdot \dfrac{x(2x-1)}{(2x+1)(x-1)} \cdot \dfrac{(5x-1)(2x+1)}{(3x+2)(2x-1)}.$$

The application of Theorem 6.1-2 reduces the fraction to

$$\dfrac{1}{x-1}.$$

31. $\dfrac{x^2+4x+3}{x^2-8x+7} \cdot \dfrac{x^2-2x-35}{x^2-7x-8} \div \dfrac{x^2+8x+15}{x^2-9x+8}$

32. $\dfrac{x^2+6x+9}{x^2-8x+7} \cdot \dfrac{x^2-2x-35}{x^2+5x+6} \div \dfrac{x^2+2x-3}{2x^2+3x-2}$

33. $\dfrac{x^2-x}{x^2-2x-3} \cdot \dfrac{x^2+2x+1}{x^2+4x} \div \dfrac{x^2-3x-4}{x^2-16}$

34. $\dfrac{x^2+11x+18}{x^2+4x-5} \cdot \dfrac{x^2-6x-7}{x^2+8x+12} \div \dfrac{x^2-7x-8}{x^2+2x-15}$

35. $\dfrac{3+5x-2x^2}{1-3x-18x^2} \cdot \dfrac{6x^2-5x+1}{2x+1} \div \dfrac{2x^2-7x+3}{1-36x^2}$

36. $\dfrac{2x^2-5x+2}{2x^2+5x+2} \cdot \dfrac{1-4x^2}{3x^2-7x+2} \div \dfrac{1-4x^2}{1-x-6x^2}$

37. $\dfrac{3x^2+10xy+3y^2}{2x^2+5xy-3y^2} \cdot \dfrac{4x^2+12xy+9y^2}{6x^2+11xy+3y^2} \div \dfrac{x+3y}{2x-y}$

38. $\dfrac{x^2+xy-2y^2}{2x^2+7xy+3y^2} \cdot \dfrac{2x^2+5xy-3y^2}{x^2+3xy+2y^2} \div \dfrac{y-2x}{x+y}$

In Problems 39–48, perform the indicated operations to find a fraction in simplest form that is equivalent to the given expression.

Example $\left(\dfrac{x}{x-y} - \dfrac{y}{x+y}\right)\left(\dfrac{x+y}{x^2+y^2}\right)$

Solution We begin by finding a fraction equivalent to the difference in the left-hand parentheses. By inspection, we determine the least common denominator to be $(x-y)(x+y)$. Hence we obtain

$$\left(\dfrac{x(x+y)}{(x-y)(x+y)} - \dfrac{y(x-y)}{(x-y)(x+y)}\right)\left(\dfrac{x+y}{x^2+y^2}\right)$$

$$= \left(\dfrac{x^2 + xy - xy + y^2}{(x-y)(x+y)}\right)\left(\dfrac{x+y}{x^2+y^2}\right)$$

$$= \dfrac{x^2+y^2}{(x-y)(x+y)} \cdot \dfrac{x+y}{x^2+y^2},$$

which reduces to $\dfrac{1}{x-y}$.

39. $\left(1 - \dfrac{2xy}{x^2+y^2}\right)\left(\dfrac{x^2+y^2}{x^2-y^2}\right)$

40. $\left(x + \dfrac{2xy}{x-2y}\right)\left(\dfrac{x^2-4y^2}{x}\right)$

41. $\left(\dfrac{x^2}{y^2} - 4\right)\left(\dfrac{y}{x+2y}\right)\left(2 - \dfrac{x}{y}\right)$

42. $\left(\dfrac{1}{4x-5} - \dfrac{1}{4x+5}\right)\left(\dfrac{4x^2+9x+5}{2}\right)$

43. $\left(\dfrac{3x}{2x+2y} + \dfrac{4}{x^2-y^2}\right)\left(\dfrac{x^2}{2x+y} - \dfrac{y^2}{2x+y}\right)$

44. $\left(\dfrac{6}{y} - \dfrac{7}{x} - \dfrac{3y}{x^2}\right)\left(\dfrac{x}{3y-2x} + \dfrac{y}{y+3x}\right)$

45. $\left(\dfrac{x^4-y^4}{x^2y^2}\right) \div \left(\dfrac{3x^2}{y^4} + \dfrac{1}{y^2} - \dfrac{2}{x^2}\right)$

46. $\left(\dfrac{3x-2}{3x} - \dfrac{2}{3x+3}\right) \div \left(\dfrac{x}{5} + \dfrac{1}{5} - \dfrac{2}{5x}\right)$

47. $\left(\dfrac{x+y}{x-y} + \dfrac{x-y}{x+y}\right) \div \left(\dfrac{x+y}{x-y} - \dfrac{x-y}{x+y}\right)$

48. $\left(2 - \dfrac{1+3x-21x^2}{1+2x-8x^2}\right) \div \left(\dfrac{1+2x}{1-2x} + \dfrac{1-3x}{1+4x}\right)$

6.4 Equations Involving Fractions

In Section 2.2 you solved some simple equations involving fractions by multiplying both members of the equation by expressions that would clear the equation of fractions. You learned that the replacement set for the variable must be restricted so that the expression used as a multiplier represents a nonzero number. In this section, we discuss the solution of such equations in greater detail.

By using the addition law and the methods for simplifying sums of fractions, you can generate an equation of the form

$$\frac{P(x)}{Q(x)} = 0 \tag{1}$$

that is equivalent to the original equation. Since division by zero is undefined, $P(x)/Q(x)$ can equal zero if and only if $P(x) = 0$ and $Q(x) \neq 0$. Thus, any solution of the given equation must be a solution of $P(x) = 0$ and cannot be a solution of $Q(x) = 0$.

Example Find the solution set of

$$\frac{6}{x} - \frac{12}{x+1} + 1 = 0.$$

Solution By inspection we note that the least common denominator is $x(x + 1)$. After changing both fractions to equivalent fractions having $x(x + 1)$ as a denominator, we combine the terms of the left-hand member into one fraction.

$$\frac{6(x+1) - 12(x) + 1(x)(x+1)}{x(x+1)} = 0$$

$$\frac{6x + 6 - 12x + x^2 + x}{x(x+1)} = 0$$

$$\frac{x^2 - 5x + 6}{x(x+1)} = 0$$

Any solution of the original equation must be a solution of $x^2 - 5x + 6 = 0$. We determine the solutions of this equation by factoring the left-hand member and equating each factor to zero.

$$(x - 2)(x - 3) = 0$$

Hence, the solutions are 2 and 3.
  Substitution shows that neither 2 nor 3 is a solution of $x(x + 1) = 0$. Therefore, the solution set of the original equation is $\{2, 3\}$.

Clearing an Equation of Fractions

An alternate method, which you may find easier to use, involves multiplying both members of the equation by the least common denominator of all of the fractions. Then, upon simplifying the products obtained, you have an equation free of fractions. However, you must remember that no zero of any denominator can be a solution of the original equation.

Example Find the solution set of

$$\frac{2}{x+2} - \frac{4}{x+1} + \frac{x}{x+2} = 0.$$

Solution Observe that the LCM of the denominators is $(x + 2)(x + 1)$. If we multiply both members of the equation by this expression, we obtain

$$\frac{2(x + 2)(x + 1)}{x + 2} - \frac{4(x + 2)(x + 1)}{x + 1} + \frac{x(x + 2)(x + 1)}{x + 2} = 0(x + 2)(x + 1).$$

Then upon reducing each fraction to lowest terms and applying Theorem 1.4-3 [$a \cdot 0 = 0$] to the right-hand member, the equation becomes

$$2(x + 1) - 4(x + 2) + x(x + 1) = 0$$
$$2x + 2 - 4x - 8 + x^2 + x = 0$$
$$x^2 - x - 6 = 0$$
$$(x - 3)(x + 2) = 0.$$

You can determine by inspection that the solution set of the final equation is $\{-2, 3\}$. However, you should note that -2 is a zero of the denominator, $(x + 2)(x + 1)$, and therefore cannot be a solution of the original equation. Hence, the required solution set is $\{3\}$.

EXERCISE 6.4

WRITTEN

Find the solution set of each equation in Problems 1–22.

Example $\quad \dfrac{4}{x} + \dfrac{10}{x + 3} = 4$

Solution $\quad \dfrac{4}{x} + \dfrac{10}{x + 3} - 4 = 0$

$$\frac{4(x + 3) + 10x - 4x(x + 3)}{x(x + 3)} = 0$$

$$\frac{4x + 12 + 10x - 4x^2 - 12x}{x(x + 3)} = 0$$

Since any solution of the equation must be a zero of the numerator, we state that

$$12 + 2x - 4x^2 = 0.$$

Then by factoring the left-hand member, the equation becomes

$$2(2 - x)(3 + 2x) = 0,$$

whose solution set is $\{-\tfrac{3}{2}, 2\}$. This is the required solution set, since neither $-\tfrac{3}{2}$ nor 2 is a solution of $x(x + 3) = 0$.

1. $\dfrac{12}{x + 2} + \dfrac{14}{x} = 10$ 2. $\dfrac{8}{x - 2} = \dfrac{4}{x} + 3$

3. $\dfrac{5}{x + 1} - \dfrac{6}{x + 4} = \dfrac{2}{3}$ 4. $\dfrac{7}{x + 2} - \dfrac{2}{x - 1} = \dfrac{2}{5}$

5. $\dfrac{4}{x+1} + \dfrac{4}{x+3} = \dfrac{32}{15}$

6. $\dfrac{2}{x-1} - \dfrac{4}{x+3} = \dfrac{2}{21}$

7. $\dfrac{8}{x+2} - \dfrac{2x}{x-1} = -\dfrac{7}{5}$

8. $\dfrac{3x}{x+1} - \dfrac{2}{x+2} = \dfrac{5}{6}$

9. $\dfrac{x}{x+2} + \dfrac{x}{2x+1} = \dfrac{2}{3}$

10. $\dfrac{x+1}{x+2} - \dfrac{x}{x+1} = \dfrac{1}{12}$

11. $\dfrac{5x-5}{x^2+1} = \dfrac{4}{x+1}$

12. $\dfrac{3x-6}{x^2+2} = \dfrac{2}{x+2}$

13. $\dfrac{6}{x-1} + \dfrac{16}{x^2-1} = 5$

14. $\dfrac{4}{x-3} - \dfrac{16}{x^2-9} = 1$

15. $\dfrac{3}{x+2} - \dfrac{1}{x^2+3x+2} = \dfrac{2}{3}$

16. $\dfrac{2}{x-1} - \dfrac{2}{x^2+2x-3} = \dfrac{4}{7}$

17. $\dfrac{x+2}{x+1} - \dfrac{32}{x^2+4x+3} = \dfrac{-4}{5}$

18. $\dfrac{6x}{x^2+5x+4} - \dfrac{5}{x+1} = -1$

Example

$\dfrac{5x}{x+1} - \dfrac{2}{x^2-1} = \dfrac{5}{x-1}$

Solution Multiply both members by $(x+1)(x-1)$, the least common denominator, and simplify the resulting products to obtain

$$5x(x-1) - 2 = 5(x+1)$$
$$5x^2 - 5x - 2 = 5x + 5$$
$$5x^2 - 10x - 7 = 0.$$

The polynomial in the left-hand member is not factorable over the integers so we use the quadratic formula to obtain

$$x = \dfrac{10 \pm \sqrt{100 + 140}}{10} = \dfrac{10 \pm \sqrt{16 \cdot 15}}{10},$$

which gives the solution set

$$\left\{ \dfrac{5 - 2\sqrt{15}}{5}, \dfrac{5 + 2\sqrt{15}}{5} \right\}.$$

Since neither member of this set is a zero of $x^2 - 1$, this is the solution set of the original equation.

19. $\dfrac{x}{x+2} - \dfrac{3}{x^2-4} = \dfrac{2}{x-2}$

20. $\dfrac{2x}{x-3} + \dfrac{x}{x^2-9} = \dfrac{60}{x+3}$

21. $\dfrac{x+1}{x-2} + \dfrac{2x-1}{x^2-5x+6} = 2$

22. $\dfrac{x-1}{x+1} - \dfrac{x+1}{x-1} = 3$

Example The sum of the reciprocals of two consecutive even integers is $\tfrac{5}{12}$. Find the integers.

Solution (1) Let x represent one of the integers; then $x + 2$ represents the next consecutive even integer.

(2) The reciprocal of x is $\dfrac{1}{x}$ and that of $x + 2$ is $\dfrac{1}{x+2}$. Thus,

$$\dfrac{1}{x} + \dfrac{1}{x+2} = \dfrac{5}{12}.$$

(3) $$\frac{12(x+2) + 12x - 5x(x+2)}{12x(x+2)} = 0$$

$$\frac{-5x^2 + 14x + 24}{12x(x+2)} = 0$$

$$5x^2 - 14x - 24 = 0$$

$$(5x + 6)(x - 4) = 0$$

The solution set of the original equation is $\{-\frac{6}{5}, 4\}$, since neither is a solution of $12x(x+2) = 0$.

(4) The first integer is 4 and the second is 6.

23. If each of two consecutive odd integers is divided by the sum of itself and 1, the sum of the two quotients is $\frac{19}{12}$. What are the odd integers?
24. If the product of two consecutive integers is divided by their sum, the result is equal to the first integer decreased by $\frac{25}{11}$. Find the integers.
25. A motorboat requires 4 hours to make a trip of 30 miles downstream and then 18 miles back upstream on a river whose current has a velocity of 3 miles per hour. Find the average rate of the boat under the same conditions if there were no current.
26. If it takes 15 hours for an airplane to fly nonstop a distance of 8400 miles against a 20 mile per hour head wind and 14 hours to fly the same distance with a 20 mile per hour tail wind, what would be the average speed if there were no wind?
27. A group of boys bought a motorboat for $250, each boy contributing the same amount. Later, one boy sold his share to the others for $40, with this cost being divided equally among the remaining boys. Each share of this additional cost was found to be $40 less than a share of the original cost. How many boys shared in the purchase of the motorboat?
28. A group of boys planned a camping trip that would cost each of them $35. However, they discovered that if more boys were added to the group, the cost per boy per day would be $2 less and they could stay 2 days longer for the same amount of $35. For how many days was the trip originally planned?

6.5 Complex Fractions

In Chapter 1, we discussed fractions of the form a/b, where a and b are integers, $b \neq 0$. In this chapter, we have broadened the concept of a fraction to include quotients of real polynomials. Thus, with suitable restrictions on the denominator, we now consider a fraction to be the quotient of any two expressions that represent real numbers. Such an expression may contain one or more terms that are fractions. A fraction that contains one or more fractions in its numerator, denominator, or both, is called a **complex fraction**

Examples (a) $\dfrac{x + \frac{1}{2}}{3}$ is a complex fraction because the numerator contains a term that is a fraction.

(b) $\dfrac{x + 2}{5 - \dfrac{1}{x}}$ is a complex fraction because the denominator contains a term that is a fraction.

(c) $\dfrac{\dfrac{x}{4} - \dfrac{2}{y}}{\dfrac{2}{x} + \dfrac{y}{3}}$ is a complex fraction in which both the numerator and denominator contain fractions.

Simplifying Complex Fractions

As we stated earlier, it is usually necessary to explain what is meant by the phrase "simplest form." To express a complex fraction in simplest form means to find a single fraction in lowest terms that is equivalent to the complex fraction. We discuss two methods of simplifying complex fractions.

One way is to first simplify the expressions in the numerator and denominator, and then divide.

Examples Simplify the complex fractions.

(a) $\dfrac{x + \frac{1}{2}}{3}$ (b) $\dfrac{x + 2}{5 - \dfrac{1}{x}}$ (c) $\dfrac{\dfrac{x}{4} - \dfrac{2}{y}}{\dfrac{2}{x} + \dfrac{y}{3}}$

Solutions (a) We first simplify the numerator.

$$x + \tfrac{1}{2} = \dfrac{2x + 1}{2}$$

Then,

$$\dfrac{2x + 1}{2} \div 3 = \dfrac{2x + 1}{2} \cdot \dfrac{1}{3} = \dfrac{2x + 1}{6}.$$

(b) We first simplify the denominator.

$$5 - \dfrac{1}{x} = \dfrac{5x - 1}{x}$$

Then,

$$(x + 2) \div \dfrac{5x - 1}{x} = (x + 2) \cdot \dfrac{x}{5x - 1} = \dfrac{x^2 + 2x}{5x - 1}.$$

(c) In this case, both the numerator and denominator must first be simplified.

$$\dfrac{x}{4} - \dfrac{2}{y} = \dfrac{xy - 8}{4y}$$

and

$$\dfrac{2}{x} + \dfrac{y}{3} = \dfrac{6 + xy}{3x}$$

Then,

$$\frac{xy-8}{4y} \div \frac{6+yx}{3x} = \frac{xy-8}{4y} \cdot \frac{3x}{6+xy}$$

$$= \frac{3x^2y - 24x}{24y + 4xy^2}.$$

The second method is to clear both the numerator and denominator of fractions by multiplying them by the same nonzero expression. The best expression to use is the LCD of all the fractions contained in the complex fraction. We illustrate this method by simplifying the same complex fractions as before.

Examples Simplify the complex fractions

(a) $\dfrac{x + \frac{1}{2}}{3}$ (b) $\dfrac{x+2}{5 - \dfrac{1}{x}}$ (c) $\dfrac{\dfrac{x}{4} - \dfrac{2}{y}}{\dfrac{2}{x} + \dfrac{y}{3}}$

Solutions (a) The only single fraction in the complex fraction is $\frac{1}{2}$. Therefore, we multiply both the numerator and denominator of the complex fraction by 2.

$$\frac{x + \frac{1}{2}}{3} = \frac{2(x + \frac{1}{2})}{2 \cdot 3} = \frac{2x + 1}{6}$$

(b) The only single fraction is $\dfrac{1}{x}$. Therefore, we multiply both the numerator and denominator of the complex fraction by x.

$$\frac{x+2}{5 - \dfrac{1}{x}} = \frac{x(x+2)}{x\left(5 - \dfrac{1}{x}\right)} = \frac{x^2 + 2x}{5x - 1}$$

(c) The complex fraction contains four single fractions whose denominators are 4, y, x, and 3, respectively. The LCD is $12xy$.

$$\frac{\dfrac{x}{4} - \dfrac{2}{y}}{\dfrac{2}{x} + \dfrac{y}{3}} = \frac{12xy\left(\dfrac{x}{4} - \dfrac{2}{y}\right)}{12xy\left(\dfrac{2}{x} + \dfrac{y}{3}\right)} = \frac{3x^2y - 24x}{24y + 4xy^2}$$

EXERCISE 6.5

WRITTEN

Reduce each complex fraction to simplest form. Use either method, or as directed by your instructor. Assume that no denominator is zero.

Example

$$\frac{x - 1 - \frac{6}{x}}{1 + \frac{2}{x} - \frac{15}{x^2}}$$

Solution METHOD 1

$$\frac{x - 1 - \frac{6}{x}}{1 + \frac{2}{x} - \frac{15}{x^2}} = \frac{\frac{x^2 - x - 6}{x}}{\frac{x^2 + 2x - 15}{x^2}}$$

$$= \frac{x^2 - x - 6}{x} \cdot \frac{x^2}{x^2 + 2x - 15}$$

$$= \frac{(x^2 - x - 6)x}{x^2 + 2x - 15}$$

$$= \frac{x(x - 3)(x + 2)}{(x - 3)(x + 5)}$$

$$= \frac{x^2 + 2x}{x + 5}$$

METHOD 2

Multiply both the numerator and denominator by x^2, the least common denominator, to obtain

$$\frac{x^3 - x^2 - 6x}{x^2 + 2x - 15}.$$

Factoring and simplifying, we obtain

$$\frac{x(x - 3)(x + 2)}{(x - 3)(x + 5)} = \frac{x^2 + 2x}{x + 5}.$$

1. $\dfrac{1 + \frac{1}{2}}{1 + \frac{3}{4}}$

2. $\dfrac{5 - \frac{2}{3}}{2 + \frac{1}{6}}$

3. $\dfrac{1 + \frac{1}{x}}{1 - \frac{1}{x}}$

4. $\dfrac{1 + \frac{1}{x}}{x - \frac{1}{x}}$

5. $\dfrac{2 - \frac{1}{x}}{\frac{1}{x}}$

6. $\dfrac{x - \frac{1}{x}}{1 - \frac{1}{x}}$

7. $\dfrac{4 - \frac{1}{x^2}}{2 + \frac{1}{x}}$

8. $\dfrac{1 - \frac{1}{x^3}}{1 - \frac{1}{x}}$

9. $\dfrac{1 - \frac{3}{x} + \frac{2}{x^2}}{\frac{1}{x} - \frac{2}{x^2}}$

10. $\dfrac{1 - \frac{1}{x} - \frac{2}{x^2}}{\frac{1}{x} + \frac{1}{x^2}}$

11. $\dfrac{1 + \frac{1}{x} - \frac{2}{x^2}}{1 + \frac{2}{x} - \frac{6}{x^2}}$

12. $\dfrac{2 - \frac{7}{x} + \frac{3}{x^2}}{2 + \frac{3}{x} - \frac{2}{x^2}}$

13. $\dfrac{1 - \frac{1}{2x - 3}}{1 + \frac{1}{2x - 3}}$

14. $\dfrac{1 - \frac{2x}{1 - x}}{2 - \frac{1 + x}{1 - x}}$

15. $\dfrac{1 - \frac{x + 3}{x + 1}}{1 + \frac{1}{x + 1}}$

16. $\dfrac{2 + \dfrac{3}{x-1}}{1 + \dfrac{x}{x-1}}$

17. $\dfrac{x + 1 + \dfrac{x+1}{x-1}}{x - \dfrac{2}{x-1}}$

18. $\dfrac{x + 2 - \dfrac{2}{x-2}}{x - 1 - \dfrac{6}{x-2}}$

19. $\dfrac{2x - 1 + \dfrac{8x}{2x-1}}{2x + 3 + \dfrac{4}{2x-1}}$

20. $\dfrac{x - \dfrac{2}{2x+3}}{x - 3 + \dfrac{10}{2x+3}}$

21. $\dfrac{1 - \dfrac{1}{2x-3}}{1 + \dfrac{1}{x-3}}$

22. $\dfrac{2 + \dfrac{3}{x-2}}{1 + \dfrac{x}{x-1}}$

23. $\dfrac{x + 2 - \dfrac{2}{x-1}}{x - 1 - \dfrac{2}{x+2}}$

24. $\dfrac{x - \dfrac{2}{2x+3}}{x + 1 - \dfrac{3}{2x+1}}$

25. $\dfrac{x - \dfrac{4x-3}{2x-1}}{2x - \dfrac{x-3}{x-2}}$

26. $\dfrac{3x + \dfrac{x-5}{x-1}}{x - \dfrac{5}{3x-2}}$

27. $\dfrac{\dfrac{x+1}{x-1} - \dfrac{x-1}{x+1}}{\dfrac{x+1}{x-1} + \dfrac{x-1}{x+1}}$

28. $\dfrac{\dfrac{3x-2}{3x} - \dfrac{2}{3x+3}}{\dfrac{3}{x+1} + \dfrac{2}{x^2+x}}$

In the calculus, you will frequently encounter expressions to be simplified that are similar to those in Problems 29–34.

Example Simplify

$$\dfrac{(x^2 + 1)^{1/2} - x^2(x^2 + 1)^{-1/2}}{x^2 + 1}.$$

Solution Multiply both the numerator and the denominator by $(x^2 + 1)^{1/2}$, which gives

$$\dfrac{(x^2 + 1)^{1/2}[(x^2 + 1)^{1/2} - x^2(x^2 + 1)^{-1/2}]}{(x^2 + 1)^{1/2}(x^2 + 1)} = \dfrac{(x^2 + 1) - x^2}{(x^2 + 1)^{3/2}}$$

$$= \dfrac{1}{(x^2 + 1)^{3/2}}.$$

*29. $\dfrac{(x + 1)^{1/2} - \tfrac{1}{2}x(x + 1)^{-1/2}}{x + 1}$

*30. $\dfrac{x(2x + 1)^{-1/2} - (2x + 1)^{1/2}}{x^2}$

*31. $\dfrac{2x(x^2 - 4)^{1/2} - x^3(x^2 - 4)^{-1/2}}{x^2 - 4}$

*32. $\dfrac{(2x - 1)^{1/2}(2x + 1)^{-1/2} - (2x - 1)^{-1/2}(2x + 1)^{1/2}}{2x - 1}$

*33. $\dfrac{x(3x - 2)^{-2/3} - (3x - 2)^{1/3}}{x^2}$ [HINT: Multiply by $(3x - 2)^{2/3}$.]

*34. $\dfrac{(x^2 + 1)^{1/3} - \tfrac{2}{3}x^2(x^2 + 1)^{-2/3}}{(x^2 + 1)^{2/3}}$

Summary

1. Two fractions of the form $\dfrac{P(x)}{Q(x)}$ and $\dfrac{S(x)}{T(x)}$, $Q(x)$ and $T(x) \neq 0$, are equal if and only if $P(x) \cdot T(x) = Q(x) \cdot S(x)$. [6.1]
2. If both the numerator and the denominator of a fraction are divided or multiplied by the same nonzero quantity, the resulting fraction is equivalent to the original fraction. A fraction is in simplest form if the numerator and the denominator are relatively prime polynomials. [6.1]
3. Any two of the three signs of a fraction can be changed without changing the value of the fraction. The sign of a polynomial is changed if the signs of an *odd* number of its factors are changed. [6.1]
4. A sum or difference of two fractions can be written as one fraction provided both fractions have the same denominator. Moreover, either one or both fractions can be written equivalently so that they do have the same denominator. [6.2]
5. The **least common multiple** (LCM) of a set of polynomials is the product of all of the individual factors of the polynomials, each factor being used the greatest number of times that it occurs in any one polynomial. [6.2]
6. The product of two or more fractions can be simplified by dividing any numerator and any denominator by a common factor. [6.3]
7. The quotient of two fractions can be simplified by inverting the divisor and simplifying the resulting product. [6.3]
8. The solution set of an equation of the form $\dfrac{P(x)}{Q(x)} = 0$ contains all solutions of the equation $P(x) = 0$ that are *not* solutions of $Q(x) = 0$. [6.4]
9. An equation can be cleared of fractions by multiplying both members of the equation by the least common denominator of the fractions. However, no zero of the LCD is a solution of the equation. [6.4]
10. A **complex fraction** can be reduced to a simple fraction either by converting it to the quotient of two fractions or by multiplying both the numerator and the denominator of the complex fraction by the least common denominator of all of the fractions contained in the complex fraction. [6.5]

REVIEW EXERCISES

In Problems 1–18, assume that the values of all denominators are not zero.

SECTION 6.1

In Problems 1 and 2, is the first fraction equal to the second?

1. $\dfrac{2x^2 - 7x - 15}{x^2 - 4x - 5}, \dfrac{2x + 3}{x + 1}$

2. $\dfrac{3x^2 + x - 2}{3x^2 - 10x - 8}, \dfrac{x + 1}{x - 4}$

Reduce each fraction in Problems 3–6 to lowest terms.

3. $\dfrac{6x^2 + 11x - 35}{3x - 5}$

4. $\dfrac{8x^3 - 27}{2x - 3}$

Review Exercises 167

5. $\dfrac{x^2 + ax + xy + ay}{x + a}$

6. $\dfrac{x^4 + x^2y^2 + y^4}{x^2 + xy + y^2}$

In Problems 7–10, find the missing numerator or denominator so that the fractions in each pair are equal.

7. $\dfrac{x}{x-2}, \dfrac{?}{x^2 - 3x + 2}$

8. $\dfrac{2}{x-y}, \dfrac{?}{x^2 - y^2}$

9. $\dfrac{x+3}{x-4}, \dfrac{x^2 - x - 12}{?}$

10. $\dfrac{x+y}{x-y}, \dfrac{x^3 + y^3}{?}$

SECTION 6.2

Find each sum or difference in Problems 11–14.

11. $\dfrac{x+4}{x^2 - x + 2} + \dfrac{2x - 3}{x^2 - x + 2}$

12. $\dfrac{x - 2y}{x + y} - \dfrac{2x - y}{x - y}$

13. $\dfrac{x}{x^2 - 16} - \dfrac{x+1}{4 - 3x - x^2}$

14. $\dfrac{x}{x^2 + x + 1} + \dfrac{x}{x^3 - 1}$

SECTION 6.3

Simplify each product or quotient in Problems 15–18.

15. $\dfrac{(x+2)(x+3)}{(x-5)(x-4)} \cdot \dfrac{(x+2)(x-5)}{(x+4)(x+3)}$

16. $\dfrac{x^2 - x - 20}{x^2 + 7x + 12} \cdot \dfrac{x^2 + 6x + 9}{x^2 - 10x + 25}$

17. $\dfrac{x^2 + x - 2}{x^2 + 2x - 3} \div \dfrac{x^2 + 7x + 10}{x^2 - 2x - 15}$

18. $\dfrac{x^2 - x}{x^2 - 2x - 3} \cdot \dfrac{x^2 + 2x + 1}{x^2 + 4x} \div \dfrac{x^2 - 3x - 4}{x^2 - 16}$

SECTION 6.4

Find the solution set of each equation in Problems 19–22.

19. $\dfrac{x^2 - 5x + 6}{x^2 - 9} = 0$

20. $\dfrac{2x - 5}{3x + 1} = \dfrac{2x - 6}{3x + 8}$

21. $\dfrac{x^2 + 9x + 20}{x + 3} = \dfrac{x^2 - x - 2}{x + 1}$

22. $\dfrac{12x^2 - 12}{x^2 - 2x - 3} = \dfrac{7x}{x - 3} + \dfrac{5x}{x + 1}$

23. If the product of two consecutive even integers is divided by their sum, the quotient is equal to the sum of one-half the smaller integer and $\tfrac{5}{11}$. Find the two integers.

SECTION 6.5

Simplify the complex fractions in Problems 24 and 25.

24. $\dfrac{x^2 + \dfrac{1}{x}}{x + \dfrac{1}{x}}$

25. $\dfrac{\dfrac{x^2 + y^2}{x^2 - y^2}}{\dfrac{x - y}{x + y} - \dfrac{x + y}{x - y}}$

Relations, Functions, and Graphs—I

Graphs, charts, and tables are all familiar to anyone who reads a daily newspaper. They are devices for making quantitative comparisons of two or more things. When two things are compared in such a manner we say that they are in **relation** to each other. Thus, we may define a relation informally as a comparison of two things. The two things may be members of two different sets, or they may be members of the same set. A more formal definition will be given later.

A relation may be a random pairing of an element of one set with an element of another set (or the same set), or it may pair the elements in a specific manner. For example, the phrase "is equal to" names a relation that pairs an element in one set with an element in the second set so that both elements represent the same thing. In this book we will be concerned primarily with relations that can be defined by open sentences.

If a relation between two members of a set or between members of two sets is to be meaningful, it is necessary to identify the set or sets being discussed. The set or sets should always be named unless the relation is so clear that no misunderstanding can arise if this is not done.

7.1 Relations and Functions

In the preceding paragraphs a relation was assumed to involve some kind of a comparison between two elements of a set (or of different sets). Let us explore this concept in more detail. For an example, apply the relation *is less than* to the set $\{1, 3, 4, 7, 10\}$. You can observe that

$$1 < 3, \quad 1 < 4, \quad 1 < 7, \quad 1 < 10,$$
$$3 < 4, \quad 3 < 7, \quad 3 < 10,$$
$$4 < 7, \quad 4 < 10,$$
$$7 < 10.$$

It is evident that the relation pairs the elements in every possible way such that the first element is less than the second. Thus, the relation can be expressed as the set of ordered pairs

$$\{(1, 3), (1, 4), (1, 7), (1, 10), (3, 4), (3, 7), (3, 10), (4, 7), (4, 10), (7, 10)\}.$$

This suggests the following.

DEFINITION 7.1-1   A **relation** is a set of ordered pairs.

Since the first components of the ordered pairs form one set and the second components form another set (not necessarily different from the first), the definition of a relation implies that in forming the set of ordered pairs, the first component of each ordered pair is an element of one set, and the second component is an element of a second set.

In Section 1.2, the Cartesian product of two sets was defined informally. We now give a more formal definition in symbolic form.

DEFINITION 7.1-2   If A and B are any two sets, not necessarily distinct, the **Cartesian product**

$$A \times B = \{(a, b) \mid a \in A, b \in B\}.$$

Any set of ordered pairs (a, b) where $a \in A$ and $b \in B$ is a relation that is a subset of $A \times B$. Any subset of $A \times B$ is said to be a relation *from* A *to* B. If $B = A$, then $A \times B = A \times A$, and the relation is said to be *on* A.

Domain and Range

The set of first components of the ordered pairs in a relation is called the **domain** of the relation. The set of second components is called the **range** or **image set** of the relation. Each element of the range is an image under the relation of an element in the domain.

Example   Specify the domain and the range of the relation

$$\{(1, 2), (2, 4), (3, 6), (4, 8)\}.$$

Solution   Domain, $\{1, 2, 3, 4\}$. Range, $\{2, 4, 6, 8\}$.

Since, by our definition, relations are sets whose elements are ordered pairs, they can be named by using set notation. Any of the methods for naming sets may be used.

(1) List method: all ordered pairs are listed.

   Example: $\{(1, 2), (2, 4), (3, 6), (4, 8)\}$

(2) Rule method: either a word statement or set-builder notation where the "rule" is stated symbolically. In set-builder notation, the "rule" is usually an equation or an inequality.

Examples: {ordered pairs (x, y) such that x is less than y}
$$\{(x, y) | x < y\}$$

The domain and range are usually not evident when a word statement is used to name a relation. In such a case the set (or sets) containing the domain and the range must be specifically stated, or else some convention must be agreed upon. In this text we shall agree that, unless otherwise specified, the domain and the range of a relation are subsets of the set of real numbers.

Notation

Relations are also named by using single letters, such as f, g, or P. When a single letter is used to name a relation, the symbols $f(x)$, $g(x)$, $P(x)$, etc., may be used to name the image of x under the particular relation. That is, if x is any element in the domain, then $f(x)$ [read "f of x"] is the corresponding element in the range.

A statement such as
$$f(x) = 2x + 3$$
defines the relation
$$f = \{[x, f(x)] | f(x) = 2x + 3\}.$$
This statement could also be written as
$$f = \{(x, y) | y = 2x + 3\},$$
where y is the image of x under f.

Restrictions

In any real number relation, the replacement set for the first component of an ordered pair, that is, for x in (x, y), must be restricted so that each image, or second component, is a real number. Recall that division by zero is undefined and that the square root of a negative number is not a real number.

Example   List the elements in
$$f = \left\{[x, f(x)] \Big| f(x) = \frac{5}{x - 4}\right\},$$
if $x \in \{1, 2, 3, 4, 5\}$ and $f(x) \in R$.

Solution $f = \{(1, -\frac{5}{3}), (2, -\frac{5}{2}), (3, -5), (5, 5)\}.$

Note that the element 4 is not a member of the domain, because $f(x)$ is not defined for $x = 4$.

Specifying the Domain and the Range of a Relation

Restrictions on the set containing the range can be determined in the manner shown in the following example.

Example Specify the domain and the range of the relation defined by
$y = \sqrt{x^2 - 9}$.

Solution Since the square root of a negative number is not real, to obtain real values for y, x must be a number such that $x^2 - 9 \geq 0$. Hence, $x^2 \geq 9$, and $x \geq 3$, or $x \leq -3$. The domain is specified as

$\{x \mid x \leq -3\} \cup \{x \mid x \geq 3\}$.

Since the square root of any positive number is defined as a positive number, the range is

$\{y \mid y \geq 0\}$.

Functions

For some relations, the rule for forming the ordered pairs is such that each first component is paired with one and only one second component. Such a relation is given the special name **function**.

DEFINITION 7.1-3 A **function** is a relation such that for each first component there is one and only one second component.

This implies that no two distinct ordered pairs in a function can have the same first component. However, such ordered pairs may have the same second component.

Examples (a) $\{(1, 2), (2, 3), (3, 4)\}$ is a function.

(b) $\{(1, 2), (2, 3), (3, 4), (5, 4)\}$ is a function even though the ordered pairs $(3, 4)$ and $(5, 4)$ have the same second component.

(c) $\{(1, 2), (1, 3), (2, 3), (3, 4)\}$ is not a function, because the ordered pairs $(1, 2)$ and $(1, 3)$ have the same first component and different second components.

EXERCISE 7.1

ORAL

Define:

1. Relation
2. Domain
3. Range
4. Cartesian product
5. Function
6. Image
7. Image set
8. $f(x)$

WRITTEN

Given $S = \{1, 2, 3, 5, 6\}$. If $x, y \in S$, write the set of ordered pairs specified in Problems 1–6.

Example $\{(x, y) | (x - y) \in S\}$

Solution $\{(2, 1), (3, 1), (6, 1), (3, 2), (5, 2), (5, 3), (6, 3), (6, 5)\}$

1. $\{(x, y) | (x + y) \in S\}$
2. $\{(x, y) | xy \in S\}$
3. $\{(x, y) | x/y \in S\}$
4. $\{(x, y) | x < y\}$
5. $\{(x, y) | x > y\}$
6. $\{(x, y) | x = y\}$

In Problems 7–16, $x \in \{-3, -2, -1, 0, 1, 2, 3\}$.

(a) Specify the domain of the relation so that each element of the range $\in R$.
(b) Write the relation as a set of ordered pairs.

Example $\left\{(x, y) \mid y = \dfrac{x}{x - 3}\right\}$

Solution (a) The domain is $\{-3, -2, -1, 0, 1, 2\}$.

(b) $\{(-3, \tfrac{1}{2}), (-2, \tfrac{2}{5}), (-1, \tfrac{1}{4}), (0, 0), (1, -\tfrac{1}{2}), (2, -2)\}$

7. $\{(x, y) | y = x - 2\}$
8. $\{(x, y) | y = x + 3\}$
9. $\{[x, f(x)] | f(x) = |x|\}$
10. $\{[x, f(x)] | f(x) = |x - 1|\}$
11. $\left\{(x, y) \mid y = \dfrac{2}{2 - |x|}\right\}$
12. $\left\{(x, y) \mid y = \dfrac{5}{|x| - 3}\right\}$
13. $\{(x, y) | y = \sqrt{4 - x^2}\}$
14. $\{(x, y) | y = \sqrt{x^2 - 4}\}$
15. $\left\{[x, g(x)] \mid g(x) = \dfrac{2}{\sqrt{4 - x^2}}\right\}$
16. $\left\{[x, P(x)] \mid P(x) = \dfrac{3}{\sqrt{x^2 - 5}}\right\}$

Specify (a) the domain and (b) the range for each relation in Problems 17–28.

Example $\{(x, y) | y = x^2 + 7\}$

Solution (a) By inspection we see that $y \in R$ for any real number replacement of x. Hence, the domain is $\{x | x \in R\}$.

(b) Since $x^2 \geq 0$ for any real number replacement of x, the range is $\{y | y \geq 7\}$.

17. $\{(x, y) | y = x^2 - 1\}$
18. $\{(x, y) | y = x^2 + 3\}$
19. $\{(x, y) | y = \sqrt{4 - x^2}\}$
20. $\{(x, y) | y = \sqrt{16 - x^2}\}$
21. $\{(x, y) | x^2 + y^2 = 9\}$
22. $\{(x, y) | x^2 + y^2 = 25\}$
23. $\left\{(x, y) \mid y = \dfrac{1}{x + 1}\right\}$
24. $\left\{(x, y) \mid y = \dfrac{3}{x - 7}\right\}$
25. $\{(x, y) | x^2 - y^2 = 16\}$
26. $\{(x, y) | x^2 - y^2 = 4\}$
27. $\left\{(x, y) \mid y = \dfrac{x}{\sqrt{4 - x^2}}\right\}$
28. $\left\{(x, y) \mid y = \dfrac{x}{\sqrt{x^2 - 4}}\right\}$

29. Which of the relations in the odd-numbered problems 17–27 are also functions?
30. Which of the relations in the even-numbered problems 18–28 are also functions?

7.2 Graphs of Relations in $R \times R$

A Cartesian product can be represented graphically by an **array** or **point lattice**. This is accomplished by using separate number lines to represent the elements of the domain and those of the range. It is customary to construct these number lines so that they meet at right angles and have a common origin.

Example   Construct a point lattice for the Cartesian product $A \times A$ if $A = \{-2, -1, 0, 1, 2\}$.

Solution $A \times A = \{(-2, -2), (-2, -1), (-2, 0), (-2, 1), (-2, 2),$
$(-1, -2), (-1, -1), (-1, 0), (-1, 1), (-1, 2),$
$(0, -2), (0, -1), (0, 0), (0, 1), (0, 2),$
$(1, -2), (1, -1), (1, 0), (1, 1), (1, 2),$
$(2, -2), (2, -1), (2, 0), (2, 1), (2, 2)\}$

The point lattice is shown in Figure 7.2-1.

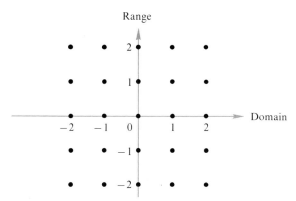

Figure 7.2-1

Coordinates

Since we have assumed a one-to-one correspondence between the set of real numbers and the set of points in a line, the example above suggests that we assume a similar correspondence between the set of ordered pairs of real numbers and the set of points in a plane. In this context, the plane is called the **coordinate plane**, a point in the plane is the **graph** of an ordered pair, and the components of the ordered pair are the **coordinates** of the point.

The first component of an ordered pair is called the **abscissa** and indicates the location of a point to the right or left of the origin. The second component is called the **ordinate** and indicates the location of a point above or below the origin. If the number lines are perpendicular to each other, they form the familiar **rectangular** or **Cartesian coordinate system**. The two number lines, called axes, separate the coordinate plane into four parts called **quadrants**

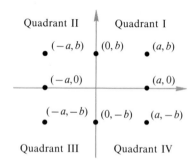

Figure 7.2-2

Graphs of ordered pairs of real numbers are located on one of the axes or in one of the four quadrants, depending upon whether one or both of the coordinates are 0, positive, or negative as shown in Figure 7.2-2 ($a, b \geq 0$). Quadrants are designated by Roman numerals.

Graphs in $U \times U$

If $A \subseteq U$ and $B \subseteq U$, where U is the universe of discourse or universal set, any relation $\{(a, b) | a \in A, b \in B\}$ is a subset of $U \times U$. You can graph the relation on the point lattice of $U \times U$ by indicating all points whose coordinate pairs are members of the relation. One such technique is shown in the following example.

Example Graph the relation $\{(a, b) | a < b,\ a, b \in U\}$ over $U \times U$, where $U = \{-2, -1, 0, 1, 2\}$.

Solution See Figure 7.2-3. Note that each point whose coordinates satisfy the condition for the relation has been circled.

Graphs in $R \times R$

If the domain of a relation is R or some subset of R such as $\{x | a \leq x \leq b, x \in R\}$, the graph is not, in general, composed of isolated points, but contains all points in a portion of the plane. This is illustrated by the graph of $\{(a, b) | a < b, a, b \in R\}$ in Figure 7.2-4. The shaded half of the coordinate plane contains all

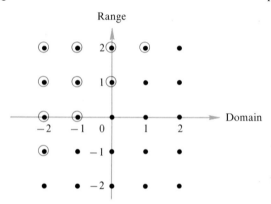

Figure 7.2-3

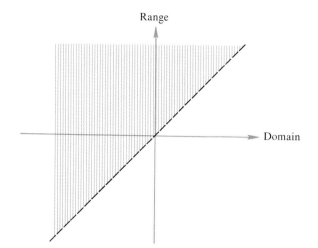

Figure 7.2-4

points whose coordinates form ordered pairs in the relation. The broken line is the graph of the relation $\{(a, b) \mid a = b, a, b \in R\}$. Graphing relations in $R \times R$ will be discussed in more detail in later sections.

EXERCISE 7.2

ORAL

Define:

1. Coordinates of a point.
2. Abscissa.
3. Ordinate.
4. Graph of a relation.

WRITTEN

Let $U = \{-5, -4, -3, -2, -1, 0, 1, 2, 3, 4, 5\}$. Graph each relation in Problems 1-10 by circling the appropriate dots on a point lattice for $U \times U$.

Example $\{(x, y) \mid y = x\}$

Solution   Write the relation as a set of ordered pairs.

$$\{(-5, -5), (-4, -4), (-3, -3), (-2, -2), (-1, -1), (0, 0),$$
$$(1, 1), (2, 2), (3, 3), (4, 4), (5, 5)\}$$

Then locate the points corresponding to these ordered pairs, as shown in Figure 7.2-5 (see page 176).

1. $\{(x, y) \mid y = x + 1\}$
2. $\{(x, y) \mid y = x - 1\}$
3. $\left\{(x, y) \mid y = \dfrac{x}{2}\right\}$
4. $\left\{(x, y) \mid y = \dfrac{x}{3}\right\}$
5. $\left\{(x, y) \mid y = \dfrac{6}{x}\right\}$
6. $\left\{(x, y) \mid y = \dfrac{10}{x}\right\}$

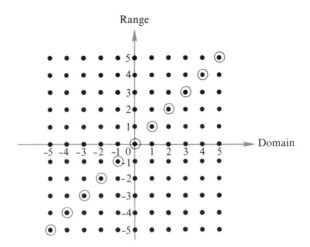

Figure 7.2-5

7. $\{(x, y) | y < x\}$
8. $\{(x, y) | y > x\}$
9. $\{(x, y) | y > x^2\}$
10. $\{(x, y) | y < x^2\}$

7.3 Linear Functions

The function f defined by

$$f = \{(x, y) | y = mx + b\},$$

where m and b are constants, $m \neq 0$, is called a **linear function**. Such a function is frequently defined by one of the equations

$$f(x) = mx + b$$

or

$$y = mx + b \qquad (m \neq 0).$$

When a number of the ordered pairs (x, y) of a linear function are plotted on a rectangular coordinate system, all the points appear to lie on the same straight line. It can be shown, but we shall not do so, that the graph of a linear function is a straight line, and that *the coordinates of every point on the line satisfy the equation.*

Graphs of Linear Functions

You learned in plane geometry that two points determine a line. That is, all lines that contain both of two given points are collinear and can be considered to be the same line. Thus, you can construct the graph of a linear function by locating two points corresponding to two distinct ordered pairs in the function and drawing a line through these points.

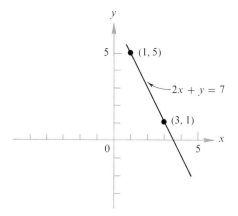

Figure 7.3-1

Example Construct the graph of $\{(x, y) | 2x + y = 7\}$.

Solution Find the coordinates of points by selecting values for x and computing the corresponding values for y. For instance, if $x = 1$, then $2 + y = 7$ and $y = 5$. If $x = 3$, then $6 + y = 7$ and $y = 1$. Thus, $(1, 5)$ and $(3, 1)$ are the coordinates of two points. The graph is completed as shown in Figure 7.3-1.

Sometimes the graph of a relation is referred to as the *graph of the equation* (or inequality). Thus, the line in Figure 7.3-1 is called the graph of $2x + y = 7$. While two points are sufficient to determine a line, it is better to find at least three points in order to check your computations of the first two.

Intercepts

The point of intersection of a line and an axis is of special interest. Since any point on the y-axis has 0 as its first component, the value of y for the point where the graph of an equation intersects the y-axis can be found by replacing x with 0. This value of y is called the **y-intercept**. The **x-intercept**, which is the x-coordinate of the intersection of the graph and the x-axis, is found by replacing y with 0. Construction of a graph by determining the intercepts is called the **intercept method**

Example Use the intercept method to construct the graph of $3x - 4y = 12$.

Solution Replacing x with 0 gives

$$0 - 4y = 12$$
$$y = -3.$$

Then, by replacing y with 0,

$$3x + 0 = 12$$
$$x = 4.$$

The graph is shown in Figure 7.3-2 (see page 178).

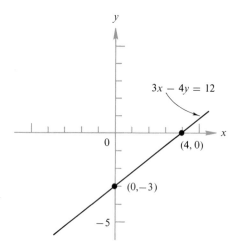

Figure 7.3-2

Lines Parallel to an Axis

If $m = 0$ in $y = mx + b$, the resulting equation, $y = b$, defines a function called a **constant function**. In such a function, every ordered pair has the same second component. The graph is the line parallel to the x-axis and passing through the point $(0, b)$.

The equation $x = a$ defines a relation in which all ordered pairs have the same first component. Such a relation is not a function, but since the graph is a line parallel to, and $|a|$ units from, the y-axis, it is called a **linear relation**.

Examples Construct the graph of each equation.

    (a) $y = 2$ (b) $x = -3$

Solutions (a) See Figure 7.3-3.   (b) See Figure 7.3-4.

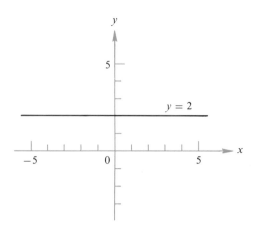

Figure 7.3-3

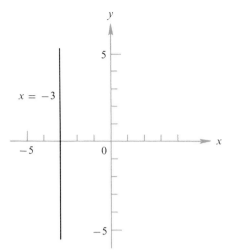

Figure 7.3-4

EXERCISE 7.3

WRITTEN

Construct the graph of each function in Problems 1–10.

1. $\{(x, y) \mid x + y = 2\}$
2. $\{(x, y) \mid x - y = 4\}$
3. $\{(x, y) \mid 4x - 2y = 3\}$
4. $\{(x, y) \mid 6x + 3y = 4\}$
5. $\{(x, y) \mid x - 4y = 2\}$
6. $\{(x, y) \mid x + 3y = 5\}$
7. $\{(x, y) \mid 2x + 3y = 9\}$
8. $\{(x, y) \mid 3x - 5y = 5\}$
9. $\{(x, y) \mid 2x + 7y = 4\}$
10. $\{(x, y) \mid 3x - 8y = 2\}$

In Problems 11–16, graph each equation by locating the intercepts.

11. $3x - 2y = 6$
12. $5x + 3y = 15$
13. $4x - 3y + 12 = 0$
14. $2x - 3y + 9 = 0$
15. $2x + 7y - 14 = 0$
16. $7x + 3y = 21$

Graph each equation in Problems 17–22.

17. $y = 4$
18. $y + 2 = 0$
19. $y - 5 = 0$
20. $y = -7$
21. $x = 2$
22. $x + 3 = 0$

7.4 Equations for Lines

In Section 7.3 we discussed construction of the graphs of linear functions. In this section we wish to discuss some properties of these graphs and some techniques for finding equations for lines that meet given conditions. Since a line can be determined by two points, let us begin by considering the segment connecting two given points. From here on, we shall name points by their coordinate pairs. For example, the point $(2, 3)$ means the point with coordinates 2 and 3.

Distance between Two Points

If you graph two numbers, say 3 and 8, on the number line as shown in Figure 7.4-1, you can observe that the distance between the two points is determined by the difference of the two coordinates, $8 - 3 = 5$. More generally, if the number

Figure 7.4-1

line corresponds to the x-axis, the distance between any two points whose coordinates are x_1 and x_2, respectively, can be found by evaluating $|x_2 - x_1|$. In a similar manner, you can determine that the distance between any two points on the y-axis is given by $|y_2 - y_1|$.

These facts are useful in finding the distance between two points that are not both on the same axis. For example, consider the distance between the points $(2, 3)$ and $(5, 7)$ (Figure 7.4-2). If perpendicular lines are drawn from each point to both coordinate axes, you see that the distance between the points along the y-axis is $7 - 3$, or 4, and the distance between the points along the x-axis is $5 - 2$, or 3.

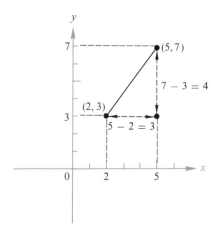

Figure 7.4-2

If we now extend a line perpendicular to the y-axis through $(2, 3)$ until it intersects the line from $(5, 7)$ perpendicular to the x-axis, a right triangle is formed whose sides are 4 and 3 units long, respectively. Then, by using the Pythagorean theorem we find the distance between the two points.

$$\begin{aligned} d &= \sqrt{4^2 + 3^2} \\ &= \sqrt{16 + 9} \\ &= \sqrt{25} \\ &= 5 \end{aligned}$$

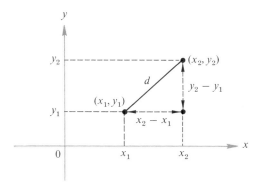

Figure 7.4-3

If we apply this technique to the more general case of the distance between the points (x_1, y_1) and (x_2, y_2), we have, from Figure 7.4-3,

$$d \sqrt{(x_2 - x_1)^2 + (y_2 - y_1)^2},$$

which is called the **distance formula**.

Slope of a Line

A line is also determined by one point on the line and the direction of the line. Ordinarily, the direction of a line is given with respect to the positive x-axis. When this is done, we refer to the direction as the **slope** of the line. If you consider a line to be the path of a point whose horizontal movement is in the direction of the positive x-axis, then the slope of the line can be defined informally as the change in vertical distance per unit of increase in the horizontal distance.

In Figure 7.4-3, if a point moves from (x_1, y_1) to (x_2, y_2), the change in vertical distance is $y_2 - y_1$ and the increase in horizontal distance is $x_2 - x_1$. Thus, the slope of the line containing the two points is given by the formula

$$m = \frac{y_2 - y_1}{x_2 - x_1}$$

Example Find the slope of the line containing $(2, 3)$ and $(5, 7)$.

Solution $y_2 = 7$, $y_1 = 3$, $x_2 = 5$, and $x_1 = 2$. Substituting the given values in the formula

$$m = \frac{y_2 - y_1}{x_2 - x_1},$$

we obtain

$$m = \frac{7 - 3}{5 - 2} = \frac{4}{3}.$$

A slope of $\frac{4}{3}$ means that if x increases 3 units, y increases 4 units. Thus, as a point moves from left to right, its distance above the x-axis increases at a constant rate. A negative slope shows that the y values decrease as x increases. If the slope is 0, there is no change in y, and the line is parallel to the x-axis.

Line Containing Two Given Points

The slope of a line is constant. Thus, the slopes of any two segments of the same line are equal. This fact can be used to find an equation for the line containing two given points.

Example Find an equation for the line containing the points (1, 3) and (3, 7).

Solution Since a line is a set of points, we let (x, y) be any other point in the line. By equating the slopes of the two segments, we form the equation

$$\frac{y-3}{x-1} = \frac{y-7}{x-3}.$$

Then, applying the theorem about equal fractions, we find

$$(x-3)(y-3) = (x-1)(y-7)$$
$$xy - 3y - 3x + 9 = xy - y - 7x + 7,$$

which is equivalent to

$$4x - 2y = -2.$$

An alternate technique for the example above is to use the segment determined by the given points. Thus, for the line containing the points (x_1, y_1) and (x_2, y_2), we have

$$\frac{y - y_1}{x - x_1} = \frac{y_2 - y_1}{x_2 - x_1},$$

which is called the **two-point form** of the equation for a line.

If the two given points are the intercepts of the line (0, b) and (a, 0), $a, b \neq 0$, the two-point form becomes

$$\frac{y - b}{x - 0} = \frac{0 - b}{a - 0}$$
$$a(y - b) = -bx$$
$$ay - ab = -bx$$
$$bx + ay = ab.$$

If both members of the last equation are divided by ab, the equation becomes

$$\frac{x}{a} + \frac{y}{b} = 1.$$

This equation is called the **intercept form** of the equation for a line, since a is the x-intercept and b is the y-intercept.

Line with Given Slope Containing a Given Point

Since a line can be determined by a point and a direction, an equation for a line can be formed if the slope of the line and the coordinates of one point in the line are known.

Example Find an equation for the line with slope $\frac{3}{5}$ and containing the point $(2, -1)$.

Solution Let (x, y) be any other point on the line. Then the slope of the segment between (x, y) and $(2, -1)$ must equal $\frac{3}{5}$.

$$\frac{y - (-1)}{x - 2} = \frac{3}{5}$$

$$5(y + 1) = 3(x - 2)$$
$$5y + 5 = 3x - 6,$$

which is equivalent to

$$3x - 5y = 11.$$

For the more general case of the line with slope m and containing the point (x_1, y_1), we use any other point (x, y) and the given point in the slope formula to obtain

$$\frac{y - y_1}{x - x_1} = m,$$

which is equivalent to

$$y - y_1 = m(x - x_1).$$

This equation is called the **point-slope form** of the equation for a line. The two-point form is easily converted into the point-slope form, since

$$\frac{y_2 - y_1}{x_2 - x_1} = m.$$

If the given point is the y-intercept $(0, b)$, the point-slope form becomes

$$y - b = m(x - 0)$$

or

$$y = mx + b,$$

which is called the **slope-intercept** form of the equation for a line. This form is the equation that was used to define a linear function.

Now consider the equation of first degree in two variables,

$$Ax + By = C.$$

If you solve this equation for y in terms of x, you obtain

$$y = -\frac{A}{B}x + \frac{C}{B}.$$

When you compare this equation with the slope intercept form you see that the slope of a line is the negative of the coefficient of x divided by the coefficient of y and that the y-intercept is the constant term divided by the coefficient of y.

EXERCISE 7.4

WRITTEN

In Problems 1–12, find (a) the distance between the points and (b) the slope of the segment connecting the points.

Example $(3, -5)$ and $(-2, 7)$

Solution   (a) Substitute the coordinates into the distance formula.

$$d = \sqrt{(-2 - 3)^2 + [7 - (-5)]^2}$$
$$= \sqrt{(-5)^2 + 12^2}$$
$$= \sqrt{25 + 144}$$
$$= \sqrt{169}$$
$$= 13$$

(b) Substitute the coordinates into the slope formula.

$$m = \frac{7 - (-5)}{-2 - 3}$$
$$= \frac{12}{-5}$$
$$= \frac{-12}{5}$$

1. $(4, 5)$ and $(8, 8)$
2. $(3, 3)$ and $(6, 7)$
3. $(-2, 1)$ and $(1, 5)$
4. $(-3, 2)$ and $(1, -1)$
5. $(3, 6)$ and $(8, -6)$
6. $(-3, -5)$ and $(2, 7)$
7. $(2, 3)$ and $(4, 7)$
8. $(3, -2)$ and $(5, -1)$
9. $(-1, 4)$ and $(5, -1)$
10. $(-1, -2)$ and $(-5, -3)$
11. $(-2, 5)$ and $(5, 5)$
12. $(-6, -2)$ and $(7, -2)$

In Problems 13–18, use the two-point form to find an equation for the line containing the two given points.

Example $(3, -5)$ and $(-2, 7)$

Solution   Replace x_1, y_1, x_2, and y_2 in the two-point form with the given values.

$$\frac{y - (-5)}{x - 3} = \frac{7 - (-5)}{-2 - 3}$$
$$\frac{y + 5}{x - 3} = \frac{12}{-5}$$
$$-5(y + 5) = 12(x - 3)$$
$$-5y - 25 = 12x - 36$$
$$12x + 5y = 11$$

13. (3, 4) and (5, 9)   14. (1, 1) and (4, 4)   15. (−3, 2) and (2, −3)
16. (−1, −4) and (3, 5)   17. (−3, 7) and (3, 7)   18. (4, −3) and (4, 2)

In Problems 19–26, find an equation for the line with the given slope and containing the given point.

Example $\frac{3}{4}$, (5, 2)

Solution   Replace x_1, y_1, and m in the point-slope form with the given values.

$$y - 2 = \tfrac{3}{4}(x - 5)$$
$$4y - 8 = 3x - 15$$
$$3x - 4y = 7$$

19. $\frac{2}{3}$, (2, 8)   20. 5, (1, 2)   21. −2, (3, −1)   22. $-\frac{1}{2}$, (3, −1)
23. $\frac{5}{2}$, (−2, −3)   24. $\frac{7}{3}$, (3, −4)   25. $-\frac{4}{5}$, (−1, 4)   26. 0, (5, 7)

27. Find an equation for the line with the same slope as the graph of $2x - y = 5$ and containing the point (3, 2).
28. Find an equation for the line with the same slope as the graph of $3x + 4y = 2$ and containing the point (1, −1).
29. Find an equation for the line whose slope is twice as great and which has the same y-intercept as the line with equation $x - 2y = 4$.
30. Find an equation for the line containing (2, 3) whose slope is the negative reciprocal of the slope of the line with equation $x - 3y = -7$.

7.5   Quadratic Functions

In Chapter 5, you found the solution sets of quadratic equations in one variable, where the equation was of the form

$$ax^2 + bx + c = 0.$$

If the left-hand member of this equation is used to form the sentence in two variables

$$f(x) = ax^2 + bx + c \quad \text{or} \quad y = ax^2 + bx + c,$$

the resulting equation defines a **quadratic function**.

The graph of a quadratic function is not a straight line but a curve called a **parabola**. The accuracy with which such a curve can be constructed depends upon the number of points located. Locating a large number of points can be quite tedious, so we will make use of some of the properties of a parabola to simplify the procedure.

Intercepts

The intercepts of the curve defined by

$$y = ax^2 + bx + c$$

are relatively easy to find. When $x = 0$, $y = c$, and when $y = 0$, the quadratic equation $ax^2 + bx + c = 0$ may be solved by any convenient method to find values for x. (See Chapter 5.) However, for $x \in R$, $ax^2 + bx + c = 0$ may have zero, one, or two real number solutions. Thus the graph of $y = ax^2 + bx + c$ may intersect the x-axis in zero, one, or two points.

Vertex of a Parabola

If the curve is constructed by locating a number of points, it becomes evident that as x increases, the values for y will either decrease for some interval and then increase, or increase and then decrease. In either case, there is a point, called the vertex, at which the curve changes direction. Such a point will be either the lowest point on the graph (minimum) or the highest point on the graph (maximum).

The coordinates of the lowest, or highest, point can be determined by completing the square.

If
$$y = ax^2 + bx + c,$$
then
$$y = a\left(x^2 + \frac{b}{a}x\right) + c$$
$$= a\left(x^2 + \frac{b}{a}x + \frac{b^2}{4a^2}\right) + c - \frac{b^2}{4a}$$
$$= a\left(x + \frac{b}{2a}\right)^2 - \frac{b^2 - 4ac}{4a}.$$

In the right-hand member of the equation, if $a > 0$, $a\left(x + \frac{b}{2a}\right)^2$ is always positive except when $x = \frac{-b}{2a}$. If x is replaced by $\frac{-b}{2a}$,
$$a\left(x + \frac{b}{2a}\right)^2 = 0 \quad \text{and} \quad y = -\frac{b^2 - 4ac}{4a}.$$

For any other value of x,
$$y > -\frac{b^2 - 4ac}{4a}.$$

Thus, at $x = \frac{-b}{2a}$, y has the least value. Therefore, if $a > 0$, the number $-\frac{b^2 - 4ac}{4a}$ is called the **minimum function value**.

If $a < 0$, $a\left(x - \frac{b}{2a}\right)^2 \leq 0$, and the number $-\frac{b^2 - 4ac}{4a}$ is then the **maximum function value**.

Example Find the coordinates of the vertex for the graph of the equation $y = 2x^2 - 8x + 12$, and state whether the y-coordinate of the point is a maximum or a minimum value of the set of all y's.

Solution $a = 2$, $b = -8$, $c = 12$

$$\frac{-b}{2a} = \frac{8}{4} = 2$$

$$-\frac{b^2 - 4ac}{4a} = -\frac{64 - 96}{8} = 4$$

The vertex is $(2, 4)$.

Since $a > 0$, the vertex is the point where the y value is a minimum.

Symmetry

A second property of a parabola that is helpful in graphing is that of **symmetry**. Two points are symmetric to a third point if and only if the third point is the bisector of the line segment connecting the other two.

In Figure 7.5-1, P_1 and P_2 are symmetric with respect to P_3 because $|P_1P_3| = |P_3P_2|$ and P_1, P_2, and P_3 are collinear. (The symbol $|P_1P_3|$ means the length of the segment from P_1 to P_3.)

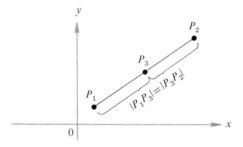

Figure 7.5-1

Two points are symmetric with respect to a line if the line is the perpendicular bisector of the segment connecting the two points (see Figure 7.5-2). Each point is called the **mirror image** of the other.

A curve is symmetric with respect to a line if for each point on the curve, except the point of intersection of the line and the curve, its mirror image with respect to the line is also on the curve.

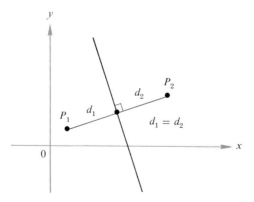

Figure 7.5-2

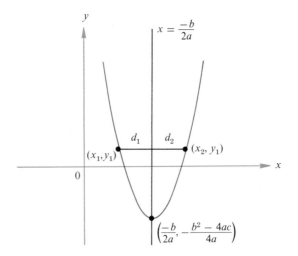

Figure 7.5-3

Axis of Symmetry

If a curve is symmetric with respect to a line, the line is called the **axis of symmetry** of the curve. Figure 7.5-3 shows the graph of an equation of the form $y = ax^2 + bx + c$, $a > 0$. The x-coordinate of the minimum point is $-b/2a$. If there is an axis of symmetry, it must pass through this point and be parallel to the y-axis. It can be shown that a parabola with equation $y = ax^2 + bx + c$ is symmetric with respect to the line $x = -b/2a$. (See Problem 29, Exercise 7.5.)

The principle of symmetry enables us to find a second point on the curve for each point whose coordinates are determined by direct substitution. The second point has the same ordinate as the first and is the same distance from the axis of symmetry, but on the opposite side.

Graphs of Quadratic Functions

We now combine the techniques discussed previously.

Example Graph the equation $y = x^2 - 8x + 12$.

Solution (1) The y-intercept is the point (0, 12).

(2) The x-intercepts are found by solving the equation

$$x^2 - 8x + 12 = 0.$$
$$(x - 2)(x - 6) = 0$$
$$x = 2, \quad x = 6$$

(3) The coordinates of the vertex are

$$x = \frac{-b}{2a} = \frac{-(-8)}{2} = 4$$

and

$$y = -\frac{b^2 - 4ac}{4a} = -\frac{64 - 48}{4} = -4.$$

Thus, the vertex, is $(4, -4)$, which is a minimum because $a > 0$.

(4) The curve is symmetric to the line $x = 4$.

If $x = 3$, $y = (3)^2 - 8(3) + 12 = -3$, and the points $(3, -3)$ and $(5, -3)$ are on the curve (both points are 1 unit from $x = 4$).

If $x = 1$, $y = (1)^2 - 8(1) + 12 = 5$. The points $(1, 5)$ and $(7, 5)$ are on the curve (both points are 3 units from $x = 4$). Thus, nine points are located, and the curve is sketched with reasonable accuracy (Figure 7.5-4).

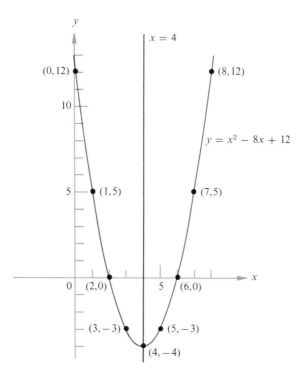

Figure 7.5-4

In constructing a graph in this manner, we are assuming the one-to-one correspondence of the set of real numbers with the points on a line and that for every permissible replacement for x within an interval, a corresponding value of y also exists. Thus, the graph of a quadratic function is constructed as a continuous smooth curve.

EXERCISE 7.5

ORAL

1. Define: (a) quadratic function, (b) vertex of a parabola.

2. What is the necessary condition for two points to be symmetric with respect to a third point? With respect to a line?
3. What is the necessary condition for a curve to be symmetric with respect to a line?

WRITTEN

In Problems 1–10, find the coordinates of points symmetric to the given points with respect to the given line.

Example $(2, 5), (3, 7), (-4, -3), (-7, 2);\quad x = -1$

Solution We illustrate this example graphically in Figure 7.5-5.

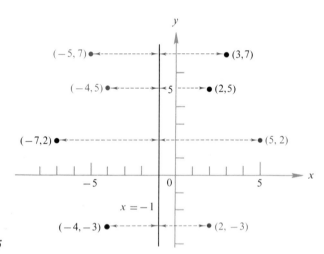

Figure 7.5-5

1. $(4, 5), (6, 2), (-3, 4), (3, -2);\ x = 0$
2. $(5, 6), (4, -4), (-1, 0), (-3, -3);\ x = 1$
3. $(-2, 0), (0, 2), (3, 1), (4, -1);\ x = -3$
4. $(6, 2), (0, 5), (-2, -1), (2, -3);\ x = 4$
5. $(2, 3), (4, 5), (-3, -2), (-6, 2);\ x = -1$
6. $(1, 2), (3, 5), (-3, 7), (-5, 1);\ x = -\frac{1}{2}$
7. $(4, 5), (6, 2), (-3, 4), (3, -2);\ y = 0$
8. $(5, 6), (4, -3), (-1, 0), (-3, -3);\ y = 1$
9. $(3, 7), (2, 5), (-4, -3), (-2, -7);\ y = -1$
10. $(1, 3), (3, 5), (-5, 1), (-3, -3);\ y = \frac{3}{2}$

For the graph of each equation in Problems 11–28, find:
(a) the x-intercepts and the y-intercept;
(b) the vertex (state whether a maximum or a minimum);
(c) the equation of the axis of symmetry.
(d) Then use the information found in (a), (b), and (c) to graph the equation.

Example $y = 6 + x - x^2$

Solution (a) To find the x-intercepts, let $y = 0$. Then,

$$6 + x - x^2 = 0$$
$$(3 - x)(2 + x) = 0$$
$$x = 3, \quad x = -2,$$

and the x-intercepts are $(3, 0)$ and $(-2, 0)$.
The y-intercept is $(0, 6)$.

(b) $a = -1, b = 1, c = 6$.

The vertex $\left(\dfrac{-b}{2a}, -\dfrac{b^2 - 4ac}{4a}\right)$ is

$$\left(\dfrac{-1}{2(-1)}, -\dfrac{1 - 4(-1)(6)}{4(-1)}\right) = \left(\dfrac{1}{2}, \dfrac{25}{4}\right).$$

Since $a < 0$, this is a maximum.

(c) The equation of the axis of symmetry is $x = \frac{1}{2}$.

(d) The mirror image of $(0, 6)$ with respect to the axis of symmetry is $(1, 6)$. If $x = -3$, $y = 6 - 3 - 9 = -6$, and the point $(-3, -6)$, as well as its mirror image $(4, -6)$, is a point on the curve.

These five points are sufficient to construct a reasonable approximation of the graph, as shown in Figure 7.5-6.

11. $y = x^2 - 4$
12. $y = x^2 - 9$
13. $y = 16 - x^2$
14. $y = 4 - x^2$
15. $y = 6 - x - x^2$
16. $y = 10 + 3x - x^2$
17. $y = x^2 + x - 6$
18. $y = x^2 - 2x - 8$
19. $y = 2x^2 - x - 15$
20. $y = 3x^2 - 10x - 8$
21. $y = 2x^2 + 3x - 14$
22. $y = 4x^2 - 4x - 15$
23. $y = x^2 + 3x - 4$
24. $y = 3 - 2x - x^2$
25. $y = x^2 + 2x - 3$
26. $y = 2 + x - x^2$
27. $y = x^2 + 4$
28. $y = x^2 + 9$

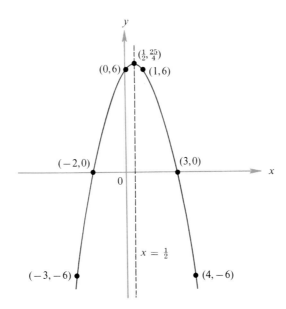

Figure 7.5-6

*29. If $f(x) = ax^2 + bx + c$ and t is any real number, show that

$$f\left(\frac{-b}{2a} - t\right) = f\left(\frac{-b}{2a} + t\right).$$

Explain how this equality proves that the graph of $f(x) = ax^2 + bx + c$ is symmetric to the line $x = -b/2a$.

Summary

1. A **relation** is a set of ordered pairs. A relation is usually defined by an equation or inequality in two variables. [7.1]
2. The set of first components of the ordered pairs in a relation is called the **domain** of the relation. The set of second components is called the **range** of the relation. [7.1]
3. When a single letter such as f, g, or P is used to name a relation, a symbol such as $f(x)$, $g(x)$, or $P(x)$ is used to represent the element in the range corresponding to x in the domain. $f(x)$ is also called the **image** of x under f. [7.1]
4. A **function** is a relation such that for each element in the domain there is one and only one element in the range. An element in the range may be paired with more than one element in the domain. [7.1]
5. We have assumed a **one-to-one correspondence** between the set of ordered pairs of real numbers and the set of points in a plane. The **coordinate plane** is separated into four **quadrants** by two number lines called **axes**, one horizontal and one vertical. The point of intersection of the axes is the **origin**. [7.2]
6. The **graph of a relation** is the set of points whose coordinates are the set of ordered pairs in the relation. [7.2]
7. The function $f = \{(x, y) \mid y = mx + b\}$ where m and b are constants, $m \neq 0$, is called a **linear function**. [7.3]
8. The graph of a linear function is a straight line. [7.3]
9. The **distance** between the points (x_1, y_1) and (x_2, y_2) is given by

$$d = \sqrt{(x_2 - x_1)^2 + (y_2 - y_1)^2}.$$ [7.4]

10. The **slope** of the line passing through the points (x_1, y_1) and (x_2, y_2) is given by

$$m = \frac{y_2 - y_1}{x_2 - x_1}.$$ [7.4]

11. The equation of a line can occur in any one of the following forms:

$\dfrac{y - y_1}{x - x_1} = \dfrac{y_2 - y_1}{x_2 - x_1}$ (Two-point form)

$\dfrac{x}{a} + \dfrac{y}{b} = 1$ (Intercept form)

$y - y_1 = m(x - x_1)$ (Point-slope form)

$$y = mx + b \quad \text{(Slope-intercept form)}$$
$$Ax + By = C \quad \text{(Standard form)} \quad [7.4]$$

12. If the standard form of a linear equation in two variables is written equivalently in the form

$$y = \frac{-A}{B}x + \frac{C}{B},$$

then $-A/B$ is the **slope** of the line and C/B is the **y-intercept**. [7.4]

13. A function defined by an equation of the form

$$y = ax^2 + bx + c$$

is called a **quadratic function** [7.5]

14. The graph of a quadratic function is a smooth curve that is called a **parabola** [7.5]

15. The **vertex** of the parabola is the point

$$\left(\frac{-b}{2a}, -\frac{b^2 - 4ac}{4a}\right).$$ [7.5]

16. If P_3 is the bisector of the line segment connecting P_1 and P_2, then the two points are **symmetric** with respect to P_3. [7.5]

17. Two points are symmetric with respect to a line if the line is the **perpendicular bisector** of the line segment connecting the two points. Each point is called the **mirror image** of the other. [7.5]

18. A parabola is symmetric with respect to its axis, which is a line through the vertex parallel to the y-axis. [7.5]

REVIEW EXERCISES

SECTION 7.1

1. Given $x, y \in S = \{1, 2, 3, 4, 5, 6\}$ and the relation $\{(x, y) | y/x \in S\}$. List: (a) the domain, (b) the range, and (c) the set of ordered pairs.
2. If $x \in \{-3, -2, -1, 0, 1, 2, 3\}$ and $y \in R$, specify the domain and list the set of ordered pairs in the relation $y = \sqrt{4 - x^2}$.
3. Which of the following relations are functions?
   (a) $\{(x, y) | y = 3x^2 - 5\}$ (b) $\{(x, y) | x^2 + y^2 = 4\}$
   (c) $\{(x, y) | x = |y|\}$ (d) $\{(x, y) | y = |x + 2|\}$

SECTION 7.2

4. Given $U = \{-4, -3, -2, -1, 0, 1, 2, 3, 4\}$ and $x, y \in U$. Graph the relation

$$\{(x, y) | y = 12/x\}$$

on the point lattice for $U \times U$.

SECTION 7.3

Construct the graph of each linear relation in Problems 5–8.

5. $\{(x, y) | x + y = 5\}$
6. $\{(x, y) | 2x - 3y = 6\}$
7. $\{(x, y) | y = -3\}$
8. $\{(x, y) | x + 2 = 0\}$

SECTION 7.4

Find the distance and the slope of the line segment between the points in Problems 9 and 10.

9. $(4, 5)$ and $(8, 2)$
10. $(-3, 2)$ and $(2, 14)$

In Problems 11–14, write an equation for the line satisfying the given conditions.

11. Containing the points $(-4, 2)$ and $(2, -4)$.
12. x-intercept $= 5$ and y-intercept $= 3$.
13. Slope $= \frac{3}{4}$ and y-intercept $= -2$.
14. Slope $= -\frac{2}{3}$ and contains the point $(-3, -4)$.
15. Find the slope and the y-intercept of the line with equation $5x - 3y = 6$.

SECTION 7.5

16. Find the coordinates of the points symmetric to each of the points $(1, 2)$, $(-3, 3)$, $(-4, -1)$, and $(2, -5)$ with respect to the line $x = -1$.
17. Find the coordinates of the points symmetric to each of the points in Problem 16 with respect to the line $y = 1$.

In Problems 18 and 19, find the coordinates of the vertex of the parabola and state whether y is a maximum or a minimum value.

18. $y = x^2 + 4x + 5$
19. $x^2 - 6x + y - 4 = 0$

20. Find the x- and y-intercepts, write the equation of the axis of symmetry, and construct the graph of $y = x^2 - x - 6$.

Conic Sections

In this chapter we discuss some special relations consisting of sets of points determined by the intersection of a plane and a right circular cone. Figure 8.0-1 shows the more important of these intersections, which are called **conic sections**. The **axis** of the cone is the line perpendicular to the base and passing through the **vertex**, which is the common intersection of all lines lying in the surface of the cone. Each of these lines is called an **element**.

If the intersecting plane is perpendicular to the axis of the cone, the intersection is called a **circle**. If the plane is not perpendicular to the axis of the cone and intersects all elements above the vertex or all those below the vertex, the intersection is called an **ellipse**. If the plane is parallel to an element, the intersection is called a **parabola**. If the plane is parallel to the axis of the cone, the intersection is called a **hyperbola**.

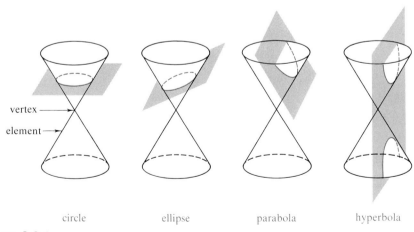

Figure 8.0-1

8.1 The Circle

DEFINITION 8.1-1 A circle is a set of points in a plane all of which are at a given distance, called the **radius**, from a given point, called the **center**

Equations for Circles

Since each point in the set is the same distance from the center, an equation for a circle can be derived by using the distance formula (p. 181).

Example Derive an equation for the circle whose center is at (3, 2) and whose radius is 4.

Solution By the definition of a circle, the distance between any point in the circle (x, y) and $(3, 2)$ equals 4. Hence, by the distance formula,

$$\sqrt{(x - 3)^2 + (y - 2)^2} = 4.$$

Squaring both members gives

$$(x - 3)^2 + (y - 2)^2 = 16.$$

The equation may be left in this form, or the binomials may be expanded to form the equivalent equations,

$$x^2 - 6x + 9 + y^2 - 4y + 4 = 16$$

or

$$x^2 + y^2 - 6x - 4y - 3 = 0.$$

More generally, if (h, k) is the center and r is the radius, then the equation for a circle is

$$(x - h)^2 + (y - k)^2 = r^2, \tag{1}$$

as illustrated in Figure 8.1-1. You should observe that if the center is at the origin, then Equation (1) becomes

$$x^2 + y^2 = r^2.$$

Standard Form of the Equation for a Circle

If we expand the binomials in Equation (1) we obtain

$$x^2 - 2hx + h^2 + y^2 - 2ky + k^2 = r^2$$

or, equivalently,

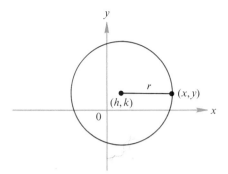

Figure 8.1-1

$$x^2 + y^2 - 2hx - 2ky + h^2 + k^2 - r^2 = 0.$$

Since h, k, and r are constants, we can replace $-2h$ with D, $-2k$ with E, and $h^2 + k^2 - r^2$ with F to obtain the general equation for a circle in standard form,

$$x^2 + y^2 + Dx + Ey + F = 0 \qquad (2)$$

Center and Radius

The center and the radius of a circle can be determined from the standard form by completing the squares in x and y and then transforming the equation into its equivalent in the form of Equation (1).

Example Find the center and the radius of the circle with equation

$$x^2 + y^2 - 4x + 10y + 13 = 0.$$

Solution We can rewrite the equation equivalently as

$$(x^2 - 4x) + (y^2 + 10y) = -13.$$

Then, by inspection, we note that 4 must be added to complete the square in x and 25 must be added to complete the square in y. However, since we are working with an equation, these numbers must be added to both members. Hence,

$$(x^2 - 4x + 4) + (y^2 + 10y + 25) = 4 + 25 - 13,$$

from which we obtain

$$(x - 2)^2 + (y + 5)^2 = 4^2.$$

Therefore, the center is $(2, -5)$ and the radius is 4.

EXERCISE 8.1

WRITTEN

Derive an equation in standard form for the circle with center and radius as given in Problems 1–12.

Example $(1, -4); r = 2$

Solution Substituting the given values into Equation (1) (p. 196) gives
$$(x - 1)^2 + (y + 4)^2 = 2^2$$
$$x^2 - 2x + 1 + y^2 + 8y + 16 = 4$$
$$x^2 + y^2 - 2x + 8y + 13 = 0.$$

1. $(2, 4); r = 5$
2. $(5, 1); r = 6$
3. $(-1, 2); r = 2$
4. $(-2, 2); r = 3$
5. $(3, -2); r = 7$
6. $(5, -4); r = 6$
7. $(-2, -2); r = 7$
8. $(-5, -4); r = 5$
9. $(3, 0); r = 8$
10. $(-4, 0); r = 3$
11. $(0, 4); r = 4$
12. $(0, -3); r = 3$

In Problems 13–24, find the center and the radius of the circle with the given equation.

Example $x^2 + y^2 - 6x - 4y - 12 = 0$

Solution Write the equation equivalently as
$$(x^2 - 6x) + (y^2 - 4y) = 12.$$

Then by completing the squares in x and y,
$$(x^2 - 6x + 9) + (y^2 - 4y + 4) = 12 + 9 + 4$$
$$(x - 3)^2 + (y - 2)^2 = 5^2.$$

Therefore, the center is $(3, 2)$ and the radius is 5.

13. $x^2 + y^2 - 8x - 2y + 16 = 0$
14. $x^2 + y^2 - 4x - 8y + 4 = 0$
15. $x^2 + y^2 + 4x - 2y - 11 = 0$
16. $x^2 + y^2 + 6x - 6y - 18 = 0$
17. $x^2 + y^2 - 4x = 0$
18. $x^2 + y^2 + 6y = 0$
19. $x^2 + y^2 + 3x - 5y - \frac{1}{2} = 0$
20. $x^2 + y^2 + x - 7y + \frac{7}{2} = 0$
21. $x^2 + y^2 - x - 4y - 7 = 0$
22. $x^2 + y^2 - 2x + 5y + 7 = 0$
23. $2x^2 + 2y^2 + 3x + 5y + 2 = 0$
24. $2x^2 + 2y^2 - 5x + 7y - 2 = 0$

(HINT: In Problems 23 and 24 divide both members by 2.)

8.2 The Parabola

In Section 7.5, on quadratic functions, you learned that the graph of a function defined by an equation of the form $y = ax^2 + bx + c, a \neq 0$, is called a *parabola*. You were also shown that a parabola is symmetric with respect to its axis. In this section, we shall discuss parabolas more generally. However, we shall limit the discussion to those parabolas whose axes are parallel to one of the coordinate axes.

DEFINITION 8.2-1 A **parabola** is a set of points in a plane such that the distance from any point in the set to a fixed line, called the **directrix**, is equal to the distance from the point to a fixed point, called the **focus**.

Figure 8.2-1 shows these equal distances and two other important features of a parabola. The **axis** of the parabola is perpendicular to the directrix and passes

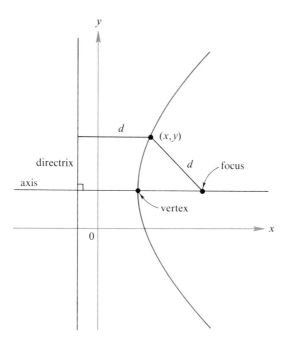

Figure 8.2-1

through the focus. The point of intersection of the parabola with its axis is the **vertex** of the parabola. It follows from the definition above that the vertex must be the midpoint of the segment of the axis that lies between the directrix and the focus.

Equation for a Parabola

An equation for a parabola is derived by equating the two distances described in Definition 8.2-1. For example, suppose a parabola has its vertex at the origin and its focus on the positive x-axis. If we designate the distance from the directrix to the focus as p, then the focus is at $(p/2, 0)$ and the equation for the directrix is $x = -p/2$. This is illustrated in Figure 8.2-2.

Since any point on the directrix has coordinates $(-p/2, y)$, the distance from any point (x, y) to $(-p/2, y)$ is

$$x - \left(\frac{-p}{2}\right) = x + \frac{p}{2}.$$

The distance from (x, y) to $(p/2, 0)$ is

$$\sqrt{\left(x - \frac{p}{2}\right)^2 + (y - 0)^2}.$$

Then by Definition 8.2-1, we have

$$\sqrt{\left(x - \frac{p}{2}\right)^2 + (y - 0)^2} = x + \frac{p}{2}.$$

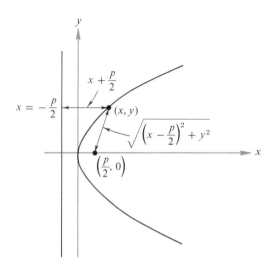

Figure 8.2-2

Now if both members of this equation are squared and the squares of the binomials under the radical are expanded, the equation becomes

$$x^2 - px + \frac{p^2}{4} + y^2 = x^2 + px + \frac{p^2}{4},$$

or, equivalently,

$$y^2 = 2px \tag{1}$$

In a similar manner, we can derive equations for parabolas with vertices at the origin and foci on the negative x-axis, the positive y-axis, or the negative y-axis. These equations are listed below in this order and are illustrated in Figure 8.2-3.

$$y^2 = -2px \quad [\text{focus: } (-p/2, 0)] \tag{2}$$

$$x^2 = 2py \quad [\text{focus: } (0, p/2)] \tag{3}$$

$$x^2 = -2py \quad [\text{focus: } (0, -p/2)] \tag{4}$$

Example Find an equation for the parabola with vertex at the origin and focus at $(3, 0)$. Also, state the equation for the directrix.

Solution Since the distance from the vertex to the focus is $p/2$, by the definition of a parabola,

$$\frac{p}{2} = \sqrt{(3-0)^2 + (0-0)^2} = 3,$$

and

$$p = 6.$$

Then, since the focus is on the positive x-axis, we replace p with 6 in Equation (1) to obtain

$$y^2 = 12x.$$

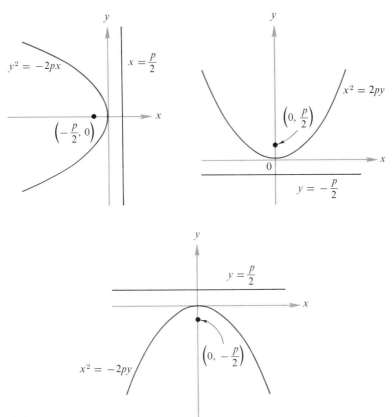

Figure 8.2-3

Also, for this parabola the directrix is the line with equation $x = -p/2$, or

$$x = -3.$$

Parabola with Vertex Not at the Origin

Now consider the parabola with vertex at $(2, 3)$ and focus at $(5, 3)$, shown in Figure 8.2-4. Since the axis passes through both the vertex and the focus, and these points have the same y-coordinate, the equation for the axis is $y = 3$. Also note that the axis is parallel to the x-axis. Thus, the directrix is perpendicular to the x-axis and is $p/2$ units to the left of the vertex. Since the distance between the vertex and the focus is also $p/2$, we have $p/2 = 3$. The x-coordinate of a point 3 units to the left of $(2, 3)$ is found by $x = 2 - 3 = -1$. Thus, the equation for the directrix is $x = -1$.

Next, we apply Definition 8.2-1, which gives

$$\sqrt{(x-5)^2 + (y-3)^2} = x - (-1).$$

Then, by squaring both members, we obtain

$$x^2 - 10x + 25 + (y-3)^2 = x^2 + 2x + 1,$$

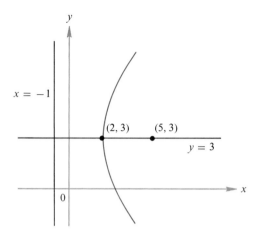

Figure 8.2-4

which is equivalent to

$$(y - 3)^2 = 12x - 24$$
$$(y - 3)^2 = 12(x - 2).$$

If you examine this equation, you will observe that the numerals within the parentheses are the coordinates of the vertex and that $12 = 2p$. This suggests that the equation for a parabola with vertex at (h, k) is of the form

$$(y - k)^2 = 2p(x - h) \tag{1a}$$

$$(y - k)^2 = -2p(x - h) \tag{2a}$$

$$(x - h)^2 = 2p(y - k) \tag{3a}$$

or

$$(x - h)^2 = -2p(y - k). \tag{4a}$$

The graphs of these equations are shown in Figure 8.2-5.

Example Find an equation for the parabola whose focus is at $(-5, 1)$ and whose directrix is the line $x = 3$.

Solution The distance from the focus to the directrix is $3 - (-5)$ or 8. Hence, $p/2 = 4$, and the vertex is at $(-1, 1)$. Then, since the focus is to the left of the directrix, we substitute these values into Equation (2a) to obtain

$$(y - 1)^2 = -16(x + 1).$$

The parabola is shown in Figure 8.2-6 (see page 204).

Properties

The vertex, focus, directrix, and axis are called the **properties** of a parabola. They can all be determined from the equation for the parabola if the equation

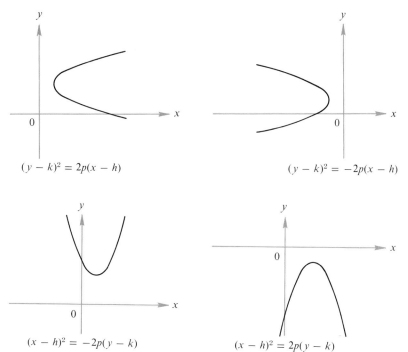

Figure 8.2-5

can be put into one of the forms we have discussed. This is usually done by completing the square in the variable that is of second degree.

Example Find the properties of the parabola with equation

$$y^2 - 8y + 18x + 52 = 0.$$

Solution By adding $-18x - 52$ to both members we form the equivalent equation

$$y^2 - 8y = -18x - 52.$$

Then, we complete the square in the left-hand member.

$$y^2 - 8y + 16 = -18x - 52 + 16$$
$$(y - 4)^2 = -18x - 36$$
$$(y - 4)^2 = -18(x + 2)$$

By comparing the last equation with Equation (2a), we observe that the vertex is at $(-2, 4)$, and that the focus is to the left of the vertex. Since $2p = 18$, $p/2 = \frac{9}{2}$, and the focus is at $(-2 - \frac{9}{2}, 4)$ or $(-\frac{13}{2}, 4)$. The directrix, being to the right of the vertex, has as its equation

$$x = -2 + \frac{9}{2} \quad \text{or} \quad x = \frac{5}{2}.$$

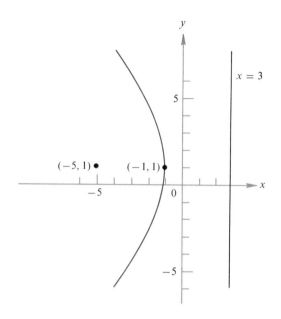

Figure 8.2-6

EXERCISE 8.2

WRITTEN

In Problems 1–24, find an equation for the parabolas satisfying the given conditions.

Example Vertex, $(0,0)$; directrix, $x = -4$

Solution The directrix is perpendicular to the x-axis and is 4 units to the left of the vertex. Hence, $p/2 = 4$, and Equation (1) is the model. Therefore,

$$y^2 = 16x$$

is the required equation.

1. Vertex, $(0,0)$; directrix, $x = -3$
2. Vertex, $(0,0)$; directrix, $x = -5$
3. Vertex, $(0,0)$; directrix, $y = -2$
4. Vertex, $(0,0)$; directrix, $y = 3$
5. Vertex, $(0,0)$; directrix, $x = 4$
6. Vertex, $(0,0)$; directrix, $y = -5$
7. Vertex, $(0,0)$; directrix, $y = 6$
8. Vertex, $(0,0)$; directrix, $x = 3$
9. Vertex, $(0,0)$; focus, $(\frac{3}{2}, 0)$
10. Vertex, $(0,0)$; focus, $(-1, 0)$
11. Vertex, $(0,0)$; focus, $(0, \frac{5}{2})$
12. Vertex, $(0,0)$; focus, $(0, -\frac{3}{4})$
13. Vertex, $(0,0)$; focus, $(-\frac{5}{4}, 0)$
14. Vertex, $(0,0)$; focus, $(0, \frac{9}{4})$
15. Vertex, $(0,0)$; focus, $(0, -2)$
16. Vertex, $(0,0)$; focus, $(\frac{13}{4}, 0)$
17. Vertex, $(2, 4)$; directrix, $x = -1$
18. Vertex, $(3, \frac{1}{2})$; directrix, $y = -\frac{3}{2}$
19. Focus, $(3, 5)$; directrix, $x = -3$
20. Focus, $(5, -2)$; directrix, $y = 2$
21. Vertex, $(6, 3)$; focus, $(-4, 3)$
22. Vertex, $(4, -5)$; focus, $(4, 0)$

23. Directrix, $x = -3$; axis, $y = 3$; $p = 2$
24. Directrix, $y = 5$; axis, $x = -2$; $p = \frac{3}{2}$

Use the distance formula to find an equation for each of the parabolas in Problems 25–28.

25. Directrix, $x = -1$; focus, $(4, 2)$
26. Directrix, $x = 3$; focus, $(2, -1)$
27. Directrix, $y = -5$; focus, $(1, 3)$
28. Directrix, $y = 10$; focus, $(3, 7)$

In Problems 29–40, find the properties of the parabolas whose equations are given.

Example $x^2 - 6x - 6y - 3 = 0$

Solution By adding $6y + 3$ to both members, we form the equivalent equation

$$x^2 - 6x = 6y + 3.$$

Completing the square in the left-hand member, we have

$$x^2 - 6x + 9 = 6y + 12$$
$$(x - 3)^2 = 6(y + 2).$$

Comparing this equation with Equation (3a) on page 202, we observe that the vertex is the point $(3, -2)$ and that $2p = 6$ or $p = 3$. Since p is positive, the directrix is below the vertex and the focus is above. Hence, the focus is $(3, -2 + \frac{3}{2})$ or $(3, -\frac{1}{2})$, and an equation for the directrix is $y = -2 - \frac{3}{2}$ or $y = -\frac{7}{2}$.

29. $y^2 = 8x$
30. $y^2 = -16x$
31. $x^2 = 4y$
32. $x^2 = -6y$
33. $(y - 2)^2 = 6(x + 1)$
34. $(y + 4)^2 = 16 - 8x$
35. $x^2 - 2x - 6y + 7 = 0$
36. $x^2 - 2x - y + 3 = 0$
37. $y^2 + 2y - x - 4 = 0$
38. $2x^2 - 4x + y + 5 = 0$
39. $4x^2 + 3x - y = 0$
40. $3y^2 - x + y = 0$

8.3 The Ellipse

DEFINITION 8.3-1 An ellipse is a set of points in a plane such that the sum of the distances from any point in the set to two fixed points, called the **foci**, is a constant.

Figure 8.3-1 shows an ellipse whose foci are the points (h_1, k_1) and (h_2, k_2). By the definition above, the sum of the distances d_1 and d_2 from any point (x, y) is a constant. The line passing through the foci is called the **principal axis** of the ellipse. The point C, the midpoint of the segment connecting the foci, is called the **center**. The points V_1 and V_2, where the ellipse intersects its principal axis, are the **vertices** of the ellipse. The distance between the vertices is called the **major axis**.

This discussion is limited to ellipses with center at the origin and one of the coordinate axes as the principal axis. One such ellipse is shown in Figure 8.3-2.

Equation for an Ellipse

Consider the ellipse with foci at $(-c, 0)$ and $(c, 0)$ and vertices at $(-a, 0)$ and $(a, 0)$ (Figure 8.3-2). Since $(a, 0)$ lies on the ellipse, the distances d_1 and d_2 from

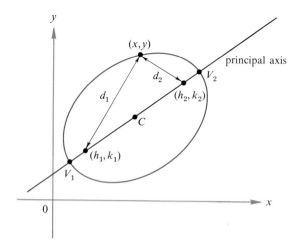

Figure 8.3-1

the foci to the point are

$$d_1 = a - (-c) = a + c \quad \text{and} \quad d_2 = a - c.$$

Therefore,

$$d_1 + d_2 = a + c + a - c = 2a.$$

Thus, the major axis is $2a$, and a is called the **semi-major axis**.

We can use the fact that $d_1 + d_2 = 2a$ along with the distance formula to derive an equation for the ellipse in Figure 8.3-2 as follows.

$$d_1 = \sqrt{[x - (-c)]^2 + (y - 0)^2} \quad \text{and} \quad d_2 = \sqrt{(x - c)^2 + (y - 0)^2}.$$

Hence,

$$d_1 + d_2 = \sqrt{(x + c)^2 + y^2} + \sqrt{(x - c)^2 + y^2} = 2a.$$

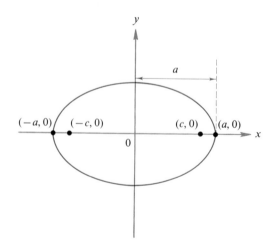

Figure 8.3-2

Adding the negative of the second radical to both members of this equation and squaring both members gives

$$(x + c)^2 + y^2 = 4a^2 - 4a\sqrt{(x - c)^2 + y^2} + (x - c)^2 + y^2$$
$$x^2 + 2cx + c^2 + y^2 = 4a^2 - 4a\sqrt{(x - c)^2 + y^2} + x^2 - 2cx + c^2 + y^2,$$

or, equivalently,

$$a\sqrt{(x - c)^2 + y^2} = a^2 - cx.$$

Squaring both members of this equation, we obtain

$$a^2[(x - c)^2 + y^2] = a^4 - 2a^2cx + c^2x^2$$
$$a^2(x^2 - 2cx + c^2 + y^2) = a^4 - 2a^2cx + c^2x^2$$
$$a^2x^2 - 2a^2cx + a^2c^2 + a^2y^2 = a^4 - 2a^2cx + c^2x^2,$$

which is equivalent to

$$a^2x^2 - c^2x^2 + a^2y^2 = a^4 - a^2c^2$$
$$(a^2 - c^2)x^2 + a^2y^2 = a^2(a^2 - c^2).$$

If $a^2 - c^2$ is replaced by b^2, the equation becomes

$$b^2x^2 + a^2y^2 = a^2b^2. \tag{1}$$

If we then divide both members of Equation (1) by a^2b^2, we obtain

$$\frac{x^2}{a^2} + \frac{y^2}{b^2} = 1. \tag{2}$$

To see the significance of b, observe in Figure 8.3-3 that the y-axis is the perpendicular bisector of the line segment connecting the foci. Now, since any point on the perpendicular bisector of a line segment is equidistant from the endpoints of the segment, we have for the point $(0, y)$, one of the points where the ellipse intersects the y-axis, $d_1 = d_2$. But $d_1 + d_2 = 2a$; therefore $d_1 = d_2 = a$, and the distance from the center to $(0, y)$ is given by the Pythagorean Theorem,

$$b^2 = a^2 - c^2, \quad \text{or} \quad b = \sqrt{a^2 - c^2}.$$

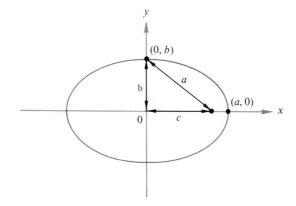

Figure 8.3-3

We can also show that the distance from the center to $(0, -y)$, the other point of intersection of the ellipse with the y-axis, is b. Thus, $2b$ is the distance between the y-intercepts and is called the **minor axis**. b is called the **semi-minor axis**

The discussion above can help us to derive an equation for an ellipse with center at the origin and foci on one of the coordinate axes provided we know values for any two of a, b, and c.

Example   Derive an equation for the ellipse with center at the origin, one focus at $(3, 0)$, and a vertex at $(5, 0)$.

Solution   From the given information we have $a = 5$ and $c = 3$. Therefore,

$$b^2 = a^2 - c^2$$
$$= 25 - 9$$
$$= 16.$$

Using these values for a^2 and b^2 in Equation (2) (p. 207) gives

$$\frac{x^2}{25} + \frac{y^2}{16} = 1.$$

By a technique similar to that used to derive Equation (2), an equation can be derived for an ellipse with center at the origin and the y-axis as principal axis (Figure 8.3-4). We find such an equation to be of the form

$$\frac{x^2}{b^2} + \frac{y^2}{a^2} = 1 \qquad (3)$$

where a and b represent the semi-major and semi-minor axes, respectively.

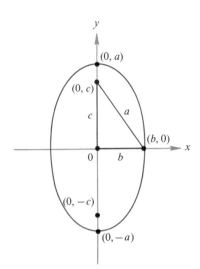

Figure 8.3-4

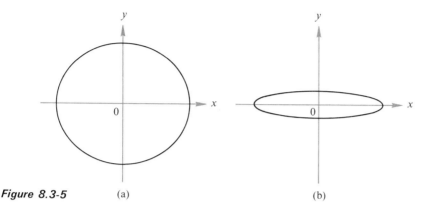

Figure 8.3-5   (a)   (b)

Eccentricity

An ellipse may be nearly round or relatively flat. Its shape is determined by its **eccentricity** e, which is defined by the equation

$$e = \frac{c}{a}.$$

Since $a > c > 0$, $0 < e < 1$. If a is kept constant and c is allowed to vary, as c approaches 0 the foci are very close together and a and b are nearly equal. Hence, the ellipse is nearly round, as shown in Figure 8.3-5a. On the other hand, if c approaches a, then b becomes quite small and the ellipse is flat, as shown in Figure 8.3-5b.

EXERCISE 8.3

WRITTEN

Find the length of the semi-major and semi-minor axes, the coordinates of the foci and vertices, and the eccentricity for each ellipse in Problems 1–10.

Example $4x^2 + 9y^2 = 36$

Solution   Transform the equation into the form of Equations (2) or (3) by dividing both members by 36. This gives

$$\frac{x^2}{9} + \frac{y^2}{4} = 1,$$

which is the equation of an ellipse with center at the origin and the x axis as principal axis. Hence, the lengths of the semi-axes are

$a = 3$ and $b = 2$.

Substituting these values into $c = \sqrt{a^2 - b^2}$ gives

$c = \sqrt{9 - 4} = \sqrt{5}.$

Therefore, the foci are at $(-\sqrt{5}, 0)$ and $(\sqrt{5}, 0)$, the vertices are at $(-3, 0)$ and $(3, 0)$, and the eccentricity is $e = \sqrt{5}/3$.

1. $16x^2 + 25y^2 = 400$
2. $9x^2 + 25y^2 = 225$
3. $9x^2 + 16y^2 = 144$
4. $x^2 + 4y^2 = 16$
5. $25x^2 + 9y^2 = 225$
6. $25x^2 + 16y^2 = 400$
7. $4x^2 + y^2 = 16$
8. $16x^2 + 9y^2 = 144$
9. $3x^2 + 2y^2 = 6$
10. $3x^2 + 4y^2 = 12$

In Problems 11–20, find an equation for the ellipse with center at the origin satisfying the given conditions.

Example   One vertex at $(5, 0)$ and eccentricity $\frac{3}{5}$.

Solution   The given vertex lies on the x-axis. Therefore, the required equation will be in the form of Equation (2). Since a is the distance from the center to the vertex, $a = 5$. Then, by substituting these values for c and a into $e = c/a$, we have

$$\frac{c}{a} = \frac{3}{5},$$

from which $c = 3$.
      We can now determine b from the equation $b = \sqrt{a^2 - c^2}$.

$$b = \sqrt{25 - 9} = \sqrt{16} = 4$$

Hence, an equation for the ellipse is

$$\frac{x^2}{25} + \frac{y^2}{16} = 1.$$

11. A vertex at $(5, 0)$ and a focus at $(4, 0)$
12. A vertex at $(6, 0)$ and a focus at $(3, 0)$
13. A vertex at $(0, 5)$ and a focus at $(0, 3)$
14. A vertex at $(0, 3)$ and a focus at $(0, 2)$
15. A vertex at $(10, 0)$ with eccentricity $\frac{4}{5}$
16. A vertex at $(4, 0)$ with eccentricity $\frac{1}{2}$
17. A vertex at $(0, 10)$ with eccentricity $\frac{3}{5}$
18. A vertex at $(0, 5)$ with eccentricity $\frac{4}{5}$
*19. Eccentricity $\frac{3}{5}$, foci on the x-axis and passing through $(4, \frac{12}{5})$
*20. Eccentricity $\frac{3}{5}$, foci on the y-axis and passing through $\left(\sqrt{7}, \frac{5\sqrt{3}}{4}\right)$

8.4   The Hyperbola

DEFINITION 8.4-1   A **hyperbola** is a set of points in a plane such that the difference of the distances from any point in the set to two fixed points is constant.

    As in the case of an ellipse, the two fixed points are the **foci**, the line passing through the foci is the **principal axis**, the midpoint of the segment connecting the foci is the **center**, and the points of intersection of the hyperbola and its principal axis are the **vertices**. These features are illustrated in Figure 8.4-1.

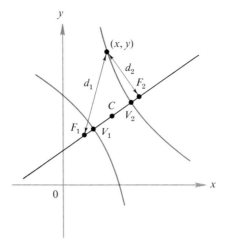

Figure 8.4-1

Figure 8.4-1 also illustrates the fact that a hyperbola consists of two branches. The right-hand branch is determined by equating $d_1 - d_2$ to a constant, and the left-hand branch is determined by equating $d_2 - d_1$ to the same constant.

If $(-c, 0)$ and $(c, 0)$ are selected as the foci, the center is at the origin. Let $(x, 0)$ be the vertex on the right (Figure 8.4-2) and $2a$ the constant. Then, for $(x, 0)$, $d_1 = x - (-c)$ and $d_2 = c - x$. By Definition 8.4-1 we have

$$[x - (-c)] - (c - x) = 2a$$
$$x + c - c + x = 2a$$
$$2x = 2a$$
$$x = a.$$

Hence, the right-hand vertex is the point $(a, 0)$. In a similar manner, it can be shown that the left-hand vertex is $(-a, 0)$. The distance between the two vertices, $2a$, is called the **transverse axis**, and a is called the **semi-transverse axis**.

Equation for a Hyperbola

We can derive an equation for the hyperbola in Figure 8.4-2 in a manner similar to that in which an equation for an ellipse was derived. If (x, y) is any point on the hyperbola, then by the distance formula we have

$$\sqrt{[x - (-c)]^2 + (y - 0)^2} - \sqrt{(x - c)^2 + (y - 0)^2} = 2a.$$

Then

$$\sqrt{(x + c)^2 + y^2} = 2a + \sqrt{(x - c)^2 + y^2},$$

which, upon squaring both members, gives

$$x^2 + 2cx + c^2 + y^2 = 4a^2 + 4a\sqrt{(x - c)^2 + y^2} + x^2 - 2cx + c^2 + y^2$$
$$4cx - 4a^2 = 4a\sqrt{(x - c)^2 + y^2}$$
$$cx - a^2 = a\sqrt{(x - c)^2 + y^2}.$$

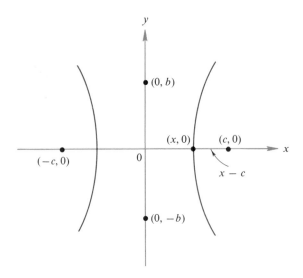

Figure 8.4-2

Then, by squaring both members again,

$$c^2x^2 - 2a^2cx + a^4 = a^2(x^2 - 2cx + c^2 + y^2)$$
$$c^2x^2 - 2a^2cx + a^4 = a^2x^2 - 2a^2cx + a^2c^2 + a^2y^2,$$

or, equivalently,

$$c^2x^2 - a^2x^2 - a^2y^2 = a^2c^2 - a^4$$
$$(c^2 - a^2)x^2 - a^2y^2 = a^2(c^2 - a^2).$$

If we now replace $c^2 - a^2$ with b^2, we obtain

$$b^2x^2 - a^2y^2 = a^2b^2.$$

Finally, dividing both members by a^2b^2 gives

$$\frac{x^2}{a^2} - \frac{y^2}{b^2} = 1 \qquad (1)$$

The geometric significance of b is not the same as for the ellipse. In that case $(0, b)$ and $(0, -b)$ were the points of intersection of the ellipse with the y-axis. However, if we replace x with 0 in Equation (1) above, we have

$$-\frac{y^2}{b^2} = 1 \quad \text{or} \quad -y^2 = b^2,$$

but since y^2 and b^2 are both positive there is no real value for y that satisfies this equation. Thus, the hyperbola does not intersect the y-axis. However, the segment of length $2b$ connecting the points $(0, -b)$ and $(0, b)$ is called the **conjugate axis** and b is called the **semi-conjugate axis**. Further geometric significance of b will be discussed in the next section.

If we use the technique above to derive an equation for a hyperbola with center at the origin and foci on the y-axis, we obtain the equation

$$\frac{y^2}{a^2} - \frac{x^2}{b^2} = 1 \qquad (2)$$

Comparison of Equations (1) and (2) reveals that the positive term in the left-hand member in either equation relates to the principal axis.

Another fact to consider is that since we used b^2 to represent $c^2 - a^2$, b can be less than a, equal to a, or greater than a.

If we know the values of a and b, we can derive an equation for a hyperbola by substituting these values into Equation (1) or Equation (2).

Example Find an equation for the hyperbola with foci at $(-5, 0)$ and $(5, 0)$ and transverse axis of length 8.

Solution Since the foci are on the x-axis, this axis is the principal axis. Thus, this hyperbola has an equation like Equation (1). The given information tells us that $c = 5$ and $a = 4$. We need to find the value of b^2. Substituting the values for c and a in $b^2 = c^2 - a^2$, we find that $b^2 = 9$. Finally, replacing a^2 and b^2 in Equation (1) with their respective values, we have as the required equation

$$\frac{x^2}{16} - \frac{y^2}{9} = 1.$$

Eccentricity

As in the case of an ellipse, the shape of a hyperbola depends on its eccentricity, which is determined by $e = c/a$. Since $c > a > 0$, the eccentricity of a hyperbola is always greater than 1, that is, $1 < e < \infty$. If e is very close to 1, then c and a are very nearly equal and b is quite small. Thus, the curve is narrow, as shown in Figure 8.4-3a. If e is very large, then c and b are very nearly equal and a is quite small. Thus, the curve is wide, as shown in Figure 8.4-3b.

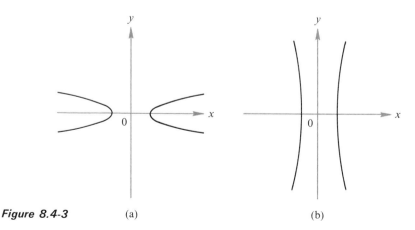

Figure 8.4-3   (a)   (b)

Example Find the eccentricity of the hyperbola with equation

$$\frac{x^2}{4} - \frac{y^2}{16} = 1.$$

Solution Since the term involving x is positive, the principal axis of the hyperbola is along the x-axis. This means that $a^2 = 4$ and $b^2 = 16$. To find the eccentricity, we must know the value of c. Substituting the values of a^2 and b^2 in $c^2 = a^2 + b^2$, we find $c^2 = 20$, or $c = 2\sqrt{5}$. Then,

$$e = \frac{c}{a} = \frac{2\sqrt{5}}{2} = \sqrt{5}.$$

EXERCISE 8.4

WRITTEN

Find the coordinates of the vertices and foci, the length of the conjugate axis, and the eccentricity of each hyperbola in Problems 1–10.

Example $4x^2 - 9y^2 = 36$

Solution Dividing both members of the given equation by 36, we obtain

$$\frac{x^2}{9} - \frac{y^2}{4} = 1,$$

which we recognize as an equation for a hyperbola with center at the origin and foci on the x-axis. Then, upon comparing this equation with Equation (1), page 212, we find

$a^2 = 9; \quad a = 3$

and

$b^2 = 4; \quad b = 2.$

Then, since $c^2 = a^2 + b^2$,

$c^2 = 9 + 4 = 13; \quad c = \sqrt{13}.$

Therefore, the vertices are $(-3, 0)$ and $(3, 0)$; the foci are $(-\sqrt{13}, 0)$ and $(\sqrt{13}, 0)$; the conjugate axis has length $2b = 4$; and the eccentricity is $e = c/a = \sqrt{13}/3$.

1. $9x^2 - 16y^2 = 144$
2. $16x^2 - 9y^2 = 144$
3. $16x^2 - 25y^2 = 400$
4. $9x^2 - 25y^2 = 225$
5. $16y^2 - 9x^2 = 144$
6. $25y^2 - 16x^2 = 400$
7. $25y^2 - 9x^2 = 225$
8. $16y^2 - 25x^2 = 400$
9. $3x^2 - 2y^2 = 6$
10. $3y^2 - 4x^2 = 12$

In Problems 11–20, find an equation for the hyperbola satisfying the given conditions.

Example One focus at $(5, 0)$ and eccentricity $\frac{5}{3}$.

Solution The given focus lies on the x-axis. Therefore, the required equation will be in the form of Equation (1). Since a focus is at $(5, 0)$ and the center is at the origin, we know that $c = 5$. Then, from $e = c/a = \frac{5}{3}$, we have $a = 3$. Next, we must find the value of b^2. Substituting the values for c and a in $b^2 = c^2 - a^2$, we find $b^2 = 16$. Finally, we replace a^2 and b^2 with their respective values in Equation (1) to obtain the required equation

$$\frac{x^2}{9} - \frac{y^2}{16} = 1.$$

11. A vertex at $(4, 0)$ and a focus at $(5, 0)$
12. A vertex at $(3, 0)$ and a focus at $(6, 0)$
13. A focus at $(0, 5)$ and a vertex at $(0, 3)$
14. A focus at $(0, 3)$ and a vertex at $(0, 2)$
15. A focus at $(2, 0)$ with eccentricity 2
16. A focus at $(0, 10)$ with eccentricity $\frac{5}{3}$
17. A vertex at $(0, 5)$ with eccentricity $\frac{5}{4}$
18. A focus at $(10, 0)$ with eccentricity $\frac{5}{4}$

*19. Eccentricity $\frac{2}{3}$, foci on the x-axis, and passing through $(5, \frac{16}{3})$
*20. Eccentricity $\frac{5}{3}$, foci on the y-axis, and passing through $(3, \frac{15}{4})$

8.5 Sketching the Conic Sections

You will encounter many situations in higher mathematics in which it is essential that you sketch the graph of a relation. The conic sections have some properties that are useful in sketching such graphs.

Circles

If you know the center and the length of the radius of a circle, you can draw the circle with a compass, so no further discussion should be necessary.

Parabolas

In Section 7.5 you saw that a parabola is symmetric with respect to its axis. That is, for every point on a parabola except the vertex, there is a point that lies on the line through the given point perpendicular to the axis and at the same distance from the axis. Since our discussion of parabolas has been limited to those with axes parallel to one of the coordinate axes, you can write the equation equivalently in the form of Equation (1a), (2a), (3a), or (4a) on page 202 and thus determine the vertex, the location of the axis, and the position of the curve. If you then locate a few points and use the property of symmetry, the curve is easily sketched.

Example Sketch the parabola with equation

$$y^2 + 4y - x + 7 = 0.$$

Solution By adding $x - 7$ to both members of the given equation, we obtain the equivalent equation

$$y^2 + 4y = x - 7.$$

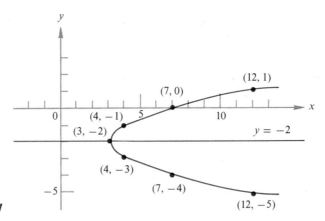

Figure 8.5-1

Then, by completing the square in the left-hand member,

$$y^2 + 4y + 4 = x - 7 + 4$$
$$(y + 2)^2 = x - 3,$$

which is in the form of Equation (1a), page 202. Hence, the vertex is the point $(3, -2)$, and the axis is the line with equation $y = -2$. Then, by solving the last equation above for y in terms of x, we have

$$y = \sqrt{x - 3} - 2.$$

We can now determine a few points by replacing x with suitable values. A few such points are $(4, -1)$, $(7, 0)$, and $(12, 1)$. The corresponding symmetric points are $(4, -3)$, $(7, -4)$, and $(12, -5)$, respectively. We now have seven points, and the parabola can be sketched with reasonable accuracy as illustrated in Figure 8.5-1.

Ellipses

If $-x$ can be substituted for x without changing an equation, the graph of the equation is symmetric with respect to the y-axis. Similarly, if $-y$ can be substituted for y, the graph is symmetric with respect to the x-axis. Since the equation of an ellipse of either of the forms we have discussed involves only x^2 and y^2, and since both $(-x)^2 = (x)^2$ and $(-y)^2 = (y)^2$, it follows that such an ellipse is symmetric with respect to both axes. This means that for every point on an ellipse, but not on one of the axes, we can locate three additional points by changing the signs of the coordinates. For example, if (s, t) is on the ellipse, then so are $(s, -t)$, $(-s, t)$, and $(-s, -t)$. The symmetric property of an ellipse is further illustrated in the following example.

Example Sketch the ellipse with equation

$$\frac{x^2}{16} + \frac{y^2}{9} = 1.$$

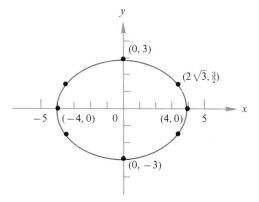

Figure 8.5-2

<u>Solution</u> By inspection, we determine that $a = 4$ and $b = 3$. Hence, the vertices are $(-4, 0)$ and $(4, 0)$, and the y-intercepts are $(0, 3)$ and $(0, -3)$. Solving the equation for y in terms of x gives

$$y = \pm \tfrac{3}{4}\sqrt{16 - x^2}.$$

Replacing x with $2\sqrt{3}$, we find that $y = \tfrac{3}{2}$ or $-\tfrac{3}{2}$.
Replacing x with $-2\sqrt{3}$, we find that $y = \tfrac{3}{2}$ or $-\tfrac{3}{2}$.
Therefore, the points $(2\sqrt{3}, \tfrac{3}{2})$ $(2\sqrt{3}, -\tfrac{3}{2})$, $(-2\sqrt{3}, \tfrac{3}{2})$ and $(-2\sqrt{3}, -\tfrac{3}{2})$ are on the ellipse. The sketch is shown in Figure 8.5-2.

Hyperbolas

A hyperbola is also symmetric with respect to its axes. However, a hyperbola is not a closed curve, and additional aids are necessary in sketching. Consider the two equations

$$\frac{x^2}{a^2} - \frac{y^2}{b^2} = 1 \qquad (1)$$

and

$$\frac{x^2}{a^2} - \frac{y^2}{b^2} = 0. \qquad (2)$$

If both of these equations are solved for y in terms of x, Equation (1) becomes

$$y = \frac{b}{a}\sqrt{x^2 - a^2} \quad \text{or} \quad y = \frac{-b}{a}\sqrt{x^2 - a^2}, \qquad (3)$$

and Equation (2) becomes

$$y = \frac{b}{a}x \quad \text{or} \quad y = \frac{-b}{a}x. \qquad (4)$$

Thus, Equation (2) is an equation for a pair of lines through the origin, and furthermore these lines contain the diagonals of the rectangle, called the **central rectangle** of the hyperbola, whose center is at the origin and whose sides are $2a$ and $2b$.

In a similar manner, we can show that

$$y = \frac{a}{b}x \quad \text{and} \quad y = \frac{-a}{b}x \tag{5}$$

are equations for the lines containing the diagonals of the central rectangle of a hyperbola whose principal axis is the y-axis.

Now compare Equations (3) and (4). Since a^2 is always positive, $x^2 - a^2$ must be less than x^2 for any replacement of x, and $\sqrt{x^2 - a^2} < x$. However, as $|x|$ becomes very large, the difference between $\sqrt{x^2 - a^2}$ and x becomes relatively small in comparison to x. Thus, a value of y in Equation (3) gets closer and closer to the corresponding value of y in Equation (4), which is to say that the hyperbola gets closer and closer to the lines. On the other hand, since $x^2 - a^2$ is always less than x^2, the hyperbola can never intersect either of the lines.

When a curve gets very close to a line without actually touching it as either $|x|$ or $|y|$ becomes very large, the line is called an **asymptote*** to the curve. Thus, a hyperbola has two asymptotes that contain the diagonals of the central rectangle of the hyperbola. This fact, together with the property of symmetry, is useful in sketching hyperbolas.

Example Sketch the hyperbola with equation

$$\frac{x^2}{16} - \frac{y^2}{9} = 1.$$

Solution Since $a = 4$ and $b = 3$, we first construct the central rectangle with vertices at $(4, 3)$, $(4, -3)$, $(-4, 3)$, and $(-4, -3)$. The diagonals of this rectangle, when extended, are the asymptotes of the hyperbola. Next, we solve the equation of the hyperbola for y in terms of x to obtain

$$y = \tfrac{3}{4}\sqrt{x^2 - 16}.$$

We use this equation to find values of y corresponding to selected values of x. For instance, if we select $x = 5$, then $y = \tfrac{9}{4}$. Hence, the point $(5, \tfrac{9}{4})$ lies on the hyperbola, and by symmetry so do the points $(5, -\tfrac{9}{4})$, $(-5, \tfrac{9}{4})$, and $(-5, -\tfrac{9}{4})$. We are now able to make a reasonably accurate sketch of the hyperbola. The right-hand branch is constructed by drawing a smooth curve through $(4, 0)$, $(5, -\tfrac{9}{4})$, and $(5, \tfrac{9}{4})$ so that as the curve is extended it gets closer and closer to the asymptotes. The left-hand branch is constructed in a similar manner. The completed hyperbola is shown in Figure 8.5-3.

EXERCISE 8.5

WRITTEN

Sketch the graph of each of the conic sections by using a suitable technique as demonstrated in the preceding examples.

---

*Asymptotes are discussed in greater detail in Section 9.5.

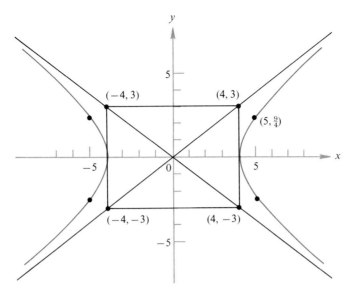

Figure 8.5-3

1. $x = y^2 - 9$
2. $x = 16 - y^2$
3. $x = 15 + 2y - y^2$
4. $x = y^2 - 2y + 24$
5. $9x^2 + 4y^2 = 36$
6. $16x^2 + 9y^2 = 144$
7. $4x^2 + 25y^2 = 100$
8. $25x^2 + 9y^2 = 225$
9. $4x^2 - 9y^2 = 36$
10. $9x^2 - 4y^2 = 36$
11. $16y^2 - 9x^2 = 144$
12. $4y^2 - 25x^2 = 100$

Summary

1. The curves called **conic sections** are the intersections of planes with a right circular cone. The conic sections are also defined as sets of points in a plane that satisfy certain conditions. [8.0]
2. A **circle** is a set of points in a plane all of which are a given distance, called the **radius**, from a fixed point, called the **center**. [8.1]
3. The general equation for the circle with center at (h, k) and radius r is

$$(x - h)^2 + (y - k)^2 = r^2,$$

which, upon expanding the binomials and combining the constants, becomes

$$x^2 + y^2 + Dx + Ey + F = 0,$$

where $D = -2h$, $E = -2k$, and $F = h^2 + k^2 - r^2$. [8.1]
4. A **parabola** is a set of points in a plane such that the distance from any point in the set to a fixed point, called the **focus**, is equal to the distance from the point to a fixed line, called the **directrix**. [8.2]
5. The equation for the parabola with vertex at the point (h, k) and directrix parallel to one of the coordinate axes is

$$(x - h)^2 = \pm 2p(y - k) \quad \text{or} \quad (y - k)^2 = \pm 2p(x - h).$$

[8.2]

6. An **ellipse** is a set of points in a plane such that the sum of the distances from any point in the set to two fixed points, called the **foci**, is constant. [8.3]
7. The equation of an ellipse with center at the origin is

$$\frac{x^2}{a^2} + \frac{y^2}{b^2} = 1 \quad \text{or} \quad \frac{x^2}{b^2} + \frac{y^2}{a^2} = 1,$$

where a and b are the **semi-major** and **semi-minor axes**, respectively. The foci are on the major axis. [8.3]
8. A **hyperbola** is a set of points in a plane such that the difference of the distances from any point in the set to two fixed points (foci) is constant. [8.4]
9. The equation of a hyperbola with center at the origin is

$$\frac{x^2}{a^2} - \frac{y^2}{b^2} = 1 \quad \text{or} \quad \frac{y^2}{a^2} - \frac{x^2}{b^2} = 1,$$

where a and b are the **semi-transverse** and **semi-conjugate axes**, respectively. The foci and the vertices lie on the transverse axis. The curve does not intersect the conjugate axis. [8.4]
10. The property of **symmetry** is helpful in sketching the conic sections. A parabola has an axis of symmetry passing through the vertex and perpendicular to the directrix. Ellipses and hyperbolas are symmetric with respect to both of their respective axes. [8.5]
11. A hyperbola has two **asymptotes**. These are lines intersecting at the center with equations

$$y = \pm \frac{b}{a} x$$

if the foci are on the x-axis, or

$$y = \pm \frac{a}{b} x$$

if the foci are on the y-axis. [8.5]

REVIEW EXERCISES

SECTION 8.1

1. Derive an equation in standard form for the circle with center at $(1, 2)$ and radius 3.
2. Find the coordinates of the center and the length of the radius of the circle with equation $x^2 + y^2 + 4x - 6y - 36 = 0$.

SECTION 8.2

Find an equation for each parabola with the properties as stated in Problems 3–5.

3. Vertex at the origin and focus at $(-4, 0)$
4. Vertex at $(3, 4)$ and directrix $x = -1$
5. Directrix $y = -2$ and focus at $(2, 5)$

6. Find the coordinates of the vertex and the equations of the directrix and the axis of symmetry of the parabola with equation $x^2 - 2x - 8y - 31 = 0$.

SECTION 8.3

Find the semi-major and semi-minor axes and the coordinates of the foci and vertices of each ellipse in Problems 7 and 8.

7. $9x^2 + 4y^2 = 36$ 8. $5x^2 + 7y^2 = 35$

In Problems 9 and 10, find an equation for each ellipse with center at the origin and other properties as given.

9. Foci on the y-axis, vertices at $(0, \pm 7)$, and intersecting the x-axis at $(\pm 3, 0)$
10. Foci at $(\pm 5, 0)$ and vertices at $(\pm 13, 0)$

SECTION 8.4

Find the coordinates of the vertices and foci of each hyperbola in Problems 11 and 12.

11. $9x^2 - 4y^2 = 36$ 12. $5y^2 - 7x^2 = 35$

In Problems 13 and 14, find an equation for each hyperbola with center at the origin and other properties as given.

13. A focus at $(0, 5)$ and semi-conjugate axis $= 3$
14. A vertex at $(12, 0)$ and a focus at $(13, 0)$

SECTION 8.5

Use such information as intercepts, symmetry, asymptotes, coordinates of vertices, etc., to sketch the graph of each equation in Problems 15–17.

15. $12x = y^2 + 12$ 16. $16x^2 + 9y^2 = 144$ 17. $9x^2 - 25y^2 = 225$

Relations, Functions, and Graphs—II

9.1 Inequalities

So far, our discussion of relations, functions, and graphs has been primarily concerned with equations. However, there are many important relations that cannot be expressed in terms of equations. For example, consider the problem of a manufacturer. In the manufacture of any article, there are costs of setting up the machinery, making dies, and so on, that do not recur. These costs are spread over the entire number of articles made, so that the actual cost of producing one article decreases as the number of articles increases. This means that the greater the number of articles made, the lower the selling price can be. With a lower price, the manufacturer can expect to sell more articles. However, if he makes too many articles, the market will be flooded, and he will be left with a large number of unsold articles. Thus, the manufacturer must make a decision as to the number of articles that can be made to give the maximum profit.

Decisions such as this can often be made by applying a technique called **linear programming** to the problem. This technique involves relations that are called **inequalities**. The basic assumptions and definitions governing inequalities are stated as follows:

(1) If $x \in \mathbf{R}$, then one and only one of the following is true:

x is positive, $-x$ is positive, or $x = 0$.

(2) If x and y are both positive, then $x + y$ and xy are both positive.
(3) $x > y$ implies that $y < x$.
(4) $x \geq y$ means that $x > y$ or $x = y$.
    $x \leq y$ means that $x < y$ or $x = y$.
(5) $x < 0$ or $0 > x$ means that x is negative.

(6) $x < a < y$ means that a is between x and y. That is, $a > x$ and $a < y$.

(7) Two inequalities are said to have the same *sense* if their order symbols are alike when read in the same direction. For example, $a < b$ and $c < d$ have the *same sense*, but $a > b$ and $c < d$ have *opposite sense*.

Other useful properties of inequalities, some of which were proved in Section 1.5, are

(1) If $a < b$ and $b < c$, then $a < c$. (Transitive property)
(2) If $a < b$, then $a + c < b + c$.
(3) If $a < b$ and $c < d$, then $a + c < b + d$.
(4) If $a < b$, then $ac < bc$ if $c > 0$, and $ac > bc$ if $c < 0$.

An inequality does not define a function. For example, the equation $y = x$ implies that for each x, there is one and only one y; but in $y < x$, for each replacement of x there are an infinite number of values of y, each of which is less than the selected value for x.

As with equations, inequalities may be of any degree. Also, they may contain any number of variables. However, for our purposes, we will consider only inequalities in two variables that are of first or second degree. In this discussion, we are concerned with specifying the solution sets of such inequalities.

Linear Inequalities

You have learned that an equation of the form

$$ax + by + c = 0$$

is a linear equation. If the equal sign is replaced by one of the order symbols ($<, >, \leq$, or $\geq$) we have a **linear inequality**.

The solution set of an inequality can be specified in a manner similar to that used for equations. If we solve

$$ax + by + c < 0$$

for y in terms of x, we have $by < -(ax + c)$, so that

$$y < \frac{-(ax + c)}{b} \text{ if } b > 0 \quad \text{and} \quad y > \frac{-(ax + c)}{b} \text{ if } b < 0.$$

The solution set is specified as

$$\left\{ (x, y) \mid y < \frac{-(ax + c)}{b} \right\} \quad \text{or as} \quad \left\{ (x, y) \mid y > \frac{(-ax + c)}{b} \right\},$$

respectively. However, in practical situations, it is better to show the solution set of an inequality by a graph.

Graphs of Linear Inequalities

You have learned that the graph of a linear equation in two variables is a line. A line separates a plane into two parts that we call **half-planes**. The graph of a linear inequality in two variables is such a half-plane. For example, if $x, y \in R$, there are three possible relations existing between the two variables. Either $x < y$, $x = y$, or $x > y$. Figure 9.1-1 shows how the graph of $x = y$ separates the coordinate plane into two half-planes. One half-plane contains all points where $x < y$, and the other contains all points where $x > y$. Each point is a **solution** of the respective inequality, because the statement is true when the coordinates are substituted for the variables.

Examples   (a) $(2, 3)$ is a solution of $x < y$ because $2 < 3$.

(b) $(7, 4)$ is a solution of $x > y$ because $7 > 4$.

To construct the graph of a linear inequality, we first construct the graph of the corresponding linear equation. This graph is a solid line if the equality is included in the original statement (denoted by $\geq$ or $\leq$) and a broken line if the equality is not included (denoted by $>$ or $<$). We then shade the portion of the coordinate plane containing the points whose coordinate pairs are solutions of the inequality.

The simplest way to determine which portion of the plane to shade is to replace both x and y with 0. If the resulting statement is true, the ordered pair $(0, 0)$ is a member of the relation and the portion of the plane containing the origin is to be shaded. If the resulting statement is not true, the ordered pair $(0, 0)$ is not a member of the relation and the portion of the plane containing the origin is left unshaded. If the graph of the corresponding equation passes through the origin, the coordinates of some point other than the origin must be used as replacements for x and y.

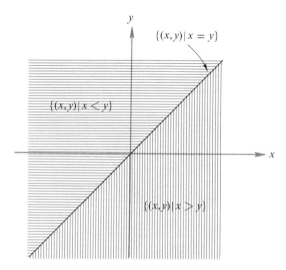

Figure 9.1-1

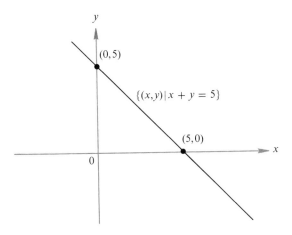

Figure 9.1-2

Example Construct the graph of the relation defined by $x + y \geq 5$.

Solution Since the equality is included in the relation, the graph of $x + y = 5$ is drawn as a solid line (see Figure 9.1-2). Since this line does not pass through the origin, zero can be substituted for both x and y in the inequality to obtain

$0 + 0 > 5$,

which is a false statement. Hence, $(0,0)$ is not a solution of the inequality, and the origin will lie in the unshaded portion of the plane. Thus, the other half-plane is shaded to complete the graph. (See Figure 9.1-3.)

Quadratic Inequalities

Relations defined by quadratic inequalities can also be represented graphically. The technique used is quite similar to that above. We construct the graph of the

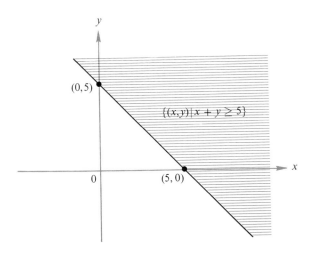

Figure 9.1-3

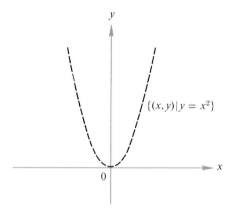

Figure 9.1-4

corresponding equation as a solid curve or a broken curve and then determine which portion of the plane to shade.

Example Graph the inequality $y < x^2$.

Solution Observe that the equation is not a part of the defining statement. Hence, the graph of $y = x^2$ is drawn as a broken curve. (See Figure 9.1-4.) Now observe that the curve passes through the origin. This means that we must use some point other than the origin to determine which portion of the plane to shade. By inspecting the graph, we see that the point $(2, 3)$ does not lie on the curve. Hence, it is suitable to use as a test point. (Any other point not on the curve can be used if desired.) If we replace x with 2 and y with 3 in $y < x^2$, we find

$$3 < 4,$$

which is a true statement. This means that $(2, 3)$ is a member of the relation, and the graph is completed by shading that portion of the plane bounded by the graph of $y = x^2$ that contains the point $(2, 3)$. (See Figure 9.1-5.)

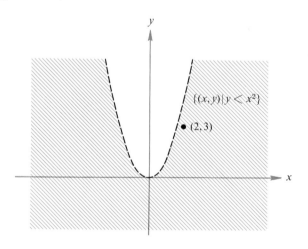

Figure 9.1-5

EXERCISE 9.1

WRITTEN

Graph each inequality in Problems 1–32.

Example $3x + y > 6$

Solution Since the statement defining the relation does not contain an equation, we draw the graph of $3x + y = 6$ as a broken line. We now note that this line does not pass through the origin, which means that we can use $(0, 0)$ as a test point. Replacing both x and y with zero in $3x + y > 6$ gives

$$0 + 0 > 6,$$

which is a false statement. Thus, the origin will lie in the unshaded half-plane. The completed graph is shown in Figure 9.1-6.

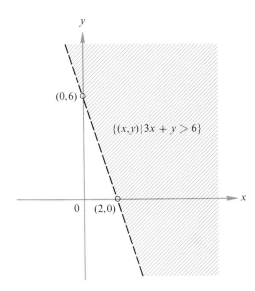

Figure 9.1-6

1. $2x + y > 4$
2. $3x + y < 6$
3. $2x - y < 3$
4. $x - 2y > 4$
5. $x + y < 1$
6. $x - y < -2$
7. $2x + 3y \geq 0$
8. $x - 3y \geq 9$
9. $x + 2y \geq 6$
10. $3x + 2y \leq 12$
11. $2x + y \leq 1$
12. $2x - 3y \leq 1$

Example $y \leq |x - 1|$

Solution The graph of $y = |x - 1|$ is V-shaped, and since the equation is part of the defining statement, the graph is solid. Since the graph does not pass through the origin, we can use $(0, 0)$ as a test point. Replacing both x and y with zero gives

$$0 < |0 - 1|,$$

which is a true statement. Hence, the portion of the plane bounded by the graph of the equation and containing the origin is shaded. (See Figure 9.1-7.)

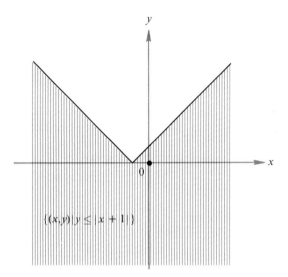

Figure 9.1-7

13. $y \geq |x + 1|$ **14.** $y \leq |x - 2|$ **15.** $y < |x| + 2$
16. $y > |x| - 3$ **17.** $y > |2x| + 1$ **18.** $y \leq |3x - 2|$

Example $\dfrac{x^2}{4} + \dfrac{y^2}{9} \leq 1$

Solution   The graph of the equation is an ellipse that does not pass through the origin. Substituting zero for both x and y gives

$$0 + 0 < 1,$$

which is a true statement, so the portion of the plane containing the origin is shaded. (See Figure 9.1-8.)

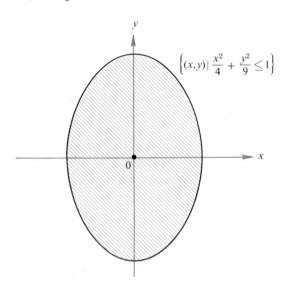

Figure 9.1-8

19. $x^2 + y^2 \leq 16$ 20. $x^2 + y^2 \geq 9$ 21. $4x^2 + 9y^2 \geq 36$
22. $16x^2 + 9y^2 \leq 144$ 23. $x^2 - y^2 \geq 4$ 24. $9x^2 - 4y^2 \leq 36$
25. $y < x^2$ 26. $y \geq x^2 + 2x$ 27. $y^2 + y < x + 2$
28. $y^2 + 3y + 2 \geq x$ 29. $4y^2 - 9x^2 \geq 36$ 30. $9y^2 - 16x^2 \leq 144$

*31. $\{(x, y) | x + y < 3\} \cap \{(x, y) | x - y > 4\}$
*32. $\{(x, y) | y \geq x^2 - 4\} \cap \{(x, y) | y \leq 4 - x^2\}$

9.2 Variation

In a study of physics you will encounter statements such as "the elongation of a spring varies directly as the force applied," or "the electrical resistance of a wire varies directly as its length and inversely as the square of its diameter." If you are to be able to solve problems involving such statements, you must learn how to interpret them.

Direct Variation

DEFINITION 9.2-1 A function defined by an equation of the form $y = kx^n$, where k is a nonzero constant and $n \in N$, is called a **direct variation**

In the definition above, k is called the **constant of variation** or the **constant of proportionality**, and y is said to vary directly as the nth power of x or to be proportional to the nth power of x. Thus, y varies directly as x means that $y = kx$; y is proportional to the square of x means that $y = kx^2$; and so on.

Examples Write each of the following statements of variation as an equation.
(a) The circumference of a circle varies directly as its diameter.
(b) The area of a circle is directly proportional to the square of its radius.
(c) The volume of a sphere varies directly as the cube of its radius.

Solutions (a) $C = kd$ (b) $A = kr^2$ (c) $V = kr^3$

Inverse Variation

DEFINITION 9.2-2 A function defined by an equation of the form $y = k/x^n$, where k is a nonzero constant and $n \in N$, is called an **inverse variation**

k is again called the constant of variation or the constant of proportionality, and y is said to vary inversely as the nth power of x or to be inversely proportional to the nth power of x. Thus, y varies inversely as x means $y = k/x$; y is inversely proportional to the square of x means $y = k/x^2$; and so on.

Examples Write each of the following statements of variation as an equation.
(a) The strength, S, of a beam varies inversely as its length, l.

(b) The resistance, R, of a wire is inversely proportional to the square of its diameter, d.

Solutions (a) $S = \dfrac{k}{l}$ (b) $R = \dfrac{k}{d^2}$

The first step in solving a variation problem is to translate the statement into an equation. Ordinarily, we are given at least one set of values of the variables so that we can find the value of the constant of variation. Then, by replacing the constant with its computed value in the equation, we have a formula that can be used to find as many sets of values as are asked for in the problem.

Example If y varies directly as x and $y = 3$ when $x = 9$, find the values of y when $x = 3$ and $x = 12$.

Solution The statement y varies directly as x is written in algebraic form as $y = kx$. We find the value of k by replacing x and y with their given values.

$$3 = k \cdot 9,$$

from which

$$k = \tfrac{1}{3}.$$

Thus, the formula for this problem is

$$y = \tfrac{1}{3}x.$$

If $x = 3$, then $y = \tfrac{1}{3}(3) = 1$.
If $x = 12$, then $y = \tfrac{1}{3}(12) = 4$.

Problems involving inverse variation are solved in the same manner.

Example If y varies inversely as x and $y = 3$ when $x = 4$, find the values of y when $x = 1$, $x = 2$, $x = 3$, and $x = 6$.

Solution The algebraic form of the statement of variation is $y = k/x$. Then, using the given values of x and y, we have

$$3 = \frac{k}{4},$$

from which

$$k = 12.$$

Thus, the formula for this problem is

$$y = \frac{12}{x}.$$

If $x = 1$, then $y = \tfrac{12}{1} = 12$.
If $x = 2$, then $y = \tfrac{12}{2} = 6$.

If $x = 3$, then $y = \frac{12}{3} = 4$.
If $x = 6$, then $y = \frac{12}{6} = 2$.

Joint Variation

Many of the formulas of geometry and physics can be thought of as variations. However, some of these formulas involve more than two variables.

DEFINITION 9.2-3 A function defined by an equation of the form $z = kxy$, where k is a constant and x or y may have an exponent other than 1, is called a **joint variation**.

The equation in Definition 9.2-3 can be stated as "z varies jointly as x and y." Note that the phrase "jointly as x and y" refers to the product of x and y.

Examples (a) The statement "the volume of a rectangular solid with a given height varies jointly as the length and the width" means $V = klw$.
(b) The statement "z varies jointly as the square of x and the cube of y" means $z = kx^2y^3$.

Combined Variation

Some problems involve both direct (or joint) variation and inverse variation. Such relations are called **combined variations**.

Example The equation $w = kx/y$ can be read as "w varies directly as x and inversely as y."

The techniques for solving problems involving joint or combined variation are the same as those used for direct or inverse variation.

For simplicity, we have discussed variation using only natural numbers as exponents. Actually, the exponents can be any positive real numbers. For example, see Problem 16 in the exercise set that follows.

EXERCISE 9.2

WRITTEN

Write each statement of variation in Problems 1–10 as an equation in which k is the constant of variation.

Example The area A of a triangle varies directly as the length of the base b.
Solution $A = kb$

1. The distance d that an object will travel at a constant velocity varies directly as the time t.

2. The volume V of a quantity of a gas varies directly as the temperature T if the pressure is constant.
3. The volume V of a quantity of a gas at constant temperature varies inversely as the pressure P.
4. The current i in amperes in an electrical circuit varies inversely as the resistance R in ohms.
5. The distance s that an object, starting from rest, will fall in a vacuum varies directly as the square of the time t in seconds.
6. The vibrating frequency f of a string varies directly as the square of the tension T in pounds.
7. The strength S of a rectangle beam varies jointly as its width w and its depth d.
8. The resistance R of a wire varies directly as its length x and inversely as the square of its diameter d.
9. The safe load L on a wooden beam supported at its ends varies directly as the product of its width w and the square of its depth d and inversely as its length x.
10. The gravitational attraction G between two bodies varies directly as the product of their masses m_1 and m_2 and inversely as the square of the distance d between them.

Example If y varies directly as x and $y = 5$ when $x = 3$, find y when $x = 18$.

Solution The statement written as an equation is

$$y = kx.$$

Then, substituting the given values,

$$5 = k \cdot 3$$
$$k = \tfrac{5}{3}.$$

Thus, the formula for this problem is

$$y = \tfrac{5}{3}x.$$

Finally, for $x = 18$,

$$y = \tfrac{5}{3} \cdot 18$$
$$y = 30.$$

Example If y varies directly as x^2 and $y = 9$ when $x = 3$, find y when $x = 4$.

Solution The algebraic form of the statement of variation is

$$y = kx^2.$$

To find k, we replace y with 9 and x with 3. This gives

$$9 = k(3)^2,$$

from which

$$k = 1.$$

Thus, the formula for this problem is $y = x^2$. If $x = 4$, then $y = 4^2 = 16$.

11. If y varies directly as x and $y = 1$ when $x = 3$, find y when $x = 5$.
12. If y varies directly as x^2 and $y = 8$ when $x = \sqrt{2}$, find y when $x = 4$.
13. If y varies inversely as x and $y = 3$ when $x = 2$, find y when $x = 4$.

14. If y varies inversely as x^2 and $y = 4$ when $x = 4$, find y when $x = 8$.
15. If y varies directly as the square of x and $y = 18$ when $x = 3$, find y when $x = 2$ and $x = 4$.
16. If y varies inversely as the square root of x and $y = 2$ when $x = 25$, find y when $x = 16$, $x = 36$, and $x = 49$.

Example If z varies jointly as x and y and $z = 36$ when $x = 2$ and $y = 3$, find z when $x = 3$ and $y = 1$.

Solution The algebraic form of the statement of variation is $z = kxy$. Replacing x, y, and z with their respective values gives

$$36 = k \cdot 2 \cdot 3,$$

from which

$$k = 6.$$

Thus, the formula for the problem is

$$z = 6xy.$$

If $x = 3$ and $y = 1$, then $z = 6(3)(1) = 18$.

17. If z varies jointly as x and y, and $z = 8$ when $x = 2$ and $y = 12$, find z when $x = 6$ and $y = 5$.
18. If r varies jointly as s and t, and $r = 60$ when $s = 2$ and $t = 3$, find r when $s = 3$ and $t = 4$.
19. z varies directly as x and inversely as the square root of y. If $z = 20$ when $x = 8$ and $y = 4$, find z when $x = 9$ and $y = 9$.
20. If z varies directly as x and inversely as y, and $z = 42$ when $x = 6$ and $y = 4$, find z when $x = 5$ and $y = 7$.
21. If an object falls 64 feet in 2 seconds, how far will it fall in 10 seconds? (See Problem 5, above.)
22. The current in an electrical circuit is 11.5 amperes when the resistance is 10 ohms. What is the strength of the current if the resistance is decreased to 5 ohms? (See Problem 4.)
23. A string vibrates 216 times per second under a tension of 3 pounds. What will be the number of vibrations per second if the tension is increased to 12 pounds? (See Problem 6.)
24. 100 feet of wire whose diameter is 0.01 inch has a resistance of 6 ohms. What is the resistance, in ohms, of 50 feet of wire of the same material with a diameter of 0.02 inch? (See Problem 8.)
*25. Consider the following table of values:

x	1	2	3	5	8
y	-1	2	5	11	20

If y varies directly as x, write an equation that defines this linear function. [HINT: Use $y - y_1 = k(x - x_1)$.]

*26. Write an equation that defines the linear function containing the following values.

x	1	2	3	5	10
y	-1	3	7	15	35

9.3 Inverse Relations

We have defined a relation from A to B as a set of ordered pairs of the form (a, b), where $a \in A$ and $b \in B$. It follows from this definition that a relation from B to A is a set of ordered pairs of the form (b, a). We have seen that the ordered pair (b, a) is generally not the same as the ordered pair (a, b). Hence, a relation from B to A is generally not the same as a relation from A to B.

The implication of the preceding statements is that if we form a relation by interchanging the components of a set of ordered pairs, the new relation is generally not the same as the original. A relation formed in this manner is called an **inverse relation**. We use the symbol $\mathcal{R}^{-1}$ to denote the inverse of the relation $\mathcal{R}$.

Since the inverse of a relation is formed by interchanging the components of each ordered pair, the domain of the relation becomes the range of the inverse relation, and the range of the relation becomes the domain of the inverse. Thus, if g is a relation between x and y, and g^{-1} is the inverse relation, then

$$y = g(x) \quad \text{implies} \quad x = g^{-1}(y).$$

Graphs of Inverse Relations

If a relation and its inverse are graphed on the same set of coordinate axes, each point representing an ordered pair (a, b) in the relation is symmetric to the corresponding point (b, a) in the inverse with respect to the line $y = x$, as shown in Figure 9.3-1.

Inverse of a Function

As with any relation, the inverse of a function is formed by interchanging the components of the ordered pairs. Generally, a function is defined by an equation in two variables and consists of an infinite set of ordered pairs. Hence, it is physically impossible to interchange all of the components. However, we can define the inverse of a function by interchanging the variables used to represent the domain

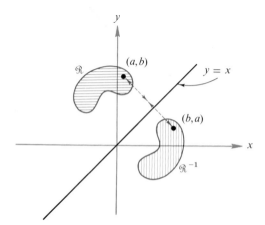

Figure 9.3-1

and the range of the function in the equation that defines the function. If the respective variables are x and y, the equation formed by interchanging the variables is solved for y in terms of x.

Example Given a function defined by the equation $y = 2x + 3$, find an equation that defines the inverse of this function.

Solution Since x represents the domain of the function, and y represents the range, we interchange them to form the equation $x = 2y + 3$. Then solving for y in terms of x

$$2y = x - 3$$
$$y = \frac{x-3}{2}.$$

If a relation is defined by an open sentence, this same technique can be used to find an open sentence that defines the inverse of the relation. However, not all such open sentences can be solved explicitly for y in terms of x.

Inverse Functions

Recall that a function is a set of ordered pairs such that no ordered pairs have the same first component and different second components. However, it may have ordered pairs with the same second component and different first components. This means that not all inverses of functions are functions. If the inverse of a function is to be a function, called an **inverse function**, then the original function must be composed of ordered pairs such that for each first component there is one and only one second component. Such a function is called a **one-to-one function**.

Functions and their inverses play an important role in mathematics, so important that we need some means of determining when the inverse of a function is an inverse function. One such method is based on the fact that the domain of a function is the range of the inverse, and vice versa. Consider the function f and its inverse g. If x represents an element in the domain of f, then $f(x)$ represents the corresponding element in the range. However, $f(x)$ is an element in the domain of g, and $g[f(x)]$ is the corresponding element in the range. On the other hand, the range of g is the domain of f, and the element in the domain of f corresponding to range of g is the domain of f, and the element in the domain of f corresponding to $f(x)$ is x. Hence, $g[f(x)] = x$. We can show in a similar way that $f[g(x)]$ is also the same as x. Thus

$$f[g(x)] = g[f(x)] = x.$$

This statement can be used to show whether or not the inverse of a function is an inverse function by replacing x in $f(x)$ with the expression that represents $g(x)$, or by replacing x in $g(x)$ with the expression representing $f(x)$.

Example Given $f(x) = 2x + 3$ and its inverse $g(x) = \frac{x-3}{2}$. Is $g(x)$ the inverse function of $f(x)$?

Solution $g[f(x)] = \dfrac{f(x) - 3}{2}$

$$= \dfrac{(2x + 3) - 3}{2}$$

$$= x$$

Therefore, $g(x)$ is the inverse function of $f(x)$.

EXERCISE 9.3

WRITTEN

1. Given the function $f = \{(0, -3), (1, 3), (2, 2), (3, 5), (4, 2)\}$. Write the ordered pairs of $\mathcal{R}^{-1}$. Is $\mathcal{R}^{-1}$ the inverse function of f?
2. Is the inverse of the function $f = \{(-2, 1), (-1, 3), (0, 5), (2, 7)\}$ the inverse function of f?

In each of Problems 3–14, write the equation that defines the inverse of the given relation.

Example $y = 5x - 7$

Solution We first interchange the variables to obtain

$x = 5y - 7$.

We then solve for y in terms of x. This gives

$y = \dfrac{x + 7}{5}$.

3. $y = 4x + 3$
4. $y = 3 - 2x$
5. $3x - y = -2$
6. $2x + 3y = 6$
7. $3x - 2y = 8$
8. $2x + 5y = 10$
9. $y = x^2$
10. $y = \sqrt{4 - x^2}$
11. $y = \sqrt{x^2 - 4}$
12. $x^2 + y^2 = 9$
13. $x^2 - y^2 = 16$
14. $4x^2 + 9y^2 = 36$

For Problems 15–20, show that the inverse of each function defined by the equations in Problems 3–8 is the inverse function.

Example $y = 5x - 7$

Solution $f(x) = 5x - 7$, and from the preceding example, we have $g(x) = \dfrac{x + 7}{5}$. We now replace x in $g(x)$ with $f(x)$, that is with $5x - 7$, to obtain

$g[f(x)] = \dfrac{f(x) + 7}{5}$

$$= \dfrac{(5x - 7) + 7}{5}$$

$$= x.$$

Therefore, $y = \dfrac{x + 7}{5}$ defines the inverse function of $y = 5x - 7$.

15. See Problem 3. 16. See Problem 4. 17. See Problem 5.
18. See Problem 6. 19. See Problem 7. 20. See Problem 8.

9.4 Graphs of Polynomial Functions

In Chapter 4 we introduced the notation $P(a)$ to represent the value of a polynomial in x when x is replaced with a. The laws of closure for real numbers with respect to the operations of addition and multiplication assure us that a polynomial with real coefficients represents a unique real number for each real number replacement of x. Thus, if $x \in R$, the set of ordered pairs $[x, P(x)]$ meets the condition of a function.

DEFINITION 9.4-1 If a_i, $x \in R$ and $n \in N$, then

$$\{(x, y) \mid y = a_0 x^n + a_1 x^{n-1} + \cdots + a_{n-1} x + a_n\}$$

is a **polynomial function**

The linear and quadratic functions that you studied in Chapter 7 are polynomial functions of first and second degree, respectively. In this section we discuss polynomial functions of higher degree, but we limit the discussion to the techniques of graphing such functions.

As with other functions, the graph of a polynomial function is constructed by first locating several points corresponding to ordered pairs in the function and then passing a smooth curve through these points. The accuracy of the graph depends upon the number of points located.

While we do not intend to discuss graphs of polynomial functions in detail, there are several theorems that are proved in more advanced courses that help us in completing such graphs. We state some of these in an informal manner.

(1) A polynomial function is a continuous function.
(2) If $P(x_1) < 0$ and $P(x_2) > 0$, then there is some number c, $x_1 < c < x_2$, such that $P(c) = 0$.
(3) A polynomial of degree n has n real zeros, not necessarily distinct, or less than n real zeros by some even natural number.
(4) The graph of a polynomial function of degree n has at most $n - 1$ turning points, or less than that number by some even natural number.

The significance of these theorems is as follows:

(1) If a function is continuous over an interval, then the function is defined for every real number replacement of the variable within the interval. This means that we can connect the points we have located with a smooth curve.
(2) If two of the located points are on opposite sides of the x-axis, then the graph must cross the x-axis.
(3) Since a real zero of a polynomial is represented graphically as an x-intercept, it follows that the graph of a polynomial function intersects

the x-axis in no more than n points. Since the number of real zeros can be less than n by an even natural number, it follows that the graph of a polynomial function of odd degree must intersect the x-axis in at least one point. Also, some of the zeros may be the same number. In this case, the graph is tangent to the x-axis if the number of repeated zeros is even.

(4) Recall that the graph of a quadratic function is a parabola, which has one turning point called the vertex. The graph of a polynomial function has an odd number of turning points $(1, 3, 5, \ldots)$ if the degree of the polynomial is even, and an even number of turning points $(0, 2, 4, \ldots)$ if the degree of the polynomial is odd. The exact location of turning points is beyond the scope of this book.

The following example illustrates how these theorems can be useful in graphing a polynomial function.

Example Construct the graph of $y = x^3 - 2x^2 - x + 2$ in the interval $-3 \leq x \leq 3$.

Solution The polynomial is of third degree. Thus, there will be either 1 or 3 zeros and 0 or 2 turning points. If you replace x successively with the integers

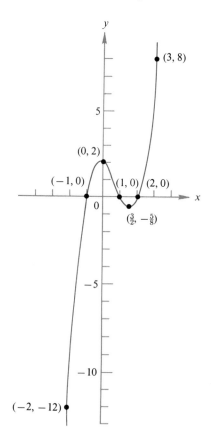

Figure 9.4-1

from -3 to 3, inclusive, you will find the following ordered pairs: $(-3, -40), (-2, -12), (-1, 0), (0, 2), (1, 0), (2, 0)$, and $(3, 8)$. Observe that $y = 0$ for both $x = 1$ and $x = 2$. In order to determine the nature of the graph between these two points, an additional ordered pair should be found. If $x = \frac{3}{2}$, then $y = -\frac{5}{8}$, which indicates that the curve is below the x-axis between 1 and 2. The eight points can now be located and connected with a smooth curve to obtain a reasonable approximation of the graph as shown in Figure 9.4-1.

EXERCISE 9.4

WRITTEN

Graph each polynomial function in the interval $-3 \leq x \leq 3$.

1. $\{(x, y) | y = x^3 + x^2 + 2x - 3\}$
2. $\{(x, y) | y = x^3 - 2x^2 - 2x + 2\}$
3. $\{(x, y) | y = 2x^3 - x^2 - 4x + 2\}$
4. $\{(x, y) | y = 2x^3 + x^2 - 6x - 3\}$
5. $\{(x, y) | y = 3x^3 + 2x^2 - 6x - 4\}$
6. $\{(x, y) | y = 3x^3 - x^2 - 9x + 3\}$
7. $\{(x, y) | y = 2x^3 - x^2 - 10x + 5\}$
8. $\{(x, y) | y = 2x^3 + x^2 - 10x - 5\}$
9. $\{(x, y) | y = x^4 - 5x^2 + 6\}$
10. $\{(x, y) | y = x^4 - 7x^2 + 10\}$

9.5 Graphs of Rational Functions

If $P(x)$ and $Q(x)$ are relatively prime polynomials (they contain no common factor), then a function defined by an equation of the form

$$f(x) = \frac{P(x)}{Q(x)} \quad \text{or} \quad y = \frac{P(x)}{Q(x)}$$

is called a **rational function**. Since we have assumed polynomial functions to be defined for every x, we may assume that a rational function is defined for all values of x except those for which $Q(x) = 0$. This means that the graph of a rational function consists of one or more smooth curves.

Whereas all of the techniques of graphing that you have previously studied are helpful, you need additional information about a rational function to construct its graph. In particular, you must consider the behavior of the value of the function as the value of the variable approaches a zero of $Q(x)$. For example, consider the function defined by

$$y = \frac{x + 1}{x - 1}.$$

By inspection, we observe that 1 is a zero of the denominator and that if $x = 1$, then y is not defined. However, if the value of x is very close to 1, but not equal to 1, then $x - 1$ is very nearly, but not quite, equal to 0, and $|y|$ is very large.

In the remainder of this discussion, the symbol $x \to c$, where c is a zero of $Q(x)$, means that $x < c$ but is almost equal to c. Similarly, the symbol $c \leftarrow x$ means that $x > c$ but is almost equal to c.

Example Discuss the behavior of y as $x \to 1$ and $1 \leftarrow x$ in the function defined by

$$y = \frac{x + 1}{x - 1}.$$

Solution If $x \to 1$, then $x + 1$ is positive, and $x - 1$ is negative but $|x - 1|$ is very small. Thus, y is negative and $|y|$ is very large. If $1 \leftarrow x$, then $x + 1$ is positive, and $x - 1$ is positive and very small. Hence, y is positive and very large. The behavior of y in the vicinity of $x = 1$ is shown in Figure 9.5-1.

Asymptotes

Since y is not defined for $x = 1$ in the example above, the graph of the function cannot intersect the line $x = 1$. However, the distance between the graph and the line can be made as small as desired by selecting values for x sufficiently close to 1. A line such as this is an **asymptote** to the graph of the function, and since this one is parallel to the y-axis, it is called a **vertical asymptote**.

The graph of a rational function may have asymptotes that are horizontal (parallel to the x-axis), oblique, or even other curves. However, this discussion is limited to vertical and horizontal asymptotes.

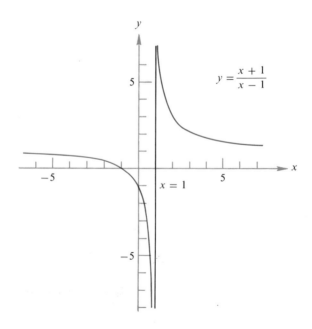

Figure 9.5-1

In Section 8.5 we saw that the hyperbola

$$\frac{x^2}{a^2} + \frac{y^2}{b^2} = 1$$

gets closer and closer to the lines

$$y = \frac{b}{a}x \quad \text{and} \quad y = \frac{-b}{a}x$$

as $|x|$ becomes very large. Since neither of these lines is parallel to one of the coordinate axes, they are called **oblique asymptotes**. Mere closeness is not sufficient to insure that a horizontal or oblique line is an asymptote to a curve. It must be shown that the curve continues to approach such a line as $|x|$ approaches infinity. This means that the distance between the curve and the line approaches zero but does not actually become zero. The proof of such a condition is beyond the scope of this course, but we can use available techniques to demonstrate, in an intuitive way, that a curve does approach an asymptote without actually touching it.

Let us again consider the equation in the preceding example. By long division, we can change the form of the equation to

$$y = 1 + \frac{2}{x-1}.$$

In this form we can observe that as $|x|$ becomes larger and larger, the value of $\frac{2}{x-1}$ gets smaller and smaller, and y is more nearly equal to 1. Thus, the curve continues to get closer to the line $y = 1$ without touching it, because no matter how large $|x|$ becomes, $\frac{2}{x-1}$ is never equal to zero. Since $y = 1$ is parallel to the x-axis, it is called a **horizontal asymptote** to the curve.

We can use a similar technique to demonstrate that a curve approaches an oblique asymptote.

The following statements can help you in finding the equations for asymptotes to the graph of

$$y = \frac{P(x)}{Q(x)}.$$

(1) The equation for a vertical asymptote is $x = c$, where c is a zero of $Q(x)$.
(2) The equation for a horizontal asymptote is $y = 0$ if the degree of $P(x)$ is less than the degree of $Q(x)$ or $y = a_n/b_n$ if the degree of $P(x)$ is equal to the degree of $Q(x)$. [a_n and b_n are the leading coefficients of $P(x)$ and $Q(x)$, respectively.] There are no horizontal asymptotes if the degree of $P(x)$ is greater than the degree of $Q(x)$.

Examples Find equations for the vertical and horizontal asymptotes to the graphs of the given equations.

(a) $y = \dfrac{x}{x^2 + x - 6}$ (b) $y = \dfrac{2x^2 + 2x + 7}{3x^2 - 5x + 2}$

Solutions   (a) $x^2 + x - 6$ may be factored as $(x + 3)(x - 2)$, which shows that the zeros of the denominator are -3 and 2. Therefore, the equations for vertical asymptotes are

$$x = -3 \quad \text{and} \quad x = 2.$$

Since the degree of the numerator is less than that of the denominator, an equation for the horizontal asymptote is

$$y = 0.$$

(b) The factored form of $3x^2 - 5x + 2$ is $(3x - 2)(x - 1)$. Hence, $\tfrac{2}{3}$ and 1 are zeros of the denominator, and the equations for vertical asymptotes are

$$x = \tfrac{2}{3} \quad \text{and} \quad x = 1.$$

Observe that the degrees of numerator and denominator are the same and that the leading coefficients are 2 and 3, respectively. Thus, the equation for the horizontal asymptote is

$$y = \tfrac{2}{3}.$$

While the graph of a rational function cannot intersect a vertical asymptote, it may cross a horizontal asymptote in cases where the asymptotic property is evident only as $|x|$ becomes very large. For example, the graph of $y = \dfrac{x - 3}{x^2 - 4}$ has

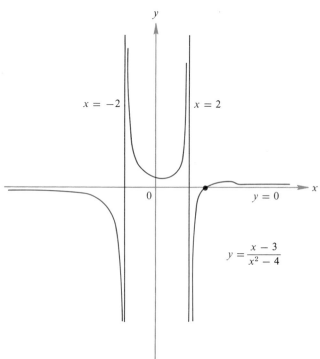

Figure 9.5-2

the graph of $y = 0$ as an asymptote; but if $x = 3$, then $y = 0$, and the graph of the function crosses the asymptote. On the other hand, as $|x| \to \infty$, $y \to 0$. (See Figure 9.5-2.)

EXERCISE 9.5

WRITTEN

For each function in Problems 1–12:
   (1) Determine equations for all vertical and horizontal asymptotes and graph them.
   (2) Find all x- and y-intercepts.
   (3) Determine coordinates of other points as necessary to construct the graph.

Example $\left\{(x, y) \mid y = \dfrac{x}{x - 3}\right\}$

Solution (1) Since $x - 3 = 0$ for $x = 3$, $x = 3$ is the equation for a vertical asymptote.

As $x \to 3$, $\dfrac{x}{x-3}$ is negative and $|y|$ is increasing rapidly. As $3 \leftarrow x$, $\dfrac{x}{x-3}$ is positive and y is increasing rapidly. $y = 1$ is a horizontal asymptote since the degrees of the numerator and denominator are equal and $a_n/b_m = 1$. $y \to 1$ as $|x|$ gets very large.

(2) If $x = 0$, $y = 0$. Hence, $(0, 0)$ is the only intercept.

(3) $(-3, \tfrac{1}{2}), (-2, \tfrac{2}{5}), (-1, \tfrac{1}{4}), (1, -\tfrac{1}{2})$, and $(2, -2)$ are coordinates of some points where $x < 3$, and $(4, 4), (5, \tfrac{5}{2}), (6, 2), (7, \tfrac{7}{4})$, and $(9, \tfrac{3}{2})$ are some points where $x > 3$.

The graph is shown in Figure 9.5-3.

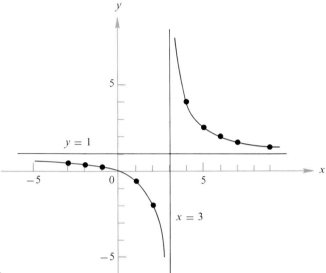

Figure 9.5-3

1. $\{(x, y) \mid y = \frac{1}{x - 2}\}$ 2. $\{(x, y) \mid y = \frac{1}{x - 3}\}$ 3. $\{(x, y) \mid y = \frac{x}{x - 1}\}$

4. $\{(x, y) \mid y = \frac{x}{x + 1}\}$ 5. $\{(x, y) \mid y = \frac{x + 1}{1 - x}\}$ 6. $\{(x, y) \mid y = \frac{3 + x}{2 - x}\}$

Example $\{(x, y) \mid y = \frac{12}{(x - 1)(x + 2)}\}$

Solution   (1) There are vertical asymptotes at $x = 1$ and $x = -2$. As $x \to -2$, both factors are negative. Thus, y is positive and increases rapidly. As $-2 \leftarrow x$, $x - 1$ is negative, but $x + 2$ is positive. Hence, y is negative and $|y|$ increases rapidly. As $x \to 1$, $x - 1$ is negative and $x + 2$ is positive. y is then negative and $|y|$ increases rapidly. As $1 \leftarrow x$, both factors are positive. Again, y is positive and increases rapidly. Since the degree of the numerator is less than that of the denominator there is a horizontal asymptote at $y = 0$. This means that $y \to 0$ as $|x| \to \infty$.

(2) If $x = 0$, $y = -6$. Hence, the y-intercept is $(0, -6)$. There is no x intercept.

(3) $(-5, \frac{2}{3})$, $(-4, \frac{6}{5})$, $(-3, 3)$ are coordinates of some points where $x < -2$. $(-1, -6)$, $(-\frac{1}{2}, -\frac{16}{3})$, $(0, -6)$, $(\frac{1}{2}, -\frac{48}{5})$ are coordinates of some points where $-2 < x < 1$. $(2, 3)$, $(3, \frac{6}{5})$, $(4, \frac{2}{3})$ are coordinates of some points where $x > 1$.

The graph is shown in Figure 9.5-4.

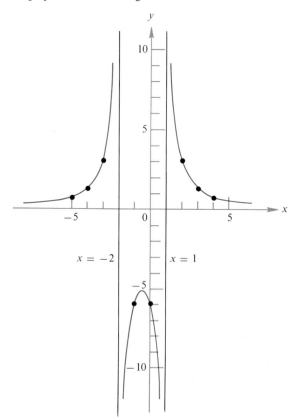

Figure 9.5-4

7. $\left\{(x, y) \mid y = \dfrac{4}{(x + 2)(x - 3)}\right\}$ 8. $\left\{(x, y) \mid y = \dfrac{8}{(x - 1)(x + 3)}\right\}$

9. $\left\{(x, y) \mid y = \dfrac{9}{x^2 - 9}\right\}$ 10. $\left\{(x, y) \mid y = \dfrac{16}{x^2 - 4}\right\}$

11. $\left\{(x, y) \mid y = \dfrac{x - 2}{x^2 - 3x - 4}\right\}$ 12. $\left\{(x, y) \mid y = \dfrac{x + 3}{x^2 - x - 6}\right\}$

Summary

1. Many relations involve inequalities rather than equations. An inequality of the form

$$ax + by + c < 0$$

(the order symbol can be $>$, $<$, $\leq$, or $\geq$) is called a **linear inequality**. [9.1]

2. The solution set of an inequality can be determined by graphing. The graph of a linear inequality is a **half-plane**. [9.1]

3. A **quadratic inequality** in two variables is an inequality that contains one or more terms of second degree. [9.1]

4. A function defined by an equation of the form $y = kx^n$, where k is a nonzero constant and $n \in N$, is called a **direct variation**. k is called the **constant of variation** or the **constant of proportionality**. [9.2]

5. A function defined by an equation of the form $y = k/x^n$, where k is a nonzero constant and $n \in N$, is called an **inverse variation**. [9.2]

6. A function defined by an equation of the form $z = kxy$, where k is a nonzero constant, is called a **joint variation**. [9.2]

7. A function that involves both direct and inverse variation is called a **combined variation**. [9.2]

8. The **inverse** of a relation can be formed by interchanging the components in each ordered pair in the relation. [9.3]

9. The inverse of a function is a relation that may or may not be a function. If $g(x)$ is the inverse of a function $f(x)$, then $g(x)$ is a function if and only if

$$f[g(x)] = g[f(x)] = x.$$ [9.3]

10. A function of the form

$$\{(x, y) \mid y = P(x)\}$$

is called a **polynomial function**. Some properties of polynomial functions that are helpful in constructing their graphs were discussed. [9.4]

11. If $P(x)$ and $Q(x)$ are relatively prime polynomials, then a function of the form

$$\left\{(x, y) \mid y = \dfrac{P(x)}{Q(x)}\right\}$$

is called a **rational function**. Some properties of rational functions that are helpful in constructing their graphs were discussed. [9.5]

REVIEW EXERCISES

SECTION 9.1

Graph the inequalities in Problems 1–3.

1. $2x - 5y < 10$
2. $y > |x| + 3$
3. $x^2 + y^2 \leq 4$

SECTION 9.2

4. If y varies directly as x^2, and $y = 4$ when $x = 7$, find y when $x = 14$.
5. If y varies inversely as the cube of x, and $y = 4$ when $x = 2$, find y when $x = 3$.
6. The weight of a sheet of metal varies jointly as its length and width. If a sheet of this metal 10 inches long and 6 inches wide weighs 90 ounces, find the weight of a sheet of the same metal that is 14 inches long and 5 inches wide.
7. If z varies directly as x and inversely as y, and $z = 12$ when $x = 8$ and $y = 2$, find z when $x = 3$ and $y = 6$.

SECTION 9.3

In Problems 8–10, find the equation that defines the inverse of each relation.

8. $y = 2x + 3$
9. $y = \sqrt{x^2 - 9}$
10. $x^2 - y^2 = 1$

11. Which of the inverse relations in Problems 8–10 are functions?

SECTION 9.4

12. Graph the function $\{(x, y) | y = x^3 - 4x^2 - 4x + 16\}$.

SECTION 9.5

Graph each of the rational functions in Problems 13 and 14.

13. $\left\{(x, y) \, y = \dfrac{1}{x - 1}\right\}$
14. $\left\{(x, y) | y = \dfrac{x - 4}{x + 1}\right\}$

Systems of Equations and Inequalities

A *system* of equations or inequalities consists of a set of two or more open sentences in two or more variables. In this discussion we consider systems containing the same number of sentences as there are variables, and we are concerned with finding the solution sets of such systems.

You saw in Chapter 7 that an equation in two variables defines a relation that consists of an infinite set of ordered pairs. This set of ordered pairs is also the solution set of the equation. The solution set of a system of equations consists of all solutions common to all equations in the system. Therefore, the solution set of a system of equations is the intersection of the solution sets of the equations.

10.1 Systems of Linear Equations—Substitution

The two equations

$$a_1x + b_1y + c_1 = 0$$
$$a_2x + b_2y + c_2 = 0$$

form a system of linear equations in two variables, and the system is in what we call **standard form**. The graph of each of these equations is a straight line, and the two lines must intersect in a single point, be parallel, or be coincident (the same line). We are primarily interested in the first case, because it is the only one of the three for which the system has a unique (one and only one) solution.

If the graphs of two linear equations are parallel, the solution sets of the equations are disjoint—that is, they have no common solution. The equations are then said to be **inconsistent**. If the graphs are coincident, the solution sets are identical, and the equations are **dependent**. If the equations are neither inconsistent nor dependent, they are **independent**. The ordered pair of coordinates of the point of intersection of the graphs is the solution of the system.

When we wish to find the solution set of a system of two linear equations in two variables, we can save time and effort by first determining whether the equations are inconsistent, dependent, or independent. We can do this by considering the slopes and y-intercepts of their graphs.

Testing for Inconsistent and Dependent Equations

Recall that when the equation $ax + by + c = 0$ is solved for y in terms of x, the result is

$$y = \frac{-a}{b}x - \frac{c}{b},$$

where $-a/b$ is the slope of the line and $-c/b$ is its y-intercept. Hence, in the system

$$a_1 x + b_1 y + c_1 = 0$$
$$a_2 x + b_2 y + c_2 = 0,$$

if $\frac{a_1}{b_1} = \frac{a_2}{b_2}$, the lines have equal slopes and are therefore parallel.

If, in addition, $\frac{c_1}{b_1} \neq \frac{c_2}{b_2}$, the lines are distinct. The equations have no common solution and are therefore *inconsistent*.

On the other hand, if $\frac{a_1}{b_1} = \frac{a_2}{b_2}$ and $\frac{c_1}{b_1} = \frac{c_2}{b_2}$, the lines have the same slope and the same y-intercept and are therefore coincident, and the equations are *dependent*.

If $\frac{a_1}{b_1} \neq \frac{a_2}{b_2}$, the two lines have different slopes and must intersect, and the equations are *independent*.

It must be pointed out that the tests given here apply only to a system of two linear equations in two variables.

Example   Are the equations in the system inconsistent, dependent, or independent?

$$3x + 4y + 7 = 0$$
$$6x + 8y + 23 = 0$$

Solution   Comparing the slopes and y-intercepts, we see that

$$\tfrac{3}{4} = \tfrac{6}{8}, \quad \text{but} \quad \tfrac{7}{4} \neq \tfrac{23}{8}.$$

Hence, the graphs of the two equations are distinct parallel lines, and the equations are *inconsistent*.

Equivalent Systems of Equations

If the solution set of a system of equations is not obvious, it is often helpful to replace the system with an equivalent system.

10.1 Systems of Linear Equations—Substitution

DEFINITION 10.1-1 Two systems of equations are **equivalent** if and only if they have the same solution set.

An equivalent system may be formed by replacing any one or all of the equations with equivalent equations. If a system of equations can be written equivalently so that one equation involves only one variable, the solution set of this equation will contain all the replacements for that variable in the ordered pairs of the solution set of the system. In this chapter we discuss two techniques for transforming a system of two linear equations in two variables into an equivalent system so that one equation is in one variable. Both of these techniques can be extended to systems involving more than two equations and variables.

Solution by Substitution

Let us first consider the system

$$3x + 2y - 7 = 0$$
$$2x - y = 0.$$

If the solution set of this system is not empty, then for the unique solution (x, y), y must have the same value in both equations. By solving the second equation for y in terms of x, we obtain

$$y = 2x.$$

Now since y must have the same value in both equations, we can replace y with $2x$ in the first equation. This gives

$$3x + 2(2x) - 7 = 0$$
$$3x + 4x - 7 = 0.$$

The solution set of this equation is $\{1\}$, and since $y = 2x$, the replacement for y is 2. Thus, the solution set of the system is $\{(1, 2)\}$.

This technique for finding the solution set of a system of equations is called **solution by substitution**.

It can be shown, but we will not do so, that if one equation in a system of two linear equations in two variables is solved for one variable in terms of the other, and the resulting expression is used to replace that variable in the other equation, the system so obtained is equivalent to the original system.

Solution Sets of Systems Involving Three Variables

A system of linear equations in more than two variables has a nonempty solution set if the system is composed of the same number of independent equations as there are variables. A solution of such a system contains a replacement for each variable and is called an **ordered n-tuple**. We will confine our discussion to methods of determining the solution sets of systems of equations involving three variables. If the three variables are x, y, and z, each solution is an **ordered triple** of the form

(x, y, z). The following example illustrates the method of finding the solution set by *substitution*.

Example Find the solution set of the system

$$x + y + z = 6 \qquad (1)$$
$$x + y - z = 0 \qquad (2)$$
$$2x - y - z = -3 \qquad (3)$$

Solution Solve any one of the equations for one variable in terms of the other two, and use the resulting expression to replace that variable in each of the other two equations. We use Equation (2) and solve for z to obtain $z = x + y$. Upon replacing z with this expression, Equation (1) becomes

$$x + y + (x + y) = 6$$

or

$$2x + 2y = 6 \qquad (4)$$

and Equation (3) becomes

$$2x - y - (x + y) = -3$$

or

$$x - 2y = -3. \qquad (5)$$

We now treat Equations (4) and (5) as a system in two variables by solving Equation (5) for x in terms of y and substituting the resulting expression for x in Equation (4).

$$x = 2y - 3$$
$$2(2y - 3) + 2y = 6$$
$$4y - 6 + 2y = 6$$
$$6y = 12$$
$$y = 2$$

Then, since $x = 2y - 3$,

$$x = 4 - 3$$
$$x = 1.$$

Finally, substituting 1 for x and 2 for y in Equation (1), we have

$$1 + 2 + z = 6$$
$$z = 3.$$

The solution set of the system is $\{(1, 2, 3)\}$. You should check this result by replacing the variables with these respective values in Equations (2) and (3).

EXERCISE 10.1

ORAL

1. Define the solution set of a system of open sentences.
2. Describe the graph of a system of two inconsistent linear equations in two variables.
3. Describe the graph of a system of two dependent linear equations in two variables.
4. How can you determine if the equations of a system of two equations in two variables are dependent without graphing or solving the system?

WRITTEN

Determine whether the equations in the systems of Problems 1–12 are inconsistent, dependent, or independent.

Examples   (a) $2x + 3y = 5$ (b) $x - 5y = 4$ (c) $2x - y = 7$
                 $4x + 6y = 8$ $3x - 15y = 12$ $x + 2y = 13$

Solutions   (a) $\frac{2}{3} = \frac{4}{6}$, but $\frac{5}{3} \neq \frac{8}{6}$.

             The equations are inconsistent.

         (b) $-\frac{1}{5} = -\frac{3}{15}$, and $-\frac{4}{5} = -\frac{12}{15}$.

             The equations are dependent.

         (c) $-\frac{2}{1} \neq \frac{1}{2}$.

             The equations are independent.

1. $x - y = 2$ 2. $6x - 5y = 14$ 3. $5x + 4y = 6$
    $3x - 3y = 6$ $7x + 2y = 32$ $10x + 8y = 13$

4. $2x + y = 5$ 5. $\frac{2}{3}x - \frac{4}{3}y = \frac{3}{5}$ 6. $\frac{3}{5}x + \frac{9}{5}y = \frac{3}{10}$
    $4x - 2y = 10$ $4x - 8y = 6$ $9x + 27y = \frac{9}{2}$

7. $2x + 3y = 10$ 8. $7x + 8y = 15$ 9. $3(x + 2y) = 9$
    $5x + 3y = 10$ $x + 8y = 15$ $6x + 12y = 18$

10. $6(x - 2y) = 5$ 11. $\dfrac{x + 3}{y} = 5$ 12. $\dfrac{x - 4}{2y} = 3$

     $3(x - y) = 8$ $2x - 10y = 9$ $2x - 12y = 8$

Use the substitution technique to find the solution set of each system of linear equations in Problems 13–30. Check each solution set.

Example $3x + 2y = 12$
            $x - y = -1$

Solution   Write the second equation equivalently as $x = y - 1$ and then replace x with this expression in the first equation.

$$3(y - 1) + 2y = 12$$
$$3y - 3 + 2y = 12$$
$$5y = 15$$
$$y = 3$$

252 SYSTEMS OF EQUATIONS AND INEQUALITIES

If y is replaced with 3 in the second equation, we obtain

$$x - 3 = -1$$
$$x = 2.$$

Hence, the solution set is $\{(2, 3)\}$.

To check, we replace x with 2 and y with 3 in the first equation. This yields

$$3(2) + 2(3) = 12,$$

which is a true statement.

13. $3x + y = 19$
    $2x - y = 1$
14. $x + y = 8$
    $y - x = 4$
15. $2x + 3y = 12$
    $3x - y = 7$
16. $6x + 5y = 22$
    $y - 4x = -6$
17. $x + 2y = 14$
    $2x + 3y = 22$
18. $2x + 5y = 4$
    $4x - 10y = 48$
19. $2(x + y) = 16$
    $3(x - y) = 6$
20. $3(x + y) = 12$
    $5(x - y) = 10$
21. $3x - y = -1$
    $x + 3y = 9$
22. $\dfrac{x + 3}{y} = 1$
    $\dfrac{y - 1}{x} = 2$
23. $\dfrac{x + 5}{2y} = 3$
    $\dfrac{6y + 2}{x} = 2$
24. $\dfrac{15}{x + y} = 3$
    $\dfrac{4}{x - y} = 2$

Example $x + y + z = 2$
$2x - y - z = 1$
$5x - y - 3z = 0$

Solution We may use any equation to express one variable in terms of the others. Let us use the third equation to express y in terms of x and z. This gives

$$y = 5x - 3z.$$

If we use this as a replacement for y in the first and second equations, we obtain the system

$$x + 5x - 3z + z = 2$$
$$2x - (5x - 3z) - z = 1.$$

The left-hand members of these equations may be simplified, which yields

$$6x - 2z = 2$$
$$-3x + 2z = 1.$$

The first of these equations may now be solved explicitly for z in terms of x.

$$z = 3x - 1.$$

Using this expression as a replacement for z, we have

$$-3x + 2(3x - 1) = 1$$
$$-3x + 6x - 2 = 1$$
$$3x = 3$$
$$x = 1.$$

Then,

$$z = 3(1) - 1 = 2$$

and
$$y = 5(1) - 3(2) = -1.$$

The solution set of the system is $\{(1, -1, 2)\}$.
Check: $1 + (-1) + 2 = 2$ and $2 - (-1) - 2 = 1$ are both true statements.

25. $x - y - z = -6$
    $2x + y + z = 0$
    $3x - 5y + 8z = 13$

26. $6x - 4y + 3z = 7$
    $5x - 3y + 2z = 5$
    $x + y - z = 0$

27. $15x - 4y + 8z = 3$
    $3x + 2y + 7z = -2$
    $x + y + z = 1$

28. $2x + 3y + 5z = 2$
    $7x - 2y + 6z = 5$
    $2x - y + 2z = 0$

29. $5x + 4y + 3z = 10$
    $8x - 7y + 5z = 3$
    $x + y + z = 3$

30. $2x + 4y + 6z = 6$
    $8x - 6y + 4z = 3$
    $2x + y - z = 1$

10.2 Systems of Linear Equations—Linear Combinations

In many cases you will find the substitution technique to be very tedious. Another way of obtaining an equivalent system of equations such that one equation is in one variable is to form a linear combination of the two equations. The resulting equation may then be paired with either of the original equations to form an equivalent system.

Linear Combinations

A **linear combination** of two equations in standard form can be made by multiplying every term of the first equation by a nonzero real number, multiplying every term of the second equation by some nonzero real number, not necessarily the same as the first, and then adding the resulting equations together, term by term. As will be seen below, the solution of the system, if one exists, must be a solution of the linear combination.

If the system
$$a_1 x + b_1 y + c_1 = 0$$
$$a_2 x + b_2 y + c_2 = 0$$
has a solution, it must be of the form (x_1, y_1). Now if s and t are nonzero real numbers, then
$$s(a_1 x + b_1 y + c_1) + t(a_2 x + b_2 y + c_2) = 0.$$
If x and y in the linear combination are replaced with x_1 and y_1, respectively, then both quantities inside the parentheses become zero. Thus,
$$s(0) + t(0) = 0$$
$$0 + 0 = 0,$$
and (x_1, y_1) is a solution by definition.

Solution of a System of Two Equations

If the equations in a system of linear equations are independent, suitable numbers can be selected for use as multipliers so that in the linear combination of the equations the coefficient of at least one, but not both, of the variables is zero. Thus, the linear combination paired with either of the original equations gives an equivalent system containing one equation in one variable, whose solution is readily obtained.

Example Find the solution set of the system of equations by forming a linear combination in which one variable has been eliminated.

$$4x + 2y - 6 = 0 \quad (1)$$
$$5x - 3y - 24 = 0 \quad (2)$$

Solution Multiply every term in Equation (1) by 3 and every term in Equation (2) by 2 to obtain

$$12x + 6y - 18 = 0$$
$$10x - 6y - 48 = 0.$$

Adding the corresponding terms of these equations gives

$$22x - 66 = 0,$$

from which

$$x = 3.$$

Thus, the system

$$4x + 2y - 6 = 0 \quad (1)$$
$$x = 3$$

is equivalent to the original system. The solution of this system can now be found by substituting 3 for x in Equation (1).

$$4(3) + 2y - 6 = 0$$
$$2y + 6 = 0$$
$$y = -3$$

The required solution set is $\{(3, -3)\}$. We could have used Equation (2) to form a new system. However, we use this equation to check our result. Substituting 3 for x and -3 for y, we have

$$5(3) - 3(-3) - 24 = 15 + 9 - 24$$
$$= 0.$$

Solution of a System of Three Equations

The technique of forming linear combinations can be used to solve a system of three independent linear equations in three variables.

Example Use linear combinations to find the solution set of the following system of equations.

10.2 Systems of Linear Equations—Linear Combinations

$$x + y - z = 0 \quad (1)$$
$$2x - y + 2z - 6 = 0 \quad (2)$$
$$3x - y - 2z + 5 = 0 \quad (3)$$

Solution Observe that if we add the corresponding terms of Equations (1) and (2) together, we form the linear combination

$$3x + z - 6 = 0. \quad (4)$$

Now if we add Equations (1) and (3) in the same way, we obtain

$$4x - 3z + 5 = 0. \quad (5)$$

Since the solution of the original system must satisfy each of these equations, it must also satisfy a linear combination of them. We can eliminate z by multiplying every term of Equation (4) by 3 and adding the results to the corresponding terms of Equation (5).

$$9x + 3z - 18 = 0$$
$$4x - 3z + 5 = 0$$
$$\overline{13x \quad\quad - 13 = 0}$$
$$x = 1$$

To find the value for z, we substitute 1 for x in Equation (4).

$$3(1) + z - 6 = 0$$
$$z - 3 = 0$$
$$z = 3$$

To find the value for y, we replace x and z in Equation (1) with the values we have found.

$$1 + y - 3 = 0$$
$$y - 2 = 0$$
$$y = 2$$

Thus, the solution set is $\{(1, 2, 3)\}$. These values should be checked in both Equations (2) and (3).

$$2(1) - 2 + 2(3) - 6 = 0$$
$$3(1) - 2 - 2(3) + 5 = 0$$

EXERCISE 10.2

WRITTEN

Find the solution set of each system of linear equations by the use of linear combination. If necessary, write any equation equivalently in the form $ax + by + c = 0$.

Example $3x + 2y - 12 = 0$
$x - y + 1 = 0$

Solution Form a linear combination in which the coefficient of y is 0 by adding 1 times the first equation to 2 times the second.

$$1(3x + 2y - 12) + 2(x - y + 1) = 0$$
$$5x - 10 = 0$$
$$x = 2$$

Then if x is replaced by 2, the first equation becomes

$$3(2) + 2y - 12 = 0,$$

from which

$$y = 3.$$

In the second equation, x is replaced by 2 and y by 3 to form the true statement

$$2 - 3 + 1 = 0.$$

Hence, $(2, 3)$ is a solution of each equation and $\{(2, 3)\}$ is the solution set of the system.

1. $3x + y - 19 = 0$
   $2x - y - 1 = 0$
2. $x + y - 8 = 0$
   $x - y + 4 = 0$
3. $2x + 3y - 12 = 0$
   $3x - y - 7 = 0$
4. $6x + 5y - 22 = 0$
   $4x - y - 6 = 0$
5. $x + 2y = 14$
   $2x + 3y = 22$
6. $2x + 5y = 4$
   $4x - 10y = 48$
7. $2x + 2y = 16$
   $3x - 3y = 6$
8. $3x + 3y = 12$
   $5x - 5y = 10$
9. $3x = y - 1$
   $x = 9 - 3y$
10. $x + 3 = y$
    $y - 1 = 2x$
11. $x + 5 = 6y$
    $6y + 2 = 2x$
12. $15 - 3x = 3y$
    $4 + 2y = 2x$
13. $2x - y = 3$
    $x = y$
14. $4x + 3y = 22$
    $2y = x$
15. $5x - 7y = 0$
    $x - y - 2 = 0$
16. $3x + y = 28$
    $2x - y + 3 = 0$
17. $3x + 2y - 5 = 0$
    $9x - 8y + 6 = 0$
18. $4x + 3y - 2 = 0$
    $8x - 9y - 9 = 0$

19. $\dfrac{1}{x} + \dfrac{1}{y} = \dfrac{5}{6}$
    $\dfrac{3}{x} - \dfrac{2}{y} = \dfrac{5}{6}$

20. $\dfrac{1}{x} + \dfrac{1}{y} = 5$
    $\dfrac{4}{x} - \dfrac{3}{y} = -1$

HINT: Solve for $1/x$ and $1/y$, then for x and y.

21. $\dfrac{5}{x} + \dfrac{2}{y} = \dfrac{19}{6}$
    $\dfrac{6}{x} - \dfrac{5}{y} = \dfrac{4}{3}$

22. $\dfrac{2}{x} + \dfrac{3}{y} = \dfrac{35}{2}$
    $\dfrac{3}{x} - \dfrac{2}{y} = 10$

23. $0.02x - 0.03y = 0.10$
    $0.05x + 0.06y = 7.00$

24. $0.03x + 0.04y = 90$
    $0.07x - 0.03y = 25$

Example $x + y + z - 2 = 0$ (1)
$2x - y + z - 5 = 0$ (2)
$5x - y - 3z = 0$ (3)

Solution Linear combinations of any two pairs of equations may be used. We shall first pair (1) and (2), then (1) and (3). If (1) and (2) are each multiplied by 1, the linear combination is

$$3x + 2z - 7 = 0. \qquad (4)$$

If (1) and (3) are each multiplied by 1, the linear combination is

$$6x - 2z - 2 = 0. \qquad (5)$$

Now form a linear combination of (4) and (5), multiplying each equation by 1 and adding to obtain

$$9x - 9 = 0,$$

from which

$$x = 1.$$

Substituting 1 for x in (4) gives

$$3(1) + 2z - 7 = 0$$
$$2z - 4 = 0$$
$$z = 2.$$

These values are then used to replace x and z in (1), which gives

$$1 + y + 2 - 2 = 0$$
$$y = -1.$$

Replacement of the variables in (2) and (3) with 1, -1, and 2, respectively, gives

$$2(1) - (-1) + 2 - 5 = 0 \quad \text{and} \quad 5(1) - (-1) - 3(2) = 0,$$

both of which are true statements.

Hence, the solution set is $\{(1, -1, 2)\}$.

25. $x - 2y + z + 1 = 0$
$3x + y - 2z - 4 = 0$
$y - z - 1 = 0$

26. $2x + 2y + z - 1 = 0$
$x - y + 6z - 21 = 0$
$3x + 2y - z + 4 = 0$

27. $4x + 8y + z + 6 = 0$
$2x - 3y + 2z = 0$
$x + 7y - 3z + 8 = 0$

28. $x + y + z = 0$
$2x - y - 4z - 15 = 0$
$x - 2y - z - 7 = 0$

29. $x + y - 2z - 2 = 0$
$3x - y + z - 5 = 0$
$3x + 3y - 5z - 7 = 0$

30. $x - 2y - 2z - 3 = 0$
$2x - 4y + 4z - 6 = 0$
$3x - 3y - 3z - 9 = 0$

31. $2x + 5z - 9 = 0$
$4x + 3y + 1 = 0$
$3y - 4z + 13 = 0$

32. $2x - 3y - 18 = 0$
$2y - z + 13 = 0$
$x + 2y + z = 0$

10.3 Systems Involving Quadratic Equations—I

Some systems of equations contain equations that are not linear. When the solution set of such a system is to be found, it is sometimes helpful to consider the possibl number of points of intersection of the graphs of the equations. In this discussion we consider systems containing one linear equation and one second-degree equation and systems containing two equations of second degree.

It can be shown that the graph of the relation

$$\{(x, y) \mid Ax^2 + Bxy + Cy^2 + Dx + Ey + F = 0\}$$

is one of the conic sections (see Chapter 8). Thus, if one equation in a system of two equations in two variables is linear and the other is of second degree, there can be no more than two points of intersection. This means that there can be no more than two real number solutions for the system. If both equations are of second degree, there can be as many as four points of intersection of their graphs, or at most four real number solutions for the system. In either case, the number of real solutions may be less.

Solution by Substitution

Certain systems involving equations of second degree can be solved by the substitution technique. If one of the equations is linear, solve this equation for one variable in terms of the other. Then use the resulting expression to replace that variable in the second-degree equation. The result will be a quadratic equation in one variable that can be solved by any of the usual methods discussed in Chapter 5.

Example Find the solution set of

$$x^2 + y^2 = 25 \qquad (1)$$
$$x + y = 7. \qquad (2)$$

Solution The graph of Equation (1) is a circle and that of Equation (2) is a straight line. Thus, there can be at most two real number solutions.

Write Equation (2) equivalently to express y in terms of x, and use this expression to replace y in Equation (1).

$$y = 7 - x$$
$$x^2 + (7 - x)^2 = 25$$
$$x^2 + 49 - 14x + x^2 = 25$$
$$2x^2 - 14x + 24 = 0$$
$$2(x - 3)(x - 4) = 0$$

and

$x = 3$ or $x = 4$.

Then, since $y = 7 - x$,

$y = 4$ or $y = 3$.

Thus, the solution set is $\{(3, 4), (4, 3)\}$.

You can also use the substitution technique in the case where one of the equations is of the form

$$bxy = k \quad (k \in R),$$

and the other contains no terms involving either x^2 or y^2.

Example Find the solution set of

$$xy + 3x - 2y - 6 = 0 \quad (1)$$
$$xy = 8. \quad (2)$$

Solution It can be shown that the graphs of both of these equations are hyperbolas and that there can be no more than four solutions. Solve Equation (2) for y in terms of x and use the resulting expression to replace y in Equation (1).

$$y = \frac{8}{x}$$

$$x\left(\frac{8}{x}\right) + 3x - 2\left(\frac{8}{x}\right) - 6 = 0$$

Then, if $x \neq 0$, both members of the equation can be multiplied by x to obtain

$$8x + 3x^2 - 16 - 6x = 0$$
$$3x^2 + 2x - 16 = 0$$
$$(3x + 8)(x - 2) = 0.$$
$$x = -\tfrac{8}{3} \quad \text{or} \quad x = 2.$$

Then, by substituting these values for x in Equation (2), the values of y are found to be

$$y = -3 \quad \text{or} \quad y = 4.$$

It can be verified that the solution set is

$$\{(-\tfrac{8}{3}, -3), (2, 4)\}.$$

In many cases, where both equations are of second degree, the substitution technique may result in an equation of third or fourth degree whose solution set cannot be determined by elementary methods. However, there are other techniques that can be applied in some cases. Two of these are discussed in Section 10.4.

EXERCISE 10.3

WRITTEN

Use the substitution technique to find the solution set of each system of equations.

Example $3x^2 + 4y^2 = 12$
$x + y = 2$

Solution The graph of the first equation is an ellipse and the graph of the second is a line. Thus, there can be no more than two solutions. Solve the second equation for y in terms of x to obtain

$$y = 2 - x.$$

Then use this expression to replace y in the first equation.

$$3x^2 + 4(2 - x)^2 = 12$$
$$3x^2 + 16 - 16x + 4x^2 = 12$$
$$7x^2 - 16x + 4 = 0$$

This quadratic equation is solved by factoring.

$$(7x - 2)(x - 2) = 0$$
$$x = \tfrac{2}{7}, \quad \text{or} \quad x = 2$$

Then, since $y = 2 - x$,

$$y = 2 - \tfrac{2}{7} = \tfrac{12}{7} \quad \text{or} \quad y = 2 - 2 = 0.$$

Thus, $(\tfrac{2}{7}, \tfrac{12}{7})$ and $(2, 0)$ are solutions of the linear equation. When these values are substituted for x and y in the second-degree equation, we find that both ordered pairs are solutions. Hence, the solution set is $\{(\tfrac{2}{7}, \tfrac{12}{7}), (2, 0)\}$.

1. $x^2 + y^2 = 25$
   $x - y = 1$
2. $x^2 + y^2 = 13$
   $x + y = 5$
3. $x^2 - y^2 = 48$
   $x - 7y = 0$
4. $x^2 + y^2 = 13$
   $2x + 3y = 13$
5. $3y^2 - 5x^2 = 7$
   $y + 2x = 7$
6. $x^2 - y^2 = 15$
   $x + y = 15$
7. $xy + 15 = 0$
   $x + y = 2$
8. $xy + 54 = 0$
   $x + y = -3$
9. $2x^2 - 2xy + y^2 = 10$
   $x + y = 7$
10. $3x^2 + 2xy + 4y^2 = 25$
    $x - y = 4$
11. $4x^2 + 5xy - 3y^2 = 2$
    $x + y = 3$
12. $3x^2 - 9xy + 4y^2 = 16$
    $2x + y = 1$

In the remaining problems, the graphs of all equations are hyperbolas.

Example $2xy + 4x - 3y - 11 = 0$ (1)
$xy = 6$ (2)

Solution Both graphs are hyperbolas so there can be at most four solutions. If Equation (2) is solved for y and the resulting expression is used to replace y in Equation (1), we have

$$y = \frac{6}{x}.$$

$$2x\left(\frac{6}{x}\right) + 4x - 3\left(\frac{6}{x}\right) - 11 = 0$$

Then, if $x \neq 0$, both members can be multiplied by x to produce

$$12x + 4x^2 - 18 - 11x = 0$$
$$4x^2 + x - 18 = 0$$
$$(4x + 9)(x - 2) = 0.$$

$x = -\frac{9}{4}$ or $x = 2$

Then, since $y = \frac{6}{x}$,

$y = -\frac{8}{3}$ or $y = 3$.

It can be shown that these pairs of values are solutions of Equation (1) and the solution set is

$\{(-\frac{9}{4}, -\frac{8}{3}), (2, 3)\}$.

13. $xy + x + y = 5$
    $xy = 2$
14. $3xy + 4x + y = 1$
    $xy + 2 = 0$
15. $3xy + x - 2y = -3$
    $xy + 4 = 0$
16. $x + y - xy = 7$
    $xy = -6$
17. $5x + 3y - 2xy = 4$
    $xy + 1 = 0$
18. $4xy - 3x - 2y + 1 = 0$
    $xy - 1 = 0$
19. $\frac{1}{x} + \frac{1}{y} = 5$
    $6xy = 1$
20. $\frac{1}{x} + \frac{1}{y} = 1$
    $xy = 4$

10.4 Systems Involving Quadratic Equations—II

In this section we discuss, very briefly, two other methods for finding solution sets of systems containing two equations of second degree.

Method 1

A system where both equations are of the form $Ax^2 + By^2 = C$ can be solved by generating an equivalent system composed of one of the original equations and a linear combination of the two equations such that the coefficient of either x^2 or y^2, but not both, is 0.

Example Find the solution set of

$$3x^2 - 4y^2 = 8$$
$$4x^2 + 9y^2 = 25.$$

Solution The graphs of these equations are a hyperbola and an ellipse, respectively. There may be as many as four real number solutions.

We multiply both members of the first equation by 9 and those of the second by 4 and then add the results, to obtain the linear combination

$$9(3x^2 - 4y^2 - 8) + 4(4x^2 + 9y^2 - 25) = 0,$$

from which

$$27x^2 - 36y^2 - 72 + 16x^2 + 36y^2 - 100 = 0.$$
$$43x^2 - 172 = 0$$
$$x^2 = 4$$
$$|x| = 2$$
$$x = 2 \quad \text{or} \quad x = -2$$

By replacing x with either 2 or -2 in the first equation, we have
$$12 - 4y^2 = 8.$$
$$-4y^2 = -4$$
$$y^2 = 1$$
$$|y| = 1$$
$$y = 1 \quad \text{or} \quad y = -1 \text{ (for either value of } x\text{)}$$

The solution set is $\{(2, 1), (2, -1), (-2, 1), (-2, -1)\}$. You should check this statement by substituting these ordered pairs for x and y in the second equation.

Method 2

If one of the equations is of the form

$$Ax^2 + Bxy + Cy^2 = 0,$$

it may be possible to factor the left-hand member into two linear factors. Each of these linear factors can be used to form a linear equation. These two equations can then be used with the other second-degree equation to form two systems whose solution sets can be found by the substitution method.

Example   Find the solution set of

$$3x^2 + 3xy - y^2 = 35 \quad (1)$$
$$x^2 - xy - 6y^2 = 0. \quad (2)$$

Solution   The graphs of both of these equations are conic sections and thus there may be as many as four solutions of the system.

Factor the left-hand member of Equation (2) to obtain

$$(x - 3y)(x + 2y) = 0.$$

By equating each of these factors to 0, we obtain two linear equations, and by pairing each of these equations with Equation (1), we form the two systems

$$3x^2 + 3xy - y^2 = 35$$
$$x - 3y = 0$$

and

$$3x^2 + 3xy - y^2 = 35$$
$$x + 2y = 0.$$

The technique of substitution can now be used to solve each system. From the first system,

$x = 3y.$
$$3(3y)^2 + 3(3y)y - y^2 = 35$$
$$27y^2 + 9y^2 - y^2 = 35$$
$$35y^2 = 35$$
$$y^2 = 1$$
$$|y| = 1$$
$y = 1$ or $y = -1.$
Since $x = 3y$, $x = 3$ or $x = -3$.

From the second system,

$x = -2y.$
$$3(-2y)^2 + 3(-2y)y - y^2 = 35$$
$$12y^2 - 6y^2 - y^2 = 35$$
$$5y^2 = 35$$
$$y^2 = 7$$
$$|y| = \sqrt{7}$$
$y = \sqrt{7}$ or $y = -\sqrt{7}.$
Since $x = -2y$, $x = 2\sqrt{7}$ or $x = -2\sqrt{7}$.

The solution set for the original system is therefore

$$\{(3, 1), (-3, -1), (2\sqrt{7}, \sqrt{7}), (-2\sqrt{7}, -\sqrt{7})\}.$$

Each of these ordered pairs should be checked in both of the original equations.

The method described above can also be used for some systems where, by forming a linear combination in which the constant term becomes 0, the left-hand member may be factorable into two linear factors. If so, you can use these factors to form equations that you can pair with either of the original equations as illustrated in the example above.

EXERCISE 10.4

WRITTEN

For each system in Problems 1–12, use the example of Method 1, page 261, to find the solution set over *R*.

264 SYSTEMS OF EQUATIONS AND INEQUALITIES

1. $9x^2 + 25y^2 = 225$
$x^2 + y^2 = 25$
2. $9x^2 + 16y^2 = 144$
$x^2 + y^2 = 16$
3. $4x^2 - 9y^2 = 36$
$x^2 + y^2 = 9$
4. $9x^2 + 4y^2 = 36$
$x^2 - y^2 = 4$
5. $9x^2 + 16y^2 = 160$
$x^2 - y^2 = 15$
6. $4x^2 - 9y^2 = 19$
$x^2 + y^2 = 34$
7. $4x^2 + y^2 = 61$
$2x^2 + 3y^2 = 93$
8. $9x^2 + 16y^2 = 288$
$x^2 + y^2 = 25$
9. $9x^2 + 16y^2 = 288$
$x^2 + y^2 = 32$
10. $x^2 + 4y^2 = 24$
$x^2 - y^2 = 9$
11. $\dfrac{1}{x^2} + \dfrac{2}{y^2} = \dfrac{9}{4}$
$\dfrac{12}{x^2} + \dfrac{9}{y^2} = 12$
12. $\dfrac{4}{x^2} + \dfrac{9}{y^2} = 13$
$\dfrac{12}{x^2} - \dfrac{18}{y^2} = 19$

(HINT: Solve for $\dfrac{1}{x}$ and $\dfrac{1}{y}$, then find x and y.)

For each system in Problems 13–20, use the example of Method 2, page 262, to find the solution set over **R**.

13. $x^2 + xy + 2y^2 = 44$
$2x^2 - 3xy + y^2 = 0$
14. $x^2 - 3xy + 2y^2 = 20$
$x^2 - 2xy - 24y^2 = 0$
15. $6x^2 + 5xy + y^2 = 35$
$x^2 - 5xy + 6y^2 = 0$
16. $4x^2 + 12xy - 4y^2 = 12$
$5x^2 - 6xy + y^2 = 0$
17. $2x^2 + 3xy + 3y^2 = 48$
$2x^2 - xy - 3y^2 = 0$
18. $3x^2 - 2xy + 4y^2 = 27$
$x^2 - 4xy + 4y^2 = 0$
19. $2x^2 + xy - y^2 = 7$
$x^2 - 2xy + y^2 = 0$
20. $3x^2 + 2xy + y^2 = 10$
$x^2 - xy - 2y^2 = 0$

In Problems 21–24, form a linear combination such that the constant is 0, then follow Method 2 as before.

*21. $x^2 - xy = 54$
$xy - y^2 = 18$
*22. $3x^2 + 3xy + 2y^2 = 8$
$x^2 - xy - 4y^2 = -4$
*23. $x^2 + 3xy - 2y^2 = 2$
$2x^2 - 5xy + 6y^2 = 3$
*24. $x^2 - xy + y^2 = 21$
$y^2 - 2xy + 15 = 0$

In Problems 25 and 26, factor the second equation in each system and then follow Method 2.

*25. $x^2 - 3xy + y^2 = 31$
$3x^2 - xy - 2y^2 + 3x + 2y = 0$
*26. $2x^2 - 4xy + 5y^2 = 11$
$x^2 - 4xy + 4y^2 + x - 2y - 2 = 0$

10.5 Systems of Inequalities

In the introduction to Chapter 9, we discussed the fact that the conditions of some problems are best stated in the form of an inequality rather than as an equation. Similarly, we find that it is sometimes better to use a system of inequalities rather than a system of equations. In this section we are concerned with specifying the solution sets of systems of open sentences involving one or more inequalities.

The usual techniques for solving a system of equations are not practical for solving a system containing inequalities. At this time, the only practical method at our disposal is the technique of graphing that was discussed on pages 224–226. All open sentences are graphed on the same set of coordinate axes, with a different kind of shading for each inequality. The solution set of the system is indicated by the region in which the separate graphs overlap, that is, the intersection of the individual solution sets.

Example By graphing, find the solution set of the system

$$2x + y = 4$$
$$x - y < 5.$$

Solution On the same set of axes, construct the graphs of $2x + y = 4$ and $x - y = 5$. The latter line should be broken because the equality is not a part of the relation. Then, substitute zero for both x and y in $x - y < 5$, to obtain the true statement $0 - 0 < 5$. Hence, the half-plane containing the origin is shaded. The solution set of the system is that part of the graph of $2x + y = 4$ that lies within the shaded region, as shown in Figure 10.5-1.

Example Graph the solution set of the system

$$3x - y \leq 6$$
$$4x + y > 8.$$

Solution Draw the graph of $3x - y = 6$ as a solid line. Since the line does not pass through the origin, replace each variable with zero. This gives the true statement $0 + 0 < 6$. Thus, the half-plane containing the origin is shaded.

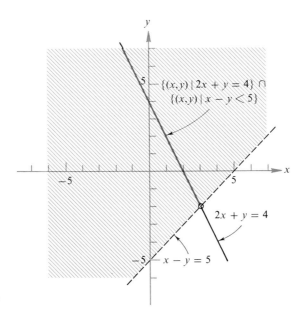

Figure 10.5-1

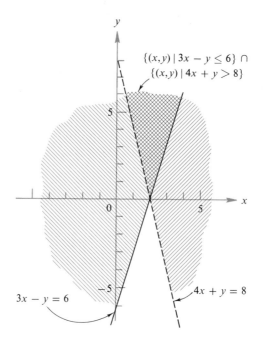

Figure 10.5-2

Draw the graph of $4x + y = 8$ as a dotted line. Since the statement $0 + 0 > 8$ is false, the half-plane that *does not* contain the origin is shaded. The intersection of the two shaded regions, including that part of the solid line within the shaded region for $4x + y > 8$, is the solution set of the system. (See Figure 10.5-2.)

Example Solve the following system graphically.

$$x^2 + y^2 < 25$$
$$x^2 - y^2 < 4$$

Solution Since neither relation contains an equation, we graph both $x^2 + y^2 = 25$ and $x^2 - y^2 = 4$ as broken curves. We now note that neither curve passes through the origin. Hence, $(0,0)$ can be used as a test point.

$$0 + 0 < 25 \quad \text{and} \quad 0 - 0 < 4$$

are both true statements. Therefore we shade the region containing the origin for both relations as shown in Figure 10.5-3.

Unlike a system of independent equations, a system of inequalities may contain more statements than variables and still have a solution. For example, a system of *three* linear inequalities in *two* variables may have a solution set whose graph is a triangular section of the coordinate plane. An example of such a system follows Problem 18 in Exercise 10.5.

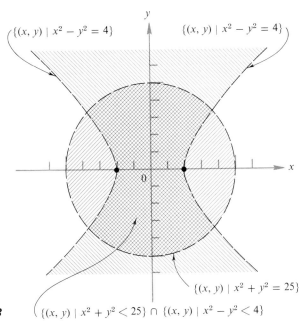

Figure 10.5-3

EXERCISE 10.5

WRITTEN

Construct the graph of the solution set of each system.

1. $6x + 5y = 30$
   $4x + y \geq 4$
2. $x + 2y \leq 14$
   $2x - 3y = 6$
3. $2x + 5y \geq 10$
   $3x + 2y = 12$
4. $3x + 4y = 12$
   $3x + 5y \leq 15$
5. $x - y > 4$
   $2x - y < 6$
6. $3x - 2y < 12$
   $4x - y > 8$
7. $2x - 5y > 10$
   $2x - y > 7$
8. $3x + 6y \leq 8$
   $4x - 5y > 10$
9. $3 - x > 0$
   $2 - x < 0$
10. $5 - y > 0$
    $3 + x \geq 0$
11. $2x - 3y \leq 0$
    $y - 4x \geq 0$
12. $x < 2y$
    $y \geq 3x$
13. $x^2 + y^2 \leq 16$
    $x + y \geq 0$
14. $y > x^2 - 2x + 1$
    $x + y < 5$
15. $y \geq x^2 - 4$
    $y \leq 4 - x^2$
16. $x^2 + y^2 \leq 25$
    $x - y^2 + 5 \geq 0$
17. $9x^2 + 4y^2 \leq 36$
    $x^2 - y^2 \geq 1$
18. $x^2 - 9y^2 \leq 1$
    $y^2 - 9x^2 \leq 1$

Example   Construct the solution set of the system

$x + y \leq 5$
$x - y \geq -3$
$y \geq -2.$

Solution   See Figure 10.5-4 on page 268.

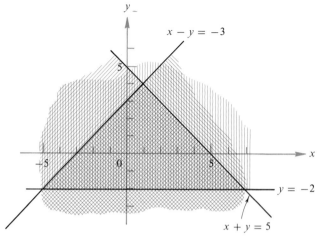

$\{(x,y) \mid x + y \leq 5\} \cap \{(x,y) \mid x - y \geq 3\} \cap \{(x,y) \mid y \geq -2\}$

Figure 10.5-4

19. $x + y \leq 4$
    $x - y \leq 4$
    $x \geq -2$
20. $x + y \geq -2$
    $x - y \geq -3$
    $x \leq 3$
21. $2x + 3y \geq 6$
    $6x + 3y \leq 18$
    $x - y \geq -2$
22. $2x - 5y \geq -10$
    $2x + 5y \geq -10$
    $4x + 2y \leq 12$
23. $x^2 + y^2 \leq 4$
    $x + y \leq 2$
    $x - y \geq -2$
24. $x^2 + y^2 \leq 25$
    $x^2 + y^2 \geq 9$
    $y + x^2 \leq 3$
25. $x^2 + y^2 \leq 25$
    $|x| \geq 2$
    $|x| \geq 3$
26. $9x^2 + 16y^2 \leq 144$
    $x^2 + y^2 \geq 9$
    $|x| \geq 2$

10.6 Word Problems

Algebraic representations of word problems can consist of equations, inequalities, or systems of either equations or inequalities. In some cases, a system of equations is more effective than a single equation for interpreting the statement of a problem. In other cases, a system involving inequalities is the only way in which a problem can be stated algebraically.

This section is devoted to the solution of problems whose algebraic representations are systems of equations. The techniques used here are those that were used in Section 2.4 except that two or more variables are used instead of one.

Example 1 Three times the sum of two integers diminished by twice their difference is 15. Find the two integers if the larger is 3 more than the smaller.

Solution Let x represent one integer, and let y represent a smaller integer. Then

$$3(x + y) - 2(x - y) = 15$$

and

$x = 3 + y.$

Equivalently,

$x + 5y = 15$
$x - y = 3.$

Multiply both members of the second equation by -1 and add the results to the first.

$6y = 12$
$y = 2$

and

$x = 5.$

The solution set of the system is $\{(5, 2)\}$. Since the solution set of the system of equations satisfies the conditions of the problem, the integers are 5 and 2.

Example 2 A bleacher section at a football field has seats for 2400 spectators, each row having the same number of seats. If the number of seats in each row were increased by 20, the same number of spectators could be seated in 10 fewer rows. Find the number of rows and the number of seats in each row in the original plan.

Solution Let x represent the number of rows, and y the number of seats in each row. Then

$xy = 2400$

and

$(x - 10)(y + 20) = 2400.$

Or equivalently,

$xy = 2400$
$xy + 20x - 10y - 200 = 2400.$

If the first of these equations is solved for y in terms of x, it becomes

$$y = \frac{2400}{x},$$

and then by substituting this expression for y the second equation becomes

$$x\left(\frac{2400}{x}\right) + 20x - 10\left(\frac{2400}{x}\right) - 200 = 2400.$$

Multiplying both sides of this equation by $\dfrac{x}{20}$ and writing the result in standard form gives

$x^2 - 10x - 1200 = 0.$
$(x + 30)(x - 40) = 0$
$x = -30 \quad \text{or} \quad x = 40$

and

$y = -80 \quad \text{or} \quad y = 60$

The solution set is

$\{(-30, -80), (40, 60)\}.$

Obviously the first of these solutions is not sensible; hence, the answer to the problem is, There are 40 rows with 60 seats each in the bleacher section.

EXERCISE 10.6

WRITTEN

Solve Problems 1-30 by using a system of equations involving two or more variables.

Example   The sum of two numbers is 25 and their difference is 13. What are the two numbers?

Solution   Let x represent one number, and let y represent a smaller number. Then

$x + y = 25$

and

$x - y = 13.$

A linear combination of these equations involving only one variable can be formed by adding the two equations together. This combination is

$2x = 38,$

from which

$x = 19.$

Replacing x with 19 in either of the original equations gives

$y = 6.$

Hence, the solution set of the system of equations is $\{(19, 6)\}$ and the two numbers are 19 and 6.

1. Find two numbers whose sum is 95 and whose difference is 15.
2. The sum of two numbers is 24, and their difference is 30. Find the two numbers.
3. The sum of three times one number and five times a second number is 94. If two times

the first number is subtracted from three times the second, the difference is 7. Find the two numbers.

4. If three times one number is added to twice a second number, the sum is 91. If three times the second number is subtracted from six times the first, the difference is 21. Find the two numbers.
5. The perimeter of a rectangle is 504 inches. If twice the width is subtracted from the length, the difference is 36 inches. Find the dimensions of the rectangle.
6. The perimeter of a rectangle is 12 feet. If three times the width is added to ten times the length, the sum is 46 feet. What are the dimensions of the rectangle?

Example The sum of the digits of a two-digit number is 6. If the digits are reversed, the number so formed exceeds the original number by 18. Find the number.

Solution Let t represent a tens digit, and u a units digit. Then

$$t + u = 6$$

and

$$10u + t - (10t + u) = 18.$$

These equations can be written equivalently as the system

$$u + t = 6$$
$$9u - 9t = 18.$$

If the second of these equations is multiplied by $\frac{1}{9}$ and the result is added to the first, we have the linear combination

$$2u = 8,$$

from which

$$u = 4.$$

Replacement of u by 4 in either equation yields the solution set of the system, $\{(2, 4)\}$.

Since $2 + 4 = 6$ and $42 - 24 = 18$, the required number is 24.

7. Find a two-digit number such that the sum of the digits is 12 and the number formed by reversing the digits exceeds the original number by 18.
8. The sum of the digits of a two-digit number is 10. If the digits are reversed, the new number is 36 less than the original number. Find the number.
9. The sum of the digits of a two-digit number is 14. If three times the tens digit is added to the number, the digits are reversed. Find the number.
10. The tens digit of a two-digit number is 2 less than the units digit. If three times the units digit is added to the number, the digits are reversed. Find the number.
11. The units digit of a two-digit number is 1 less than three times the tens digit. If the digits are reversed, the resulting number is 2 more than twice the original number. Find the number.
12. Two times the units digit of a two-digit number is 10 more than the tens digit. If the digits are reversed, the resulting number is 27 more than the original. Find the number.

Example An alloy containing 60% copper is to be made by melting together two alloys that are 25% copper and 75% copper, respectively. How many pounds of each alloy must be used to produce 200 pounds of the 60% alloy?

Solution Let x represent the number of pounds of the 25% alloy, and y represent the number of pounds of the 75% alloy. Then

$$x + y = 200. \qquad (1)$$

Since the total number of pounds of copper in the quantities of the two original alloys used must be the same as in the final alloy

$$0.25x + 0.75y = 0.60(200)$$

or

$$25x + 75y = 12{,}000. \qquad (2)$$

If Equation (1) is multiplied by -25 and the result is added to Equation (2), the linear combination will be free of x.

$$50y = 7000$$
$$y = 140$$

y is then replaced with 140 in Equation (1).

$$x + 140 = 200$$
$$x = 60$$

The solution set of the system of equations is $\{(60, 140)\}$.
  If x is replaced with 60 and y with 140 in Equation (2),

$$25(60) + 75(140) = 1500 + 10{,}500 = 12{,}000.$$

Hence, 60 pounds of the 25% alloy and 140 pounds of the 75% alloy must be used.

13. 12 ounces of a solution that is 36% alcohol are to be prepared by mixing two solutions that are 18% alcohol and 45% alcohol, respectively. How many ounces of each of the two solutions must be used?

14. How many gallons of milk that is 4% butterfat and how many gallons of cream that is 40% butterfat must be used to make 30 gallons of a mixture that is 20% butterfat?

15. One brand of lawn fertilizer is 10% nitrogen by weight and another brand is 19% nitrogen. How many pounds of each brand must be used to make 450 pounds of a mixture that is 14% nitrogen?

16. How many gallons each of two solutions that are 20% acid and 30% acid, respectively, must be mixed to prepare 50 gallons of a solution that is 26% acid?

17. 120 pounds of candy to be sold at 56 cents per pound are to be prepared by mixing two kinds that are respectively worth 60 cents and 50 cents per pound. How many pounds of each kind should be used?

18. How many pounds of nuts worth 50 cents per pound and how many pounds of nuts worth 30 cents per pound must be used to prepare 100 pounds of mixed nuts worth 35 cents per pound?

19. A collection of 76 coins, all dimes and nickels, has a value of $6.40. How many dimes and how many nickels are there in the collection?

20. The paid admissions at a football game numbered 5442 and total receipts were $8923.50. The admission prices were $1.00 for students and $2.50 for nonstudents. How many tickets were sold at each price?

Example The sum of the digits of a three-digit number is 18. The tens digit is twice the hundreds digit and the units digit is 3 more than the tens digit. Find the number.

Solution Let h represent a hundreds digit, t represent a tens digit, and u represent a units digit. A mathematical model for this problem is

$$h + t + u = 18$$
$$t = 2h$$
$$u = t + 3,$$

or, equivalently,

$$h + t + u = 18 \quad (1)$$
$$-2h + t = 0 \quad (2)$$
$$ -t + u = 3. \quad (3)$$

A system of two equations in which t is eliminated can be formed by making linear combinations of Equations (1) and (3) and Equations (2) and (3). This system is

$$h + 2u = 21 \quad (4)$$
$$-2h + u = 3. \quad (5)$$

The linear combination of two times Equation (4) added to Equation (5) is

$$5u = 45$$
$$u = 9.$$

If u is replaced with 9 in Equation (3), we have

$$-t + 9 = 3$$
$$-t = -6$$
$$t = 6.$$

If t is replaced with 6 in Equation (2), we have

$$-2h + 6 = 0$$
$$-2h = -6$$
$$h = 3.$$

Replacing the variables with the respective values in Equation (1) produces

$$3 + 6 + 9 = 18,$$

which is a true statement. Hence, the solution set of the system of equations is $\{(3, 6, 9)\}$, and the required number is 369.

21. The sum of the digits of a three-digit number is 12. The hundreds digit is 2 more than the tens digit and the units digit is 5 less than the tens digit. Find the number.
22. The sum of the digits of a three-digit number is 6. If the hundreds digit and the tens digit are interchanged, the resulting number is 90 more than the original number. If the tens digit and the units digit are interchanged, the resulting number is 9 more than the original number. Find the number. (HINT: A three-digit number may be written as $100h + 10t + u$.)

23. The perimeter of a triangle is 30 inches. The longest side is 1 inch longer than the second longest side and 2 inches shorter than three times the length of the shortest side. Find the lengths of the sides of the triangle.
24. The perimeter of a triangle whose sides, in inches, are x, y, and z is 20 inches. The sum of x and y is 13 inches. x is 4 inches less than the sum of y and z. Find the length of each side.
25. A collection of coins composed of pennies, nickels, and dimes has a value of $1.85. There are 6 more dimes than nickels and 3 less nickels than pennies. How many coins of each kind are there?
26. A man purchased 68 stamps of 5¢, 6¢, and 10¢ denominations for $4.47. He bought 6 more 5¢ stamps than 10¢ stamps. The number of 6¢ stamps was 4 less than the combined number of the other two kinds. How many stamps of each denomination did he buy?

Example The difference of two positive integers is 2, and the sum of their squares is 74. Find the integers.

Solution Let x represent a positive integer and y a second positive integer. Then, a mathematical model for the problem is the system of equations

$$x - y = 2 \quad (1)$$
$$x^2 + y^2 = 74. \quad (2)$$

If the first equation is written equivalently to express x in terms of y, it becomes

$$x = y + 2.$$

Now replace x with $y + 2$ in the second equation to produce

$$(y + 2)^2 + y^2 = 74.$$
$$y^2 + 4y + 4 + y^2 - 74 = 0$$
$$2y^2 + 4y - 70 = 0$$
$$2(y + 7)(y - 5) = 0$$

The solution set of the last equation is $\{-7, 5\}$. However, the problem specified positive integers, so nothing is to be gained by using -7. If y is replaced by 5 in Equation (1), the result is $x = 7$. Then,

$$(7)^2 + (5)^2 = 49 + 25 = 74$$

is a true statement, and the two required integers are 7 and 5.

27. The sum of two integers is 11, and the sum of their squares is 73. Find the two integers.
28. Find two positive integers such that their sum is 25 and the difference of their squares is 25.
29. Find two positive integers such that twice the square of the first plus the square of the second is 211, and the square of the first plus twice the square of the second is 179.
30. The sum of the digits of a two-digit number is 10, and the product of the digits is 21. Find the number.

Example The sum of two numbers is 10, and the sum of their reciprocals is $\frac{5}{12}$. Find the two numbers.

Solution   Let x represent one number and y the other. An algebraic representation of this problem is the system of equations

$$x + y = 10 \qquad (1)$$

$$\frac{1}{x} + \frac{1}{y} = \frac{5}{12}. \qquad (2)$$

If neither x nor y is zero, Equation (2) can be multiplied by $12xy$ to obtain

$$12y + 12x = 5xy. \qquad (3)$$

Equation (1) can be written equivalently to express y in terms of x as

$$y = 10 - x.$$

If $10 - x$ is used to replace y, Equation (3) becomes

$$12(10 - x) + 12x = 5x(10 - x)$$
$$120 - 12x + 12x = 50x - 5x^2$$
$$5x^2 - 50x + 120 = 0$$
$$5(x - 4)(x - 6) = 0.$$

The solution set of this equation is $\{4, 6\}$. Substitution of these numbers for x in Equations (1) and (2) shows the solution set of the system to be $\{(4, 6), (6, 4)\}$. Hence, one of the required numbers is 4 and the other is 6.

31. The sum of two numbers is 24, and the sum of their reciprocals is $\frac{2}{9}$. Find the numbers.
32. The difference of two numbers is 12. If the reciprocal of the first is subtracted from the reciprocal of the second, the difference is $\frac{1}{9}$. Find the numbers.
33. The speed of a boat in still water is 4 times the speed of the current in a certain river. Find the speed of the boat in still water and the speed of the current if the boat can travel a distance of 12 miles upstream in one hour more than it can travel a distance of 15 miles downstream.
34. A certain airplane requires one hour longer to fly a distance of 3990 miles against a headwind whose speed is $\frac{1}{20}$ of the airspeed of the airplane than to make the return flight at the same airspeed but with no wind. Find the airspeed of the airplane and the speed of the wind in miles per hour.
35. The sum of the lengths of the two legs of a right triangle is 28 inches, and the area is 96 square inches. Find the hypotenuse of the right triangle.
36. The perimeter of a rectangle is 36 feet. If both the length and the width of the rectangle are increased by 5 feet, the area of the resulting rectangle is 15 square feet less than three times the area of the original rectangle. Find the dimensions of the original rectangle.

In Problems 37 and 38, find a value for k for which the graph of the linear equation will be tangent to the conic section. (HINT: For a line to be tangent to a curve, there will be one and only one point of intersection. Hence, the solution set of each system will consist of a single ordered pair.)

*37. $y = x + k$
    $y = x^2 - 5x + 2$

*38. $y = x + k$
    $9x^2 + 4y^2 = 36$

Summary

1. The two equations

$$a_1x + b_1y + c_1 = 0$$
$$a_2x + b_2y + c_2 = 0$$

   form a **system of linear equations** in two variables. We call this the **standard form** of such a system. [10.1]
2. The **solution set** of a system of equations is the intersection of the solution sets of the individual equations. If the equations have no common solution, they are **inconsistent** equations. If two equations have the same solution set, they are **dependent** equations. [10.1]
3. The equations in the system above are inconsistent if

$$\frac{a_1}{b_1} = \frac{a_2}{b_2} \quad \text{and} \quad \frac{c_1}{b_1} \neq \frac{c_2}{b_2},$$

   dependent if

$$\frac{a_1}{b_1} = \frac{a_2}{b_2} \quad \text{and} \quad \frac{c_1}{b_1} = \frac{c_2}{b_2},$$

   and **linearly independent** if

$$\frac{a_1}{b_1} \neq \frac{a_2}{b_2}.$$

   [10.1]

4. Two systems of equations are **equivalent systems** if they have the same solution set. [10.1]
5. One method of solving a system of equations is by **substitution**. [10.1]
6. If every term in each of two equations is multiplied by a nonzero constant, and the corresponding terms of the two equations are then added, the resulting equation is called a **linear combination** of the two equations. [10.2]
7. The solution set of a system of equations can be determined by first forming a linear combination of the equations such that the coefficient of one, but not both, of the variables is zero. [10.2]
8. The substitution technique can be used to find the solution set of some systems that contain one or more quadratic equations. [10.3]
9. If both equations in a system of two equations are of the form

$$Ax^2 + By^2 = C,$$

   the solution set can be found by using a linear combination. [10.4]
10. Techniques for solving systems of equations are not practical for solving systems of inequalities. The solution set of such a system can be shown by graphing all of the inequalities on the same set of axes. [10.5]
11. The algebraic representations of some word problems are systems of equations or inequalities. [10.6]

REVIEW EXERCISES

SECTION 10.1

In Problems 1–3, determine whether the equations in each system are inconsistent, dependent, or independent.

1. $3x + 4y = 60$
   $\dfrac{x}{4} + \dfrac{y}{3} = 5$

2. $2x - 5y = 8$
   $2x + 5y = 8$

3. $3x + 5y = 16$
   $\dfrac{15}{2}x + \dfrac{25}{2}y = 8$

In Problems 4–7, find the solution set of each system by the substitution technique.

4. $x + 7y = 15$
   $3x - y = 1$

5. $3x + 2y = 5$
   $6x - 2y = 1$

6. $4x + 3y = 14$
   $6x - 5y = 2$

7. $x + y + 2z = 0$
   $2x - 2y + z = 8$
   $3x + 2y + z = 2$

SECTION 10.2

In Problems 8–11, find the solution set of each system by using a linear combination.

8. $x + y = 1$
   $x - y = 5$

9. $6x + 7y = 20$
   $5x - 3y = -1$

10. $\dfrac{1}{x} + \dfrac{1}{y} = 7$
    $\dfrac{2}{x} + \dfrac{3}{y} = 16$

11. $x - 2y + 3z = 4$
    $2x - y + z = 3$
    $3x + 3y + 4z = 7$

SECTION 10.3

In Problems 12–14, use the substitution technique to find the solution set of each system.

12. $x^2 + y^2 = 25$
    $2x - y = 2$

13. $x + y = 1$
    $xy = -12$

14. $2x^2 + xy + y^2 = 9$
    $x - 3y + 9 = 0$

SECTION 10.4

In Problems 15–17, use a linear combination to find the solution set of each system.

15. $4x^2 + y^2 = 25$
    $x^2 - y^2 = -5$

16. $x^2 + 4y^2 = 17$
    $3x^2 - y^2 = -1$

17. $3x^2 + 4y^2 = 16$
    $x^2 - y^2 = 3$

In Problems 18 and 19, factor the left-hand member of the first equation and use each factor to form a linear equation. Use the substitution technique to solve the systems in which each linear equation is paired with the second equation.

18. $x^2 - xy - 6y^2 = 0$
    $3x^2 + 3xy - y^2 = 35$

19. $x^2 + 2xy - 8y^2 = 0$
    $x^2 - xy + y^2 = 21$

SECTION 10.5

Construct the graph of the solution set of each system of inequalities in Problems 20 and 21.

20. $x^2 + y^2 < 9$
    $y > x^2$
    $x - 2y > -2$

21. $x^2 + y^2 \leq 9$
    $x^2 + y^2 \geq 4$
    $9x^2 - 4y^2 \geq 36$

SECTION 10.6

22. A rectangle whose area is 48 square inches is inscribed within a circle of radius 5 inches. Find the dimensions of the rectangle.

23. The annual income from an investment is $120. If the amount of the investment were increased by $600 and the rate of interest decreased by $\frac{1}{2}$%, the annual income would be $135. Find the amount and the rate of interest of the investment.

Complex Numbers

Is there a solution for the equation

$$x^2 + 4 = 0?$$

If there is a solution, it must be a number such that its square is -4. However, we know from previous discussions that the square of any nonzero real number is a positive number. (Theorem 1.5-7: If $a \in \mathbf{R}$ and $a \neq 0$, then $a^2 > 0$.) Therefore, the solution of this equation does not exist in the set of real numbers.

In the past we have found it convenient to extend a number system to include additional elements so that the operations with numbers might be defined more completely. If we continue with this concept, it seems logical to extend the number system beyond the set of real numbers to include appropriate elements so that every real number has a square root.

11.1 An Extension of the Real Number System

The development of the system of real numbers can be demonstrated by the use of simple equations.

(1) Since an equation such as $x + 3 = 2$ has no solution in the set of whole numbers, the integers were introduced.
(2) Since an equation such as $3x = 5$ has no solution in the integers, the rational numbers were introduced.
(3) Since an equation such as $x^2 = 2$ has no solution in the rational numbers, the irrational numbers were introduced.

All of these sets of numbers were then combined into a single set, called the real numbers, so that none of the properties assumed for any of the numbers were lost.

Since an equation such as $x^2 = -1$ has no solution in the set of real numbers, we need to define a different kind of number so that a solution does exist.

DEFINITION 11.1-1 The symbol i names a number with the property
$$i \cdot i = i^2 = -1.$$

If the number i is to be included in the new set, and the properties of the real numbers with respect to the multiplication and addition operations are to be maintained, the new set must be closed with respect to these operations. Thus, numbers named by symbols such as $3i$, i^3, $2 + 4i$, $3 - 2i$, etc., must be included. For the purpose of this discussion, we call any number of the form bi, $b \in \mathbf{R}$, an **imaginary number** and designate the set of all such numbers by $\mathbf{I}$.

The set of all numbers of the form $a + bi$, $a, b \in \mathbf{R}$, is called the set of **complex numbers** and is designated by $\mathbf{C}$. If both a and b are different from 0, $a + bi$ is partly a real number and partly an imaginary number. However, if $b = 0$, $a + 0i = a + 0 = a$, and if $a = 0$, $0 + bi = bi$. These statements indicate that every real number and every imaginary number can be written in the form $a + bi$. Thus, we can think of both $\mathbf{R}$ and $\mathbf{I}$ as being proper subsets of $\mathbf{C}$.

Powers of i

If we assume that $-i$ means $-1(i)$, then any power of i can be written equivalently as i, -1, $-i$, or 1.

Examples
$i = i$
$i^2 = -1$ by Definition 11.1-1
$i^3 = i^2 \cdot i = -1 \cdot i = -i$
$i^4 = i^2 \cdot i^2 = -1 \cdot -1 = 1$

One way to determine which of i, -1, $-i$, or 1 is the equivalent of some power of i is to divide the exponent of the power by 4 and then consider the remainder. If the remainder is 1, then $i^n = i$; if it is 2, $i^n = -1$; if it is 3, $i^n = -i$; if it is 0, $i^n = 1$.

Example Reduce i^{125} to one of the basic forms.

Solution $125 \div 4 = 31$ with a remainder of 1. Hence,
$$i^{125} = (i^4)^{31} \cdot i = 1^{31} \cdot i = i.$$

Square Roots of Negative Numbers

In Section 1.8, $\sqrt{a}$, $a > 0$, was defined to be a non-negative number. It follows from our informal definition that $i\sqrt{a}$ is an imaginary number. Now recall that to place a factor of a number under the square root sign, the factor must be squared. Then,
$$i\sqrt{a} = \sqrt{i^2 \cdot a} = \sqrt{-1 \cdot a} = \sqrt{-a},$$

11.1 An Extension of the Real Number System

or by the symmetric property of equality,

$$\sqrt{-a} = i\sqrt{a}.$$

Examples (a) $\sqrt{-3} = i\sqrt{3}$ (b) $\sqrt{-xy} = i\sqrt{xy}$
(c) $\sqrt{-4x} = 2i\sqrt{x}$ (d) $\sqrt{-8x^3} = 2xi\sqrt{2x}$

A product such as $\sqrt{-a}\sqrt{-b}$, $a, b > 0$, does not satisfy the conditions of Theorem 3.5-1 [If $a, b > 0$, then $\sqrt{a} \cdot \sqrt{b} = \sqrt{ab}$]. In fact, as a consequence of the informal definition of the square root of a negative number,

$$\sqrt{-a} \cdot \sqrt{-b} = i\sqrt{a}(i\sqrt{b})$$
$$= i^2\sqrt{ab}$$
$$= -\sqrt{ab}.$$

Examples (a) $\sqrt{-2}\sqrt{-3} = -\sqrt{6}$ (b) $\sqrt{-5}\sqrt{-15} = -\sqrt{75} = -5\sqrt{3}$

Equality of Complex Numbers

Before we can define addition and multiplication of complex numbers, we must establish the conditions for the equality of two complex numbers. Recall that the equals relation means that we are using two different names for the same number. However, a complex number is made up of two parts, a real number and an imaginary number. Hence, for two complex numbers to be equal, the real parts must be equal and the imaginary parts must be equal.

DEFINITION 11.1-2 If $a, b, c, d \in \mathbf{R}$, then

$$a + bi = c + di$$

if and only if $a = c$ and $b = d$.

Examples (a) $5 + 6i = (2 + 3) + (2 \cdot 3)i$
(b) $x + yi = 3 - 4i$ if and only if $x = 3$ and $y = -4$.

EXERCISE 11.1

WRITTEN

Write each power of i in Problems 1–12 as one of the basic forms $i, -1, -i,$ or 1.

Example i^7

Solution $i^7 = i^4 \cdot i^3 = 1 \cdot (-i) = -i$

1. i^5
2. i^6
3. $2i^4$
4. $5i^9$
5. $-i^{11}$
6. $-i^{10}$
7. $6i^2$
8. $10i^5$
9. $-3i^{14}$
10. $-7i^{12}$
11. $-i^{123}$
12. i^{346}

282 COMPLEX NUMBERS

In Problems 13–22, write each polynomial in the form $a + bi$.

Example $\quad 3i^7 - 2i^6 + 5i^5 + 3i^4 - 8i^3 + 4i^2 - 6i - 10$

Solution $\quad 3i^7 = 3 \cdot i^4 \cdot i^3 = 3 \cdot 1 \cdot (-i) = -3i$
$-2i^6 = -2 \cdot i^4 \cdot i^2 = -2 \cdot 1 \cdot (-1) = 2$
$5i^5 = 5 \cdot i^4 \cdot i = 5 \cdot 1 \cdot i = 5i$
$3i^4 = 3 \cdot 1 = 3$
$-8i^3 = -8 \cdot (-i) = 8i$
$4i^2 = 4 \cdot (-1) = -4$

Hence, the polynomial may be written as

$-3i + 2 + 5i + 3 + 8i - 4 - 6i - 10$
$= 2 + 3 - 4 - 10 - 3i + 5i + 8i - 6i$
$= -9 + 4i.$

13. $2i^6 + 3i^5 - 4i^3 + 10$
14. $5i^9 + 7i^8 - 2i^6 + 4i^3$
15. $7i^{14} - 8i^{13} - 2i^8 + i^7$
16. $3i^5 - 2i^6 + 8i^9 - 5i^{10}$
17. $3i^7 + 3i^5 - 2i^2 + 7$
18. $4i^8 + 2i^7 + 4i^2 - 3i$
19. $2i^9 - i^8 - 3i^7 + i^6 - 5i^5 + 4i^4 + 2i^2$
20. $4i^{13} + 5i^{12} + 2i^{11} + 3i^{10} - 2i^9 + 3i^6$

In Problems 21–32, write each square root in simplest radical form such that no radicand is negative.

Examples (a) $\sqrt{-24}$ (b) $\sqrt{\dfrac{3}{-2}}$ (c) $\sqrt{-2}\sqrt{-6}$

Solutions (a) $\sqrt{-24} = i\sqrt{24} = i\sqrt{4 \cdot 6} = 2i\sqrt{6}$

(b) $\sqrt{\dfrac{3}{-2}} = \sqrt{\dfrac{-3}{2}} = i\sqrt{\dfrac{3}{2}} = i\sqrt{\dfrac{6}{4}} = \dfrac{i\sqrt{6}}{2}$

(c) $\sqrt{-2}\sqrt{-6} = i\sqrt{2} \cdot i\sqrt{6} = i^2 \sqrt{2}\sqrt{6} = -1\sqrt{12} = -2\sqrt{3}$

21. $\sqrt{-18}$
22. $\sqrt{-27}$
23. $\sqrt{-54}$
24. $\sqrt{-125}$
25. $\sqrt{\dfrac{-2}{5}}$
26. $\sqrt{\dfrac{-3}{7}}$
27. $\sqrt{\dfrac{8}{-3}}$
28. $\sqrt{\dfrac{32}{-5}}$
29. $\sqrt{-3}\sqrt{-5}$
30. $\sqrt{-6}\sqrt{-7}$
31. $\sqrt{-3}\sqrt{-15}$
32. $\sqrt{-8}\sqrt{-12}$

In Problems 33–48, find values for x and y such that the statement is true.

Examples (a) $x + yi = 2 + 5i$ (b) $2x + 3yi = 8 - 9i$

Solutions (a) $x = 2$ and $y = 5$ (b) $2x = 8$, hence $x = 4$.
$3y = -9$, hence $y = -3$.

33. $2x + yi = 4 - 2i$
34. $5x + yi = 15 + 4i$
35. $x - 3yi = 2 + 9i$
36. $x + 5yi = 2 + 10i$
37. $3x + 4yi = 6 - 8i$
38. $4x - 3yi = 8 + 12i$

39. $x + 3 - yi = 4 + 5i$
40. $2x - 7 + yi = 3 - 2i$
41. $x - 5i = 4 + yi$
42. $2x + 8i = 4 - 2yi$
43. $x + y + 3i = 7 + xi - yi$
44. $x - y + 8i = 5 + xi + yi$
45. $x + 2y + 4i = 6 + 3xi - yi$
46. $3x - 2y - 2i = 1 + xi - 3yi$
47. $x^2 + x - 6i = 6 + y^2i + 5yi$
48. $2x^2 - x - i = 3 + 3y^2i + 4yi$

11.2 Sums and Differences

The sum of two complex numbers is defined by

DEFINITION 11.2-1 If $a + bi$ and $c + di$ are complex numbers, then
$$(a + bi) + (c + di) = (a + c) + (b + d)i.$$

Example Express the sum $(2 + 3i) + (5 - 2i)$ in the form $a + bi$.

Solution By applying Definition 11.2-1, we have
$$(2 + 3i) + (5 - 2i) = (2 + 5) + (3 - 2)i$$
$$= 7 + i.$$

Negative of a Complex Number

In the set of real numbers, the negative of a is defined to be the number $-a$ such that $a + (-a) = 0$. It is evident that the real number 0 corresponds to the complex number $0 + 0i$. Hence, if $c + di$ is the negative of $a + bi$, then
$$(a + bi) + (c + di) = 0 + 0i.$$
If we apply Definition 11.2-1 to this sum, we obtain
$$(a + c) + (b + d)i = 0 + 0i,$$
and by Definition 11.1-2
$$a + c = 0 \quad \text{and} \quad b + d = 0.$$
Then, since a, b, c, and d are all real numbers, $c = -a$ and $d = -b$. Hence, the negative of $a + bi$ is $-a + (-b)i$, or $-a - bi$.

Difference of Two Complex Numbers

Now that we have established the meaning of the negative of a complex number, we can define a difference of two complex numbers in a manner similar to that for real numbers.

DEFINITION 11.2-2 If $a + bi$ and $c + di$ are complex numbers, then
$$(a + bi) - (c + di) = (a + bi) + [-(c + di)]$$
$$= (a - c) + (b - d)i.$$

Example Express the difference $(4 + 5i) - (2 + 6i)$ in the form $a + bi$.

Solution By applying Definition 11.2-2, we have
$$(4 + 5i) - (2 + 6i) = (4 - 2) + (5 - 6)i$$
$$= 2 - i.$$

EXERCISE 11.2

WRITTEN

Express each sum or difference in the form $a + bi$.

Examples (a) $(2 + 3i) + (4 - i)$ (b) $(3 + i\sqrt{7}) - (\sqrt{2} + 3i)$

Solutions (a) $6 + 2i$ (b) $3 - \sqrt{2} + (\sqrt{7} - 3)i$

1. $(4 + 2i) + (6 - 3i)$
2. $(3 - 7i) + (-1 + 4i)$
3. $(-5 - 2i) + (3 - 4i)$
4. $(-6 - 3i) + (-1 - 4i)$
5. $(2 + yi) + (x + 4i)$
6. $(x - 3i) + (-2 + yi)$
7. $(2x + yi) + (3 - 2i)$
8. $(x - 2yi) + (4 + 3i)$
9. $(4 + \sqrt{-9}) + (6 - \sqrt{-4})$
10. $(2 - \sqrt{-16}) + (-3 + \sqrt{-25})$
11. $(\sqrt{2} + \sqrt{-3}) + (\sqrt{2} + 2\sqrt{-3})$
12. $(\sqrt{5} - \sqrt{-7}) + (2\sqrt{5} - 3\sqrt{-7})$
13. $(5 + 3i) - (2 + i)$
14. $(3 - 2i) - (4 - i)$
15. $(-6 + 2i) - (-5 + 3i)$
16. $(-7 - 5i) - (-8 - 4i)$
17. $(3 + yi) - (x + 2i)$
18. $(x - 5i) - (-7 + yi)$
19. $(x + yi) - (4 - 3i)$
20. $(2x + 3yi) - (-5 - 3i)$
21. $(3 + \sqrt{-4}) - (1 - \sqrt{-9})$
22. $(-2 + \sqrt{-16}) - (-3 - \sqrt{-25})$
23. $(\sqrt{5} + \sqrt{-3}) - (2\sqrt{5} - 2\sqrt{-3})$
24. $(3\sqrt{2} + \sqrt{-3}) - (\sqrt{2} + 2\sqrt{-3})$

11.3 Products

The product of two complex numbers can be found in a manner similar to that of finding the product of two binomials. For example, consider the product
$$(2 + 3i)(4 + i).$$
If we assume that the distributive law holds, we have
$$(2 + 3i)4 + (2 + 3i)i, \quad \text{or} \quad 8 + 12i + 2i + 3i^2.$$
Then, combining the coefficients of i and recalling that $i^2 = -1$, we have
$$(8 - 3) + (12 + 2)i = 5 + 14i.$$

For the more general case, if $a + bi$ and $c + di$ are any two complex numbers, then
$$(a + bi)(c + di) = (ac - bd) + (ad + bc)i.$$

Example Express the product $(3 - 7i)(2 + i)$ in the form $a + bi$.

Solution For this pair of complex numbers, $a = 3$, $b = -7$, $c = 2$, and $d = 1$. Hence,

$$(3 - 7i)(2 + i) = [6 - (-7)] + (3 - 14)i = 13 - 11i.$$

Multiplicative Inverses

In the set of real numbers, every number except zero has a multiplicative inverse; that is, for $a \neq 0$, $a^{-1} \cdot a = 1$. If this property is retained in the set of complex numbers, then for every nonzero complex number, there should be another number such that the product of the two equals the complex number $1 + 0i$, or simply 1.

Consider the complex number $2 + 3i$. If this number has a multiplicative inverse, it must be a number $x + yi$ such that

$$(2 + 3i)(x + yi) = 1 + 0i$$
$$(2x - 3y) + (3x + 2y)i = 1 + 0i.$$

Then, by Definition 11.1-2,

$$2x - 3y = 1$$

and

$$3x + 2y = 0.$$

Solving this system in any convenient way gives

$$x = \tfrac{2}{13} \quad \text{and} \quad y = -\tfrac{3}{13}.$$

Hence, the multiplicative inverse of $2 + 3i$ is $\tfrac{2}{13} - \tfrac{3}{13}i$.

In a similar way you can show that

$$(a + bi)^{-1} = \frac{a}{a^2 + b^2} - \frac{bi}{a^2 + b^2}. \tag{1}$$

Example Use Equation (1) to find the multiplicative inverse of $3 - 4i$, and then show that the product of the two equals 1.

Solution By Equation (1), the multiplicative inverse of $3 - 4i$ is

$$\frac{3}{25} + \frac{4i}{25}.$$

Then

$$(3 - 4i)\left(\frac{3}{25} + \frac{4i}{25}\right) = \left(\frac{9}{25} + \frac{16}{25}\right) + \left(\frac{-12}{25} + \frac{12}{25}\right)i$$
$$= \tfrac{25}{25} + 0i$$
$$= 1$$

EXERCISE 11.3

WRITTEN

In Problems 1–18, express each product in the form $a + bi$.

Examples (a) $(2 + 5i)(3 - i)$ (b) $(2 + yi)(x + 4i)$ (c) $(3 + \sqrt{-2})(2 - \sqrt{-3})$

Solutions (a) $(2 + 5i)(3 - i) = (6 + 5) + (15 - 2)i$
$= 11 + 13i$

(b) $(2 + yi)(x + 4i) = (2x - 4y) + (xy + 8)i$

(c) $(3 + \sqrt{-2})(2 - \sqrt{-3}) = (3 + i\sqrt{2})(2 - i\sqrt{3})$
$= (6 + \sqrt{6}) + (2\sqrt{2} - 3\sqrt{3})i$

1. $(5 + i)(2 + 6i)$
2. $(4 + 2i)(6 + 3i)$
3. $(2 - 3i)(8 + 7i)$
4. $(3 - 7i)(1 + 4i)$
5. $(6 - 4i)(6 + 4i)$
6. $(-4 - 7i)(-4 + 7i)$
7. $(4 + yi)(x - 3i)$
8. $(3 + yi)(x + 2i)$
9. $(x - 5i)(-3 + yi)$
10. $(x - 2i)(3 - yi)$
11. $(x + yi)(3 - 4i)$
12. $(2x - yi)(5 - 7i)$
13. $(3 + \sqrt{-2})(4 - \sqrt{-2})$
14. $(5 + \sqrt{-3})(2 + \sqrt{-3})$
15. $(3 + \sqrt{-5})(3 - \sqrt{-5})$
16. $(\sqrt{3} - \sqrt{-2})(\sqrt{3} + \sqrt{-2})$
17. $(\sqrt{5} + \sqrt{-2})(\sqrt{5} + \sqrt{-3})$
18. $(\sqrt{7} - \sqrt{-3})(\sqrt{2} + \sqrt{-3})$

In Problems 19–30, use Equation (1) (p. 285), to find the multiplicative inverse of each complex number, and then show that the product of each number and its multiplicative inverse is equal to 1.

Examples (a) $1 + i$ (b) $2 + \sqrt{-3}$

Solutions (a) The multiplicative inverse of $1 + i$ is $\dfrac{1}{2} - \dfrac{i}{2}$.

$$(1 + i)\left(\frac{1}{2} - \frac{i}{2}\right) = \left(\frac{1}{2} + \frac{1}{2}\right) + \left(\frac{i}{2} - \frac{i}{2}\right) = 1 + 0i = 1$$

(b) The multiplicative inverse of $2 + \sqrt{-3}$, or $2 + i\sqrt{3}$, is $\dfrac{2}{7} - \dfrac{i\sqrt{3}}{7}$.

$$(2 + i\sqrt{3})\left(\frac{2}{7} - \frac{i\sqrt{3}}{7}\right) = \left(\frac{4}{7} + \frac{3}{7}\right) + \left(\frac{-2i\sqrt{3}}{7} + \frac{2i\sqrt{3}}{7}\right) = 1 + 0i = 1$$

19. $1 + 5i$
20. $2 + i$
21. $-2 + 4i$
22. $-3 + 2i$
23. $3 + \sqrt{-5}$
24. $2 - \sqrt{-5}$
25. $\sqrt{2} - \sqrt{-3}$
26. $\sqrt{3} + \sqrt{-2}$
27. $\sqrt{5} + 2i$
28. $\sqrt{7} - i$
29. $x - 2i$
30. $x + yi$

11.4 Quotients

The quotient $\dfrac{a + bi}{c + di}$, c, d not both 0, can be expressed as a complex number. If the basic definition of the quotient of two real numbers is extended to complex numbers, then

$$\frac{a+bi}{c+di} = x+yi$$

such that
$$(c+di)(x+yi) = a+bi.$$

The use of the basic definition to simplify the quotient of two complex numbers involves solving a system of equations.

Example Use the basic definition of a quotient to express $\frac{2+3i}{1-i}$ in the form $a+bi$.

Solution By definition, there is a complex number $x+yi$ such that
$$(1-i)(x+yi) = 2+3i$$
$$(x+y) + (-x+y)i = 2+3i,$$
from which
$$x+y=2$$
$$-x+y=3.$$
Upon solving this system, we find
$$x = -\tfrac{1}{2} \text{ and } y = \tfrac{5}{2}.$$
Hence, the quotient can be written as $-\tfrac{1}{2} + \tfrac{5}{2}i$.

Multiplicative Inverse

A quotient of two complex numbers also can be simplified by using the multiplicative inverse of the divisor. Thus,
$$\frac{a+bi}{c+di} = (a+bi)(c+di)^{-1}, \qquad c, d \text{ not both } 0$$

Example Use the multiplicative inverse of the divisor to express $\frac{2+3i}{1-i}$ in the form $a+bi$.

Solution The multiplicative inverse of $1-i$ is $\frac{1}{2} + \frac{i}{2}$. Then
$$\frac{2+3i}{1-i} = (2+3i)\left(\frac{1}{2} + \frac{i}{2}\right) = \left(\frac{2}{2} - \frac{3}{2}\right) + \left(\frac{2}{2} + \frac{3}{2}\right)i$$
$$= -\frac{1}{2} + \frac{5}{2}i.$$

Conjugates

In Section 3.6, you learned that certain products of irrational numbers are rational, and more specifically that the product
$$(\sqrt{a} + \sqrt{b})(\sqrt{a} - \sqrt{b}) = a - b,$$
in which the two binomials are called *conjugates* of each other.

288 COMPLEX NUMBERS

A similar situation exists in the field of complex numbers. Consider the product
$$(a + bi)(a - bi) = (a^2 + b^2) + (-ab + ab)i$$
$$= a^2 + b^2 + 0i$$
$$= a^2 + b^2.$$

In this context, $a + bi$ and $a - bi$ are called **complex conjugates**. It can be shown that every complex number has a conjugate and that the two numbers differ only in the sign of the coefficient of the imaginary part.

Examples The following pairs are conjugates.

(a) $2 + 3i$; $2 - 3i$ (b) $4 - 2i$; $4 + 2i$
(c) $-1 - 4i$; $-1 + 4i$ (d) $0 + 3i$; $0 - 3i$

Quotients as Fractions

Since the quotient $\dfrac{a + bi}{c + di}$, c, d not both 0, is in the form of a fraction, when both the numerator and the denominator are multiplied by the same nonzero number the resulting fraction is equivalent to the original.

This fact and the fact that the product of two conjugate complex numbers corresponds to a real number afford us a third method of simplifying quotients.

Example Express the quotient $\dfrac{2 + 3i}{1 - i}$ in the form $a + bi$ by multiplying both the numerator and the denominator by the conjugate of $1 - i$.

Solution The conjugate of $1 - i$ is $1 + i$. Then,
$$\frac{2 + 3i}{1 - i} = \frac{(2 + 3i)(1 + i)}{(1 - i)(1 + i)}$$
$$= \frac{(2 - 3) + (2 + 3)i}{1 + 1}$$
$$= -\frac{1}{2} + \frac{5}{2}i.$$

EXERCISE 11.4

ORAL

Name the conjugate of each number.

1. $3 + 4i$ 2. $5 - 7i$ 3. $-1 + 2i$
4. $-5 - 3i$ 5. $x - yi$ 6. $2x + 3yi$
7. $\frac{3}{2} + \frac{5}{2}i$ 8. $\frac{4}{3} - \frac{2}{3}i$ 9. $3 + \sqrt{-2}$
10. $4 - \sqrt{-5}$ 11. $\sqrt{7} + \sqrt{-3}$ 12. $-2\sqrt{2} - \sqrt{-6}$

WRITTEN

In Problems 1–8, use the basic definition to express each quotient in the form $a + bi$.

Example $\dfrac{3 - 5i}{3 + 2i}$

Solution Let the quotient equal $x + yi$. Then

$$(3 + 2i)(x + yi) = 3 - 5i$$
$$(3x - 2y) + (3y + 2x)i = 3 - 5i$$
$$3x - 2y = 3$$
$$2x + 3y = -5,$$

from which, $x = -\tfrac{1}{13}$ and $y = -\tfrac{21}{13}$. Hence,

$$\frac{3 - 5i}{3 + 2i} = \frac{-1}{13} - \frac{21}{13}i.$$

1. $\dfrac{2 + 5i}{1 + 2i}$ 2. $\dfrac{3 + 5i}{2 + 2i}$ 3. $\dfrac{4 + 2i}{3 + 7i}$ 4. $\dfrac{1 + i}{2 + 2i}$

5. $\dfrac{2 + 2i}{1 + i}$ 6. $\dfrac{-3 - 4i}{2 - 5i}$ 7. $\dfrac{-1 - \sqrt{-3}}{4 - \sqrt{-3}}$ 8. $\dfrac{2 + \sqrt{-5}}{2 - \sqrt{-5}}$

In Problems 9–16, use the multiplicative inverse technique to express each quotient in the form $a + bi$.

Example $\dfrac{3 - 5i}{3 + 2i}$

Solution The multiplicative inverse of $3 + 2i$ is $\tfrac{3}{13} - \tfrac{2}{13}i$. Then,

$$\frac{3 - 5i}{3 + 2i} = (3 - 5i)\left(\frac{3}{13} - \frac{2}{13}i\right)$$
$$= \left(\frac{9}{13} - \frac{10}{13}\right) + \left(\frac{-6}{13} - \frac{15}{13}\right)i$$
$$= \frac{-1}{13} - \frac{21}{13}i.$$

9. $\dfrac{2 + 2i}{3 + 5i}$ 10. $\dfrac{3 + 7i}{4 + 2i}$ 11. $\dfrac{2 - 5i}{3 - 4i}$ 12. $\dfrac{4 - 3i}{1 - 3i}$

13. $\dfrac{-2 - 3i}{6 + i}$ 14. $\dfrac{-3 - i}{4 - 3i}$ 15. $\dfrac{2 + \sqrt{-3}}{5 - \sqrt{-3}}$ 16. $\dfrac{2 - \sqrt{-5}}{2 + \sqrt{-5}}$

In Problems 17–24, express each quotient in the form $a + bi$ by multiplying both the numerator and denominator by the conjugate of the denominator.

Example $\dfrac{3 - 5i}{3 + 2i}$

Solution The conjugate of $3 + 2i$ is $3 - 2i$. (solution continued on next page)

$$\frac{3-5i}{3+2i} = \frac{(3-5i)(3-2i)}{(3+2i)(3-2i)}$$
$$= \frac{(9-10)+(-6-15)i}{9+4}$$
$$= \frac{-1}{13} - \frac{21}{13}i.$$

17. $\dfrac{4+2i}{2-i}$ 18. $\dfrac{3-4i}{3+5i}$ 19. $\dfrac{2-5i}{4-3i}$ 20. $\dfrac{6+i}{3-i}$

21. $\dfrac{-3-3i}{1+i}$ 22. $\dfrac{1+i}{-3-3i}$ 23. $\dfrac{4+\sqrt{-5}}{2-\sqrt{-5}}$ 24. $\dfrac{3+\sqrt{-2}}{5-\sqrt{-2}}$

11.5 Solutions of Quadratic Equations

In Section 5.2 you learned that for the equation $ax^2 + bx + c = 0$, if $b^2 - 4ac < 0$, then the equation has no solution in the set of real numbers. However, since the square root of a negative number has now been defined as an imaginary number, any quadratic equation has one or two solutions in the set of complex numbers.

In this section we discuss solutions of quadratic equations with integral coefficients and some of the properties of the solutions.

The Quadratic Formula

Recall from Section 5.2 that the quadratic formula

$$x = \frac{-b \pm \sqrt{b^2 - 4ac}}{2a}$$

can be used to determine the solution set of a quadratic equation in one variable.

Example Find the solution set of $x^2 - 4x + 5 = 0$.

Solution $a = 1$, $b = -4$, and $c = 5$. Hence, by the quadratic formula,
$$x = \frac{4 \pm \sqrt{16-20}}{2}$$
$$= \frac{4 \pm \sqrt{-4}}{2}$$
$$= \frac{4 \pm 2i}{2}$$
$$= 2 \pm i.$$

Therefore, the solution set is $\{2+i, 2-i\}$.

Conjugate Solutions

An inspection of the members of the solution set of the equation in the preceding example shows that the two numbers are conjugates. In fact, it can be shown that for any quadratic equation of the form $ax^2 + bx + c = 0$, $a, b, c \in R$, if one solution is a complex number, then its conjugate is also a solution. You can use this fact to generate a quadratic equation with real number coefficients with a given complex number as a solution.

Example Find a quadratic equation with real number coefficients that has $1 + i$ as a solution.

Solution If $1 + i$ is a solution, then $1 - i$ is also a solution. Thus, $x - (1 + i) = 0$, $x - (1 - i) = 0$, and

$$[x - (1 + i)][x - (1 - i)] = 0$$
$$x^2 - x - ix - x + ix + 1 + 1 = 0$$
$$x^2 - 2x + 2 = 0,$$

which is the required equation.

Sum and Product of the Solutions

You have seen that the product of two conjugate complex numbers can be expressed as a real number. Now consider the sum of two such numbers.

$$(a + bi) + (a - bi) = (a + a) + (bi - bi) = 2a$$

Thus, the sum of two complex conjugates can be expressed as a real number, since the sum of the imaginary parts is zero.

Now consider the sum of the two solutions of any quadratic equation with real coefficients.

$$\frac{-b + \sqrt{b^2 - 4ac}}{2a} + \frac{-b - \sqrt{b^2 - 4ac}}{2a} = \frac{-2b}{2a} = \frac{-b}{a}.$$

Example For the equation $3x^2 - 4x + 5 = 0$, the sum of the two solutions is $\frac{-(-4)}{3} = \frac{4}{3}$.

The product of the two solutions of a quadratic equation is

$$\left(\frac{-b + \sqrt{b^2 - 4ac}}{2a}\right)\left(\frac{-b - \sqrt{b^2 - 4ac}}{2a}\right) = \frac{(-b)^2 - (\sqrt{b^2 - 4ac})^2}{4a^2}$$
$$= \frac{b^2 - b^2 + 4ac}{4a^2},$$
$$= \frac{c}{a}.$$

Example The product of the two solutions of $3x^2 - 4x + 5 = 0$ is $\frac{5}{3}$.

EXERCISE 11.5

WRITTEN

In Problems 1–12, find the solution set of each quadratic equation.

Example $x^2 + 2x + 2 = 0$

Solution $a = 1$, $b = 2$, and $c = 2$. Then by the quadratic formula,

$$x = \frac{-2 \pm \sqrt{4 - 8}}{2}$$

$$= \frac{-2 \pm \sqrt{-4}}{2}$$

$$= \frac{-2 \pm 2i}{2}$$

and the solution set is $\{-1 + i, -1 - i\}$.

1. $x^2 - 2x + 5 = 0$
2. $x^2 - 2x + 10 = 0$
3. $x^2 - 4x + 8 = 0$
4. $x^2 + 4x + 13 = 0$
5. $2x^2 - 2x + 1 = 0$
6. $4x^2 - 8x + 5 = 0$
7. $x^2 + 2\sqrt{2}x + 3 = 0$
8. $x^2 - \sqrt{3}x + 3 = 0$
9. $2x^2 + \sqrt{7}x + 2 = 0$
10. $3x^2 - \sqrt{15}x + 2 = 0$
11. $x^2 - 3x + 8 = 0$
12. $x^2 + x + 9 = 0$

In Problems 13–20, form a quadratic equation with real number coefficients that contains the given complex number in its solution set.

Example $2i$

Solution Since $2i$ is a solution, $-2i$ is also a solution. Hence, $x - 2i = 0$, $x + 2i = 0$, and

$$(x - 2i)(x + 2i) = 0$$
$$x^2 - 4i^2 = 0$$
$$x^2 + 4 = 0$$

is the required equation.

13. i
14. $-3i$
15. $1 - i$
16. $2 + i$
17. $2 + 3i$
18. $3 - 2i$
19. $2 + \sqrt{-5}$
20. $1 - 2\sqrt{-3}$

In Problems 21–26, find the sum and the product of the solutions of each equation.

Example $4x^2 - 7x + 3 = 0$

Solution $a = 4$, $b = -7$, $c = 3$. Hence, the sum of the solutions is $\frac{-b}{a} = \frac{7}{4}$, and the product of the solutions is $\frac{c}{a} = \frac{3}{4}$.

21. $2x^2 - 4x + 5 = 0$ 22. $3x^2 - 6x + 1 = 0$ 23. $4x^2 + 3x - 8 = 0$
24. $6x^2 - 5x - 12 = 0$ 25. $7x^2 - 3x - 4 = 0$ 26. $5x^2 + 7x - 9 = 0$

Summary

1. The real number system was extended so that equations of the form $x^2 = -a$, $a > 0$, have solutions. [11.1]
2. The symbol i names a number with the property
$$i \cdot i = i^2 = -1.$$ [11.1]
Any number of the form bi, $b \in \mathbf{R}$, is called an **imaginary number**
3. The set of numbers of the form $a + bi$, $a, b \in \mathbf{R}$, is called the set of **complex numbers**, which is designated by $\mathbf{C}$. [11.1]
4. The definition of i is used to express the root of a negative number as an imaginary number. If $a > 0$, $\sqrt{-a} = i\sqrt{a}$. [11.1]
5. The two complex numbers $a + bi$ and $c + di$ are equal if and only if $a = c$ and $b = d$. [11.1]
6. If $a + bi$ and $c + di$ are any two complex numbers, then
$$(a + bi) + (c + di) = (a + c) + (b + d)i$$
and
$$(a + bi) - (c + di) = (a - c) + (b - d)i.$$ [11.2]
7. The product of two complex numbers is
$$(a + bi)(c + di) = (ac - bd) + (ad + bc)i.$$ [11.3]
8. If a and b are not both zero, then
$$(a + bi)^{-1} = \frac{a}{a^2 + b^2} - \frac{bi}{a^2 + b^2}.$$ [11.3]
9. The quotient of two complex numbers can be simplified by multiplying the dividend by the multiplicative inverse of the divisor.
$$(a + bi) \div (c + di) = (a + bi)(c + di)^{-1}$$ [11.4]
10. The numbers $a + bi$ and $a - bi$ are called **complex conjugates** and have the property
$$(a + bi)(a - bi) = a^2 + b^2.$$ [11.4]
11. If the quotient of two complex numbers is expressed as a fraction, it can be simplified by multiplying both the numerator and the denominator by the conjugate of the denominator. [11.4]
12. If $a, b, c \in \mathbf{R}$, and if one solution of the equation $ax^2 + bx + c = 0$ is a complex number, then the conjugate of the complex number is also a solution. [11.5]
13. The sum of the two solutions of $ax^2 + bx + c = 0$ is $\frac{-b}{a}$, and the product of the two solutions is $\frac{c}{a}$. [11.5]

294 COMPLEX NUMBERS

REVIEW EXERCISES

SECTION 11.1

Write each expression in Problems 1 and 2 in the form $a + bi$.

1. $3i^5 - 2i^4 + 5i^3 + 3i^2 + 4i + 8$
2. $2i^7 - 3i^6 + 2i^5 + 3i^4 + 4i^3 - i^2 + 4i - 5$

In Problems 3–6, write each expression in simplest radical form with no negative radicands.

3. $\sqrt{-27}$ 4. $\sqrt{-48}$ 5. $\sqrt{-2}\sqrt{-8}$ 6. $\sqrt{-3}\sqrt{-12}$

Solve for x and y in Problems 7 and 8.

7. $3x + 8i = 9 - yi$ 8. $x + 2y + 4i = 5 + 2xi + yi$

SECTION 11.2

In Problems 9–12, write each sum or difference in the form $a + bi$.

9. $(2 + 3i) + (5 - 7i)$ 10. $(5 - 3i) + (3 + 4i)$
11. $(3 - 7i) - (4 - 5i)$ 12. $(6 + 2i) - (4 + i)$

SECTION 11.3

In Problems 13–16, write each product in the form $a + bi$.

13. $(2 + 3i)(5 - 7i)$ 14. $(\sqrt{3} + i\sqrt{5})(\sqrt{2} - i\sqrt{3})$
15. $(6 + 2i)(6 - 2i)$ 16. $(\sqrt{2} - \sqrt{-3})(\sqrt{2} + \sqrt{-3})$

Find the multiplicative inverse of each complex number in Problems 17 and 18.

17. $-2 + 3i$ 18. $\sqrt{5} - \sqrt{-3}$

SECTION 11.4

19. Simplify the quotient $(3 + 4i) \div (5 - 2i)$ by solving $(x + yi)(5 - 2i) = 3 + 4i$.
20. Simplify the quotient $(2 - 4i) \div (1 + 3i)$ by using the multiplicative inverse of $1 + 3i$.
21. Simplify the quotient $\dfrac{2 - \sqrt{-3}}{4 + \sqrt{-3}}$ by using the conjugate of $4 + \sqrt{-3}$.

SECTION 11.5

Find the solution set of each equation in Problems 22–25.

22. $x^2 - 2x + 2 = 0$ 23. $x^2 - 6x + 13$
24. $x^2 - 3x + 9 = 0$ 25. $2x^2 - x + 2 = 0$

In Problems 26–29, find the sum and the product of the solutions without solving the equations.

26. $x^2 + 4x + 8 = 0$ 27. $x^2 - 9x + 13 = 0$
28. $3x^2 + 6x - 10 = 0$ 29. $2x^2 - 5x + 7 = 0$

Vectors

Many physical quantities, such as length, volume, mass, and temperature, can be identified by use of a single real number or by a point on a linear scale. Such quantities are called **scalar quantities**. On the other hand, there are other physical quantities, such as force, displacement, velocity, and acceleration, that cannot be represented by a single real number because they involve both magnitude and direction. Quantities that must be represented in terms of both a magnitude and a direction are called **vector quantities** or, more simply, **vectors**.

In arithmetic, you learned to perform operations with numbers without thinking in terms of specific quantities. In this chapter, we discuss operations with vectors without connecting them with any specific physical quantity.

12.1 Vectors in the Plane

Since a vector involves both magnitude and direction, we can interpret a vector as a **displacement**, or **translation**, of a point in the plane from one position to a new position. A position on the coordinate plane is denoted by an ordered pair of numbers. Hence, the translation of a point generally involves changing both of the coordinates. Thus, there is one displacement parallel to the x axis and another parallel to the y axis. This means that we can represent a vector by an ordered pair of numbers in which the first component represents the displacement parallel to the x axis and the second component represents the displacement parallel to the y axis.

A vector can also be represented by a line segment connecting the original position of the point with the new position. In this context, the original position is called the **initial point** of the vector, and the new position is called the **terminal point**. An arrowhead is placed at the terminal point to indicate the direction of the displacement. This representation of a vector, as illustrated in Figure 12.1-1,

296 VECTORS

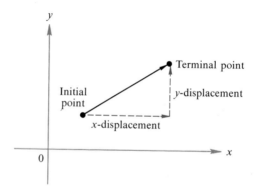

Figure 12.1-1

is called a **geometric vector**. As will be seen shortly, the geometric representation of a vector is not unique.

Symbolic Representation

When a vector is represented as an ordered pair, we use brackets [ ] to enclose the components to distinguish a vector from a point. Thus, [3, 4] names a vector that represents a displacement from an initial point to a terminal point 3 units to the right and 4 units above the initial point. Figure 12.1-2 shows several geometric vectors all corresponding to [3, 4].

For a general discussion of vectors, it is necessary to name vectors without referring to a specific vector. This is done on the printed page by using lowercase letters in boldface type, such as **v**, **u**, **w**. However, for hand-written work, a vector is named by placing an arrow over the letter, such as $\vec{v}$, $\vec{u}$, $\vec{w}$. The arrow is also used to help name a vector from one point to another. If A and B are two points, then $\overrightarrow{AB}$ names the vector from A to B. ($\overrightarrow{BA}$ or $\overleftarrow{AB}$ names the vector from B to A.)

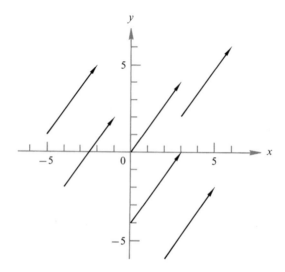

Figure 12.1-2

Equality of Vectors

Figure 12.1-2 implies an equality of vectors. The equals relation in the set of vectors in the plane is stated as follows.

DEFINITION 12.2-1 If $[a, b]$ and $[c, d]$ are two vectors in the plane, then $[a, b] = [c, d]$ if and only if $a = c$ and $b = d$.

Since geometric vectors may occupy different positions in the coordinate plane, we usually refer to two geometric vectors as **equivalent** if they correspond to the same vector. In many applications, a geometric vector can be replaced by an equivalent geometric vector.

Magnitude and Direction

A vector has both magnitude and direction. Since we have considered a vector to be a displacement from one point to another, we can use the distance formula to find the magnitude, or **norm**, of a vector.

DEFINITION 12.1-2 The norm of the vector $\mathbf{v} = [a, b]$ is
$$\|\mathbf{v}\| = \sqrt{a^2 + b^2}.$$

Example Find the norm of $\mathbf{v} = [3, 4]$.

Solution $\|\mathbf{v}\| = \sqrt{3^2 + 4^2} = \sqrt{9 + 16} = \sqrt{25} = 5$

Example Find the norm of $\overrightarrow{AB}$ where $A = (1, 3)$ and $B = (6, 15)$.

Solution $\overrightarrow{AB} = [6 - 1, 15 - 3] = [5, 12]$, then
$$\|\overrightarrow{AB}\| = \sqrt{5^2 + 12^2} = \sqrt{25 + 144} = \sqrt{169} = 13.$$

The **direction** of a vector can be specified as the same direction as that of the geometric vector from the origin to the point whose coordinates are the same as the respective components of the vector. Thus, the vector $\overrightarrow{AB} = [5, 12]$, from $(1, 3)$ to $(6, 15)$, has the same direction as the vector from $(0, 0)$ to $(5, 12)$. Two geometric vectors that have the same (or opposite) direction are parallel. No direction is specified for the zero vector $[0, 0]$.

Scalar Multiplication

As indicated in the opening paragraph of this chapter, length is a scalar quantity. You have seen that the norm of a vector is the length of a corresponding geometric vector. Now consider a vector $[x, y]$ having the same direction as $[3, 4]$ and a norm twice as great.

If we represent $[3, 4]$ and $[x, y]$ as shown in Figure 12.1-3 and drop perpen-

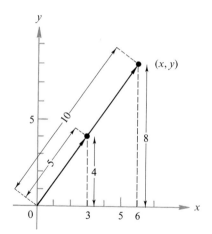

Figure 12.1-3

diculars from $(3, 4)$ and (x, y) to the x-axis, by similar triangles we find $x = 6$ and $y = 8$. This indicates that if the norm of a vector is multiplied by some scalar, the components of the resulting vector are the respective components of the original vector multiplied by the same scalar. This leads us to state

DEFINITION 12.1-3 If $\mathbf{v} = [a, b]$ and $c \in \mathbf{R}$, then

$$c\mathbf{v} = c[a, b] = [ca, cb],$$

and

$$\|c\mathbf{v}\| = |c|\,\|\mathbf{v}\|.$$

The vector $c\mathbf{v}$ is called a **scalar multiple** of $\mathbf{v}$. If $c > 0$, then $c\mathbf{v}$ has the same direction as $\mathbf{v}$. If $c < 0$, then the direction of $c\mathbf{v}$ is opposite to $\mathbf{v}$. See Figure 12.1-4.
If $c = -1$ and $\mathbf{v} = [a, b]$, then $-1(\mathbf{v})$, or $-\mathbf{v}$, equals $-1[a, b] = [-a, -b]$. $\mathbf{v}$ and $-\mathbf{v}$ have the same norm but opposite directions.

Unit Vectors

A **unit vector** is a vector whose norm is 1.

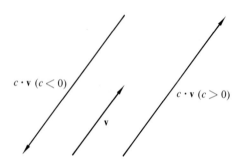

Figure 12.1-4

DEFINITION 12.1-4 If **v** is any nonzero vector, then a unit vector in the same direction as **v** is $\dfrac{\mathbf{v}}{\|\mathbf{v}\|}$.

Example Find a unit vector in the direction of $\mathbf{v} = [3, 4]$.

Solution
$$\frac{\mathbf{v}}{\|\mathbf{v}\|} = \frac{[3,4]}{\sqrt{3^2 + 4^2}} = \frac{[3,4]}{5} = \frac{1}{5} \cdot [3,4] = \left[\frac{3}{5}, \frac{4}{5}\right].$$

$$\text{Check: } \left\|\left[\frac{3}{5}, \frac{4}{5}\right]\right\| = \sqrt{\left(\frac{3}{5}\right)^2 + \left(\frac{4}{5}\right)^2} = \sqrt{\frac{25}{25}} = 1.$$

It follows from Definition 12.1-4 that $\dfrac{-\mathbf{v}}{\|\mathbf{v}\|}$ is a unit vector in the direction opposite to **v**.

Vector as a Sum

The vector [1, 0] is a unit vector and corresponds to a geometric vector of length 1 along the positive x-axis. This vector is frequently represented by **i**. Then, if a is a scalar, $a[1, 0] = [a, 0]$, and it follows that any vector of the form $[a, 0]$ can be represented by $a\mathbf{i}$. This symbol also represents a displacement of a units along the x-axis.

In a similar manner, the vector [0, 1] corresponds to a unit vector, called **j**, along the positive y-axis, and any vector of the form [0, b], or displacement along the y-axis, can be represented by $b\mathbf{j}$. These vectors are shown in Figure 12.1-5.

Since any vector represents a displacement parallel to the x-axis and a displacement parallel to the y-axis,

$$\mathbf{v} = [a, b] = a\mathbf{i} + b\mathbf{j},$$

where $a\mathbf{i}$ is called the **horizontal component** and $b\mathbf{j}$ is called the **vertical component** of **v**, as shown in Figure 12.1-6.

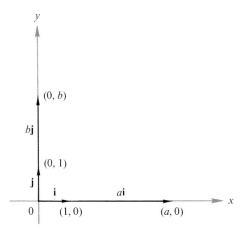

Figure 12.1-5

300 VECTORS

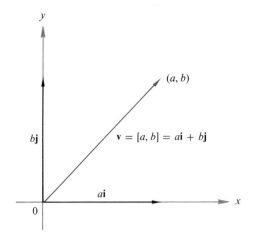

Figure 12.1-6

EXERCISE 12.1

ORAL

Define or explain each of the following:

1. Vector
2. Geometric vector
3. Equal vectors
4. Equivalent geometric vectors
5. Norm of a vector
6. Scalar multiple
7. Unit vector
8. Horizontal component
9. Vertical component

WRITTEN

Find the norm of each vector **v** in Problems 1–8.

Example $\mathbf{v} = [2, -3]$
Solution $\|\mathbf{v}\| = \sqrt{(2)^2 + (-3)^2} = \sqrt{4 + 9} = \sqrt{13}$

1. $\mathbf{v} = [1, 2]$
2. $\mathbf{v} = [3, 1]$
3. $\mathbf{v} = [\sqrt{2}, -\sqrt{7}]$
4. $\mathbf{v} = [-\sqrt{3}, \sqrt{6}]$
5. $\mathbf{v} = [-4, -3]$
6. $\mathbf{v} = [-5, -12]$
7. $\mathbf{v} = [3c, 4c]$
8. $\mathbf{v} = [d, -d\sqrt{3}]$

Find the norm of each vector $\overrightarrow{AB}$ where A and B are as given in Problems 9–14.

Example $A(2, -1), B(5, 3)$
Solution $\overrightarrow{AB} = [5 - 2, 3 - (-1)] = [3, 4]$
$\|\overrightarrow{AB}\| = \sqrt{3^2 + 4^2} = \sqrt{25} = 5$

9. $A(2, 4), B(-2, 1)$
10. $A(-3, -7), B(2, 5)$
11. $A(-3, -4), B(3, 4)$
12. $A(4, 0), B(0, 3)$
13. $A(3, 0), B(3, 5)$
14. $A(4, 2), B(7, 2)$

In Problems 15–22, construct geometric vectors with initial point at the origin corresponding to the given vector and scalar multiples.

Example $\mathbf{v} = [1, 2]$; $c_1 = 3$; $c_2 = -2$

Solution See Figure 12.1-7.

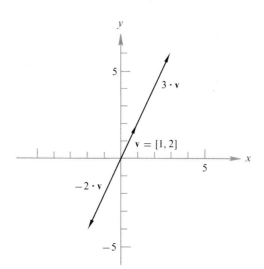

Figure 12.1-7

15. $\mathbf{v} = [3, 1]$; $c_1 = 2$; $c_2 = -2$
16. $\mathbf{v} = [2, 3]$; $c_1 = 4$; $c_2 = -3$
17. $\mathbf{v} = [3, 0]$; $c_1 = 4$; $c_2 = -1$
18. $\mathbf{v} = [0, 2]$; $c_1 = 3$; $c_2 = -2$

19. $\overrightarrow{AB}$ for $A(1, 2)$, $B(3, 5)$; $c_1 = 3$; $c_2 = -3$
20. $\overrightarrow{AB}$ for $A(3, 3)$, $B(1, 2)$; $c_1 = 2$; $c_2 = -\frac{1}{2}$
21. $\overrightarrow{AB}$ for $A(5, 7)$, $B(0, 5)$; $c_1 = \frac{3}{2}$; $c_2 = -\frac{3}{2}$
22. $\overrightarrow{AB}$ for $A(6, 4)$, $B(3, 1)$; $c_1 = \frac{5}{3}$; $c_2 = -\frac{2}{3}$

In Problems 23–32, find a unit vector in the direction of the given vector.

Example $\mathbf{v} = [2, 3]$

Solution $\|\mathbf{v}\| = \sqrt{(2)^2 + (3)^2} = \sqrt{13}$. Hence, the required unit vector is $[2/\sqrt{13}, 3/\sqrt{13}]$.

23. $\mathbf{v} = [3, 1]$
24. $\mathbf{v} = [2, 1]$
25. $\mathbf{v} = [-2, 3]$
26. $\mathbf{v} = [4, -1]$
27. $\mathbf{v} = [-\sqrt{2}, \sqrt{2}]$
28. $\mathbf{v} = [\sqrt{2}, -\sqrt{7}]$
29. $\mathbf{v} = [-3, -4]$
30. $\mathbf{v} = [-1, -1]$
31. $\overrightarrow{AB}$ for $A(3, 2)$, $B(6, 2)$
32. $\overrightarrow{AB}$ for $A(4, 5)$, $B(4, 7)$

12.2 Operations with Vectors

Before we can solve problems involving vector quantities, it is necessary for us to define some operations with vectors.

Sums

The use of the number line in explaining addition of real numbers suggests a technique for defining the sum of two vectors. For example, in Figure 12.2-1 the sum $3 + 2$ is interpreted as moving three units from the origin and then an additional two units. This movement along the number line is a displacement, and if we consider the number line to be the x-axis, the number 3 corresponds to the vector $[3, 0]$ and the number 2 corresponds to the vector $[2, 0]$. Thus, the sum $3 + 2$ corresponds to the vector $[3 + 2, 0] = [5, 0]$.

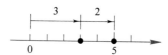

Figure 12.2-1

Now consider two vectors $[2, 3]$ and $[5, 2]$. The sum $[2, 3] + [5, 2]$ can be interpreted as a displacement from $(0, 0)$ to $(2, 3)$ and then additional displacements of 5 units horizontally and 2 units vertically. If we construct geometric vectors to represent these displacements, as in Figure 12.2-2, the terminal point of the first geometric vector is the initial point of the second. By inspection, we see that the terminal point of the second geometric vector is $(7, 5)$.

This concept of one displacement followed by additional horizontal and vertical displacements suggests the following definition.

DEFINITION 12.2-1   If $\mathbf{v} = [a, b]$ and $\mathbf{u} = [c, d]$, then

$$\mathbf{v} + \mathbf{u} = [a, b] + [c, d] = [a + c, b + d].$$

Thus, the sum of two vectors can be found by adding the corresponding components. If the $\mathbf{i}, \mathbf{j}$ notation is used, we have

If $\mathbf{v} = a\mathbf{i} + b\mathbf{j}$ and $\mathbf{u} = c\mathbf{i} + d\mathbf{j}$, then

$$\mathbf{v} + \mathbf{u} = (a\mathbf{i} + b\mathbf{j}) + (c\mathbf{i} + d\mathbf{j}) = (a + c)\mathbf{i} + (b + d)\mathbf{j}.$$

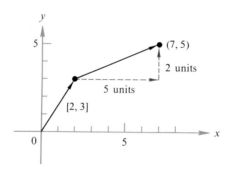

Figure 12.2-2

Examples (a) $[3, -2] + [2, 5] = [3 + 2, -2 + 5]$
$= [5, 3]$

(b) $(2\mathbf{i} - 4\mathbf{j}) + (3\mathbf{i} + 2\mathbf{j}) = (2 + 3)\mathbf{i} + (-4 + 2)\mathbf{j}$
$= 5\mathbf{i} - 2\mathbf{j}$

Differences

DEFINITION 12.2-2 The difference of two vectors **v** and **u** is

$$\mathbf{v} - \mathbf{u} = \mathbf{v} + (-\mathbf{u}).$$

Examples (a) $[5, 4] - [2, 3] = [5, 4] + [-2, -3]$
$= [5 - 2, 4 - 3]$
$= [3, 1]$

(b) $(3\mathbf{i} + 4\mathbf{j}) - (\mathbf{i} + 2\mathbf{j}) = (3\mathbf{i} + 4\mathbf{j}) + (-\mathbf{i} - 2\mathbf{j})$
$= (3 - 1)\mathbf{i} + (4 - 2)\mathbf{j}$
$= 2\mathbf{i} + 2\mathbf{j}$

EXERCISE 12.2

WRITTEN

Write each vector sum in Problems 1–12 as a vector, and construct a geometric vector as shown in the examples below.

Examples (a) $[2, -4] + [4, 1]$ (b) $(3\mathbf{i} + 2\mathbf{j}) + (\mathbf{i} - 3\mathbf{j})$
Solutions (a) $[2, -4] + [4, 1] = [6, -3]$

(See Figure 12.2-3.)

(b) $(3\mathbf{i} + 2\mathbf{j}) + (\mathbf{i} - 3\mathbf{j}) = (3 + 1)\mathbf{i} + (2 - 3)\mathbf{j}$
$= 4\mathbf{i} - \mathbf{j}$

(See Figure 12.2-4, page 304.)

1. $[3, 5] + [2, 1]$ 2. $[4, -3] + [-1, 2]$
3. $[-2, 5] + [4, -2]$ 4. $[-2, -4] + [2, 7]$

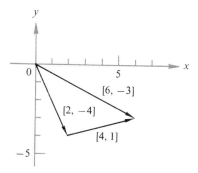

Figure 12.2-3

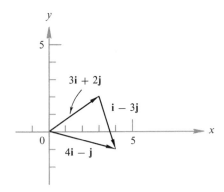

Figure 12.2-4

5. $[-3, -4] + [-2, -2]$
6. $[-5, -3] + [-1, 4]$
7. $(2i + 4j) + (3i + 2j)$
8. $(2i - j) + (i - 3j)$
9. $(-4i + 2j) + (3i - 7j)$
10. $(-4i + 7j) + (6i - 2j)$
11. $(-3i - 6j) + (i - 2j)$
12. $(5i + 7j) + (-2i - 5j)$

Write each vector difference in Problems 13–20 as a vector, and construct geometric vectors as shown in the following examples.

Examples   (a) $[2, -4] - [5, -2]$ (b) $(3i + 7j) - (-2i + 3j)$

Solutions   (a) $[2, -4] - [5, -2] = [2, -4] + [-5, 2]$
$= [-3, -2]$

(See Figure 12.2-5.)

(b) $(3i + 7j) - (-2i + 3j) = (3i + 7j) + (2i - 3j)$
$= 5i + 4j$

(See Figure 12.2-6.)

13. $[5, 6] - [3, 2]$
14. $[2, 4] - [-2, 1]$
15. $[3, -1] - [-3, 2]$
16. $[-2, 1] - [2, 4]$

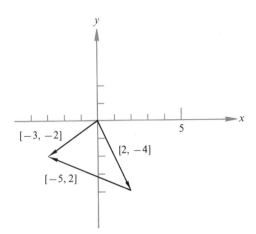

Figure 12.2-5

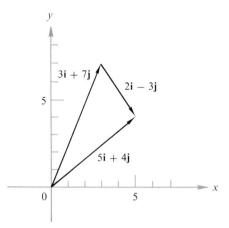

Figure 12.2-6

17. $(3i + 2j) - (5i + 6j)$
18. $(-3i + 2j) - (i - 3j)$
19. $(-4i - 6j) - (3i - 3j)$
20. $(-2i - 3j) - (5i - 7j)$

12.3 The Parallelogram Law—Applications

Consider the vector sum $[5, 2] + [1, 6] = [6, 8]$, which is shown graphically in Figure 12.3-1.

If line segments are drawn connecting the terminal points of the given vectors with that of the vector sum, and the slopes of these line segments are computed, we have

$$\frac{8-6}{6-1} = \frac{2}{5} \quad \text{and} \quad \frac{8-2}{6-5} = \frac{6}{1}.$$

Thus, the line segment connecting the terminal point of $[1, 6]$ with that of $[6, 8]$ has the same slope and is parallel to $[5, 2]$, and the line segment connecting the

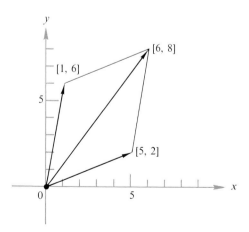

Figure 12.3-1

terminal point of [5, 2] with that of [6, 8] has the same slope and is parallel to [1, 6]. This means that the figure formed by the two given vectors and the line segments is a parallelogram and that the vector sum is a diagonal of this parallelogram.

The Parallelogram Law

This example leads to a theorem known as the **parallelogram law**, which we state without proof.

THEOREM 12.3-1 The vector sum of two noncollinear vectors is the diagonal of the parallelogram with the two given vectors as adjacent sides which has the same initial point as the given vectors.

In many cases, vector quantities are expressed in terms of directions and magnitudes rather than in terms of ordered pairs of real numbers. The direction is usually stated as the measure of an angle called the **direction angle**. The Greek letter θ (theta) is frequently used as a symbol to name a direction angle. The magnitude is often given in physical units of some kind, such as *pounds* for forces and *miles per hour* for wind and water currents.

When vector quantities are represented by geometric vectors, the direction angle is usually measured from the positive x-axis in a counterclockwise direction for positive angles and in a clockwise direction for negative angles. An exception to this procedure arises in problems of navigation involving compass headings. In this latter case, north is 0° and angles are measured in a clockwise direction.

Example Construct geometric vectors with direction angles and magnitudes as specified.

$\theta_1 = 30°, \|v_1\| = 6; \quad \theta_2 = 120°, \|v_2\| = 8; \quad \theta_3 = 250°, \|v_3\| = 5$

Solution See Figure 12.3-2.

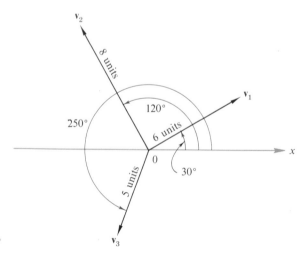

Figure 12.3-2

12.3 The Parallelogram Law—Applications

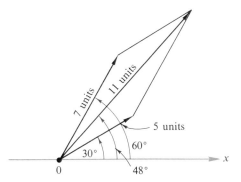

Figure 12.3-3

Sums

The parallelogram law can be used to solve problems involving the sum, or **resultant**, of two vector quantities. An analytical solution of such a problem involves techniques from plane trigonometry that are beyond the scope of this book. However, if the directions of the vector quantities are given as angles whose measures in degrees are integers, approximate solutions can be obtained graphically.

Construct geometric vectors with initial points at the origin and direction angles measured from the positive x-axis. Show magnitudes in suitable linear units. Then complete a parallelogram by drawing lines through the terminal point of each vector parallel to the other vector as shown in Figure 12.3-3. You can then determine the resultant by measuring both the direction angle and the length of the diagonal.

In Figure 12.3-3 we show the resultant of two vectors whose direction angles are 30° and 60°, respectively, and whose magnitudes are 5 and 7 units. Measurement shows the resultant to be a vector whose direction angle is approximately 48° and whose length is approximately 11 units.

Resultant of Forces

One important use of the parallelogram law is to find the resultant of two or more forces that are acting on an object, usually in different directions. Any quantity, such as a force, that has both magnitude and direction is a vector quantity and can be represented as a geometric vector. Vector v_1 in Figure 12.3-2 may be the representation of a force of 6 pounds acting in a direction of 30°.

The following example illustrates the use of the parallelogram law in solving a problem involving forces.

Example   Find the magnitude and direction of the resultant of a force of 5 pounds acting on an object in a direction of 30° and a second force of 8 pounds acting in a direction of 80°.

Solution   Construct the parallelogram that has the two vectors representing the forces as adjacent sides. Draw the diagonal (Figure 12.3-4). The direction and length of the diagonal can then be measured.

         By actual measurement we determine the direction to be approximately 61° and the magnitude to be approximately 12 pounds.

308 VECTORS

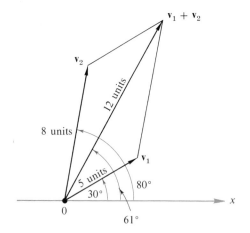

Figure 12.3-4

Differences

Since the difference of two vectors may be stated as

$$\mathbf{v}_1 - \mathbf{v}_2 = \mathbf{v}_1 + (-\mathbf{v}_2),$$

the parallelogram law may be used to find the difference of two geometric vectors by finding the diagonal of the parallelogram whose adjacent sides are the first vector and the negative of the second. The direction angle of the negative of a vector is $\theta + 180°$ if $0° \leq \theta < 180°$ and it is $\theta - 180°$ if $180° \leq \theta \leq 360°$.

Example Find the vector difference $\mathbf{v}_1 - \mathbf{v}_2$ if $\mathbf{v}_1$ has a direction angle of 20° and a magnitude of 5 and $\mathbf{v}_2$ has a direction angle of 110° and a magnitude of 4.

Solution The direction of $-\mathbf{v}_2$ is $110° + 180°$ or 290°. The magnitude is the same as that of $\mathbf{v}_2$. The parallelogram is shown in Figure 12.3-5. Measurement

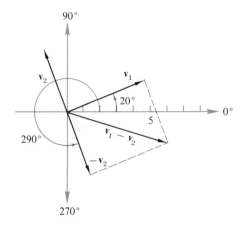

Figure 12.3-5

shows the direction of $v_1 - v_2$ to be approximately 341° and the magnitude to be approximately 6.25 units.

EXERCISE 12.3

WRITTEN

Find the direction angle to the nearest degree and the length to the nearest unit of the sum of each pair of vectors in Problems 1–10 by using the parallelogram law.

Example $\theta_1 = 40°$, $\|v_1\| = 6$; $\theta_2 = 75°$, $\|v_2\| = 4$

Solution   See Figure 12.3-6. The direction angle is 54°, the length is 10.

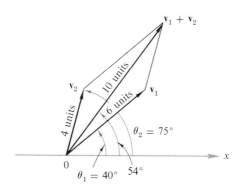

Figure 12.3-6

1. $\theta_1 = 30°$, $\|v_1\| = 5$; $\theta_2 = 60°$, $\|v_2\| = 5$
2. $\theta_1 = 20°$, $\|v_1\| = 6$; $\theta_2 = 70°$, $\|v_2\| = 3$
3. $\theta_1 = 50°$, $\|v_1\| = 8$; $\theta_2 = 110°$, $\|v_2\| = 2$
4. $\theta_1 = 45°$, $\|v_1\| = 7$; $\theta_2 = 135°$, $\|v_2\| = 7$
5. $\theta_1 = 10°$, $\|v_1\| = 3$; $\theta_2 = 260°$, $\|v_2\| = 5$
6. $\theta_1 = 25°$, $\|v_1\| = 2$; $\theta_2 = 245°$, $\|v_2\| = 8$
7. $\theta_1 = 30°$, $\|v_1\| = 5$; $\theta_2 = 330°$, $\|v_2\| = 5$
8. $\theta_1 = 45°$, $\|v_1\| = 4$; $\theta_2 = 345°$, $\|v_2\| = 6$
9. $\theta_1 = 0°$, $\|v_1\| = 12$; $\theta_2 = 90°$, $\|v_2\| = 5$
10. $\theta_1 = 180°$, $\|v_1\| = 6$; $\theta_2 = 270°$, $\|v_2\| = 8$

In Problems 11–18, use the parallelogram law to find the direction angle to the nearest degree and the length to the nearest unit of each difference $v_1 - v_2$.

Example $\theta_1 = 40°$, $\|v_1\| = 6$; $\theta_2 = 75°$, $\|v_2\| = 4$

Solution   Since $0° \leq \theta_2 \leq 180°$, the direction angle of $-v_2$ is $75° + 180° = 255°$. The direction angle of $v_1 - v_2$ as measured in Figure 12.3-7 is 0° to the nearest degree, and the length is 4 to the nearest unit.

11. $\theta_1 = 20°$, $\|v_1\| = 6$; $\theta_2 = 80°$, $\|v_2\| = 5$
12. $\theta_1 = 15°$, $\|v_1\| = 8$; $\theta_2 = 75°$, $\|v_2\| = 7$
13. $\theta_1 = 45°$, $\|v_1\| = 4$; $\theta_2 = 170°$, $\|v_2\| = 3$

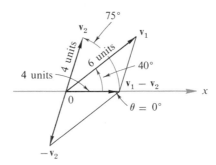

Figure 12.3-7

14. $\theta_1 = 60°$, $\|v_1\| = 3$; $\theta_2 = 150°$, $\|v_2\| = 6$
15. $\theta_1 = 120°$, $\|v_1\| = 5$; $\theta_2 = 200°$, $\|v_2\| = 8$
16. $\theta_1 = 150°$, $\|v_1\| = 7$; $\theta_2 = 210°$, $\|v_2\| = 4$
17. $\theta_1 = 240°$, $\|v_1\| = 2$; $\theta_2 = 330°$, $\|v_2\| = .5$
18. $\theta_1 = 270°$, $\|v_1\| = 12$; $\theta_2 = 180°$, $\|v_2\| = 5$

Use the parallelogram law to find the direction angle to the nearest degree and the length to the nearest unit for the sum of the vectors in Problems 19–26.

Example $\theta_1 = 30°$, $\|v_1\| = 4$; $\theta_2 = 65°$, $\|v_2\| = 6$; $\theta_3 = 110°$, $\|v_3\| = 5$

Solution The associative property of addition of vectors permits us to find the sum of two of the vectors and then the sum of this vector and the third. We shall find the sum of v_1 and v_2 first and then find the sum of this vector and v_3 (Figure 12.3-8). The direction angle is 74° and the length is 13 units.

19. $\theta_1 = 20°$, $\|v_1\| = 6$; $\theta_2 = 80°$, $\|v_2\| = 5$; $\theta_3 = 135°$, $\|v_3\| = 8$
20. $\theta_1 = 15°$, $\|v_1\| = 8$; $\theta_2 = 75°$, $\|v_2\| = 6$; $\theta_3 = 150°$, $\|v_3\| = 8$
21. $\theta_1 = 45°$, $\|v_1\| = 6$; $\theta_2 = 110°$, $\|v_2\| = 6$; $\theta_3 = 200°$, $\|v_3\| = 6$

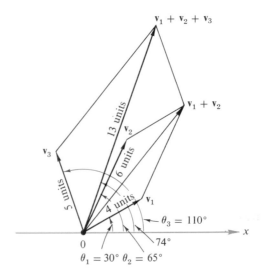

Figure 12.3-8

22. $\theta_1 = 30°$, $\|v_1\| = 5$; $\theta_2 = 100°$, $\|v_2\| = 5$; $\theta_3 = 190°$, $\|v_3\| = 5$
23. $\theta_1 = 135°$, $\|v_1\| = 5$; $\theta_2 = 180°$, $\|v_2\| = 6$; $\theta_3 = 260°$, $\|v_3\| = 7$
24. $\theta_1 = 150°$, $\|v_1\| = 4$; $\theta_2 = 210°$, $\|v_2\| = 4$; $\theta_3 = 270°$, $\|v_3\| = 8$
25. $\theta_1 = 225°$, $\|v_1\| = 8$; $\theta_2 = 300°$, $\|v_2\| = 2$; $\theta_3 = 120°$, $\|v_3\| = 5$
26. $\theta_1 = 260°$, $\|v_1\| = 5$; $\theta_2 = 330°$, $\|v_2\| = 7$; $\theta_3 = 150°$, $\|v_3\| = 4$

27. A corner fence post is pulled eastward ($\theta = 0°$) with a force of 20 pounds and south by a force of 15 pounds. Find the direction angle and magnitude of the resultant force.
28. The weight of an object in pounds is a force acting downward in a vertical direction. If a boy weighing 119 pounds is in a swing and is held to one side so that the rope makes an angle of 30° with the vertical, find, to the nearest pound, the force exerted on the rope and the horizontal force necessary to hold the boy in that position. [HINT: The force due to gravity (weight) is the resultant of the other two forces.]
29. A tent pole has four forces acting upon it, 3 pounds north, 4 pounds east, 2 pounds south, and 6 pounds west. Find the direction of the resultant of these forces and its magnitude to the nearest pound.
30. An object has a force of 5 pounds acting upon it in a direction of 60° and a second force of 5 pounds acting in a direction of 240°. In what direction will the object move?

Summary

1. A **vector** is a displacement of a point from one position to a new position, and as such it can be represented by an ordered pair of real numbers [a, b], where the first component is the displacement parallel to the x-axis and the second component is the displacement parallel to the y-axis. [12.1]
2. A vector can also be represented by a **directed line segment** from the first position to the second position. The directed line segment is called a **geometric vector**. [12.1]
3. If [a, b] and [c, d] are two vectors in the plane, then [a, b] = [c, d] if and only if $a = c$ and $b = d$. [12.1]
4. The norm (magnitude) of the vector $v = [a, b]$ is
$$\|v\| = \sqrt{a^2 + b^2}.$$
[12.1]
5. The direction of a vector is the same as that of the geometric vector from the origin to the point whose coordinates are the same as the components of the vector. [12.1]
6. If $v = [a, b]$ and $c \in R$, then
$$cv = c[a, b] = [ca, cb],$$
and
$$\|cv\| = |c|\,\|v\|.$$
[12.1]
7. A **unit vector** is a vector whose norm is 1. A unit vector with the same direction as v is $v/\|v\|$. [12.1]
8. **i** is the unit vector [1, 0] and **j** is the unit vector [0, 1]. Any vector [a, b] can be expressed as the sum $ai + bj$. [12.1]

9. If $v = [a, b]$ and $u = [c, d]$, then
$$v + u = [a, b] + [c, d] = [a + c, b + d],$$
or
$$(a\mathbf{i} + b\mathbf{j}) + (c\mathbf{i} + d\mathbf{j}) = (a + c)\mathbf{i} + (b + d)\mathbf{j}. \qquad [12.2]$$

10. The **parallelogram law** permits us to find sums and differences of vectors when the vectors are specified by their norms and direction angles. [12.3]
11. The sum of two vectors is also called the **resultant** of the vectors. [12.3]

REVIEW EXERCISES

SECTION 12.1

Find the norm of each vector in Problems 1–4.

1. $[4, 2]$ 2. $[-2, 3]$ 3. $3\mathbf{i} + 4\mathbf{j}$ 4. $-5\mathbf{i} - 2\mathbf{j}$

Find a unit vector with the same direction as the vector in each of Problems 5–8.

5. $[4, -3]$ 6. $[3, 3]$ 7. $-5\mathbf{i} + 12\mathbf{j}$ 8. $2\mathbf{i} - 7\mathbf{j}$

SECTION 12.2

In Problems 9–12, write each expression as a single vector.

9. $[2, 3] + [5, -7]$ 10. $(3\mathbf{i} - 7\mathbf{j}) + (4\mathbf{i} + 5\mathbf{j})$
11. $3[2, -1] + 5[1, 2] - 4[2, -2]$ 12. $4(2\mathbf{i} + \mathbf{j}) - 2(3\mathbf{i} - 2\mathbf{j}) + 6(\mathbf{i} + \mathbf{j})$

SECTION 12.3

Use the parallelogram law in Problems 13–15 to find each sum or difference to the nearest unit.

13. $v_1 + v_2$, where $\|v_1\| = 3$, $\theta_1 = 0°$, $\|v_2\| = 4$, and $\theta_2 = 90°$
14. $v_1 - v_2$, where $\|v_1\| = 8$, $\theta_1 = 30°$, $\|v_2\| = 6$, and $\theta_2 = 300°$
15. $v_1 + v_2 + v_3$, where $\theta_1 = 25°$, $\|v_1\| = 4$; $\theta_2 = 115°$, $\|v_2\| = 3$; and $\theta_3 = 205°$, $\|v_3\| = 12$

Introduction to Matrices and Determinants

The rapid growth of technology has emphasized the need for time-saving techniques for the solution of problems involving large volumes of data or lengthy computations. Matrix algebra is one such technique, particularly useful in work involving electronic computers.

The words *matrix* and its plural, *matrices,* are probably unfamiliar to you, but the notion of matrices has been in existence for over one hundred years. In 1858, the English mathematician Arthur Cayley introduced matrix notation as a means of expressing systems of linear equations in abbreviated form.

In this chapter we discuss some of the properties of matrices and some elementary operations. However, the application of matrices to systems of equations and transformations of vector spaces involve concepts beyond the level of this book.

13.1 Basic Definitions

There are many instances in which quantities of data are given in the form of a table. For example, consider the weekly production data of an assembly crew at a television factory.

Name	Units assembled				
	M	T	W	T	F
Archer	7	10	8	11	5
Evans	9	8	12	6	6
Jones	4	0	5	7	10
Smith	8	7	9	10	7

An arrangement such as this is called a **rectangular array**, and each individual number is called an **entry**. In an array the horizontal lines of entries are called **rows**, and the vertical lines of entries are called **columns**.

If the names of the assemblers in the table above are kept in the same order, and the days of the week are assumed in natural order, there should be no loss of clarity when the production data are presented in the form

$$\begin{bmatrix} 7 & 10 & 8 & 11 & 5 \\ 9 & 8 & 12 & 6 & 6 \\ 4 & 0 & 5 & 7 & 10 \\ 8 & 7 & 9 & 10 & 7 \end{bmatrix}.$$

This form is called a **matrix** of numbers.

Dimension or Order

We note that the matrix just displayed has 4 rows and 5 columns. The number of rows and the number of columns of a matrix are called its **dimension** or **order**. In reading the dimension or order of a matrix, the number of rows is given first. Thus, the order of the matrix above is 4×5 (read "four by five"). More generally, the number of rows is designated by m and the number of columns by n.

Entries and Addresses

Letters of the alphabet are frequently used to name the entries in a matrix, in combination with a system of double subscripts. For example, the symbol a_{23} names the entry in the *second row* and the *third column*. The symbol a_{ij} is used to represent the entry in the ith row and the jth column. The double subscript is called the **address** of the entry, because it determines the location of the entry. Note that the row is always specified first.

Examples Given the matrix

$$\begin{bmatrix} 2 & -1 & 5 & 0 \\ \sqrt{3} & \tfrac{1}{2} & -7 & 4 \\ \tfrac{3}{4} & \sqrt{5} & 3 & -2 \\ 8 & 1 & 0 & -3 \end{bmatrix},$$

(a) give the addresses of the entries $\sqrt{3}$, -2, 1, and 5;

(b) name the entries whose addresses are 14, 23, 42, and 31.

Solutions (a) 21; 34; 42; 13   (b) 0; -7; 1; $\tfrac{3}{4}$

Matrix Notation

There are several different forms of matrix notation. Some authors prefer to enclose the array within parentheses, and others use a pair of double vertical lines.

We will represent an $m \times n$ matrix by one of the following symbols,

$$A_{m \times n} = \begin{bmatrix} a_{11} & a_{12} & \cdots & a_{1n} \\ a_{21} & a_{22} & \cdots & a_{2n} \\ \cdot & \cdot & & \cdot \\ a_{m1} & a_{m2} & \cdots & a_{mn} \end{bmatrix} = [a_{ij}]_{m \times n},$$

and use a capital letter for a matrix whose order is not specified.

A matrix may contain any number of rows and columns. One that consists of a single row of entries ($1 \times n$) is called a **row matrix**. If all of the entries are in a single column ($m \times 1$), the matrix is called a **column matrix**. Any matrix that has the same number of rows as it has columns is called a **square matrix**. In a square matrix, the entries whose first and second subscripts are equal ($i = j$) form the **principal diagonal**. Only square matrices have principal diagonals.

Example Name the entries on the principal diagonal of

$$\begin{bmatrix} 4 & 3 & -1 & 7 \\ 2 & \sqrt{2} & -5 & 0 \\ 6 & 1 & 0 & -4 \\ 3 & 2 & -5 & -2 \end{bmatrix}.$$

Solution $4, \sqrt{2}, 0, -2$.

The entries in a matrix may be quantitative or nonquantitative. However, in this discussion we shall consider only matrices whose entries are real numbers.

Equal Matrices

We define the equals relation for matrices as follows.

DEFINITION 13.1-1 Two matrices A and B are said to be equal if and only if they are of the same order and all their corresponding entries are equal; that is, $a_{ij} = b_{ij}$ for all i and j.

Examples (a)
$$A = \begin{bmatrix} 2 & 1 & 0 \\ 3 & -1 & 6 \\ -2 & 2 & 2 \end{bmatrix}; \quad B = \begin{bmatrix} 2 & 1 & 0 \\ 3 & -1 & 6 \\ -2 & 2 & 2 \end{bmatrix}; \quad A = B.$$

(b) $A = \begin{bmatrix} x & 3 \\ 4 & y \end{bmatrix}; \quad B = \begin{bmatrix} 1 & 3 \\ 4 & 2 \end{bmatrix}; \quad A = B$ if and only if $x = 1$ and $y = 2$.

(c) $A = \begin{bmatrix} 1 & 2 & 3 \\ -2 & 0 & 1 \end{bmatrix}$; $B = \begin{bmatrix} 1 & -2 \\ 2 & 0 \\ 3 & 1 \end{bmatrix}$; $A \neq B$, because they are not of the same order.

EXERCISE 13.1

ORAL

Give the meaning of each of the following.

1. Dimension of a matrix
2. Order
3. Square matrix
4. Principal diagonal
5. $m \times n$
6. a_{ij}

In Problems 7–14, state the address of the element of the matrix

$$\begin{bmatrix} 3 & 2 & -1 \\ 5 & 0 & 4 \\ -2 & \sqrt{2} & \frac{2}{3} \end{bmatrix}.$$

7. -1
8. $\sqrt{2}$
9. 5
10. $\frac{2}{3}$
11. 3
12. -2
13. 0
14. 4

In Problems 15–20, name the entry of the matrix

$$\begin{bmatrix} \sqrt{3} & -1 & 0 \\ 4 & 2 & -3 \\ 5 & 7 & 8 \end{bmatrix}$$

whose address is given.

15. 23
16. 33
17. 21
18. 11
19. 31
20. 13

WRITTEN

In Problems 1–16, find the replacement sets for the variables such that the matrices in each pair are equal.

Examples (a) $\begin{bmatrix} x & 3 \\ 1 & -7 \end{bmatrix} = \begin{bmatrix} 2 & 3 \\ -1 & -7 \end{bmatrix}$ (b) $\begin{bmatrix} x+y & 2 \\ 4 & x-y \end{bmatrix} = \begin{bmatrix} 5 & 2 \\ 4 & 1 \end{bmatrix}$

Solutions (a) $\{2\}$ (b) $x + y = 5$
$x - y = 1$
$2x = 6$
$x = 3$
$\{(3, 2)\}$

1. $\begin{bmatrix} 3 & 5 \\ x & -1 \end{bmatrix} = \begin{bmatrix} 3 & 5 \\ 4 & -1 \end{bmatrix}$

2. $\begin{bmatrix} x & -3 \\ 6 & -2 \end{bmatrix} = \begin{bmatrix} 5 & -3 \\ 6 & -2 \end{bmatrix}$

3. $\begin{bmatrix} 4 & y+1 \\ 1 & -1 \end{bmatrix} = \begin{bmatrix} 4 & 2 \\ 1 & -1 \end{bmatrix}$

4. $\begin{bmatrix} x-2 & 2 \\ 3 & -1 \end{bmatrix} = \begin{bmatrix} 4 & 2 \\ 3 & -1 \end{bmatrix}$

5. $\begin{bmatrix} x & 3 \\ 2 & y \\ 4 & 0 \end{bmatrix} = \begin{bmatrix} 3 & 3 \\ 2 & 6 \\ 4 & 0 \end{bmatrix}$

6. $\begin{bmatrix} -1 & 1 \\ x & y \\ 4 & 0 \end{bmatrix} = \begin{bmatrix} -1 & 1 \\ 3 & 4 \\ 4 & 0 \end{bmatrix}$

7. $\begin{bmatrix} x+1 & 4 \\ 2 & 3 \end{bmatrix} = \begin{bmatrix} 5 & 4 \\ 2 & y-1 \end{bmatrix}$

8. $\begin{bmatrix} -3 & x-1 \\ 4 & -2 \end{bmatrix} = \begin{bmatrix} -3 & 4 \\ 5 & -2 \end{bmatrix}$

9. $\begin{bmatrix} x & 2 & y \\ 3 & z & 0 \end{bmatrix} = \begin{bmatrix} -1 & 2 & -3 \\ 3 & 2 & 0 \end{bmatrix}$

10. $\begin{bmatrix} 3 & x & -4 \\ y & 2 & z \end{bmatrix} = \begin{bmatrix} 3 & 5 & -4 \\ -2 & 2 & 3 \end{bmatrix}$

11. $\begin{bmatrix} x+2 & 4 & -1 \\ 2 & y-1 & 7 \\ 5 & 0 & z+1 \end{bmatrix} = \begin{bmatrix} 3 & 4 & -1 \\ 2 & 3 & 7 \\ 5 & 0 & 3 \end{bmatrix}$

12. $\begin{bmatrix} x-2 & 2 & 3 \\ y+3 & 4 & 5 \\ z+1 & 6 & 7 \end{bmatrix} = \begin{bmatrix} 1 & 2 & 3 \\ 1 & 4 & 5 \\ 1 & 6 & 7 \end{bmatrix}$

13. $\begin{bmatrix} x+y & 2 \\ 3 & x-y \end{bmatrix} = \begin{bmatrix} 6 & 2 \\ 3 & 2 \end{bmatrix}$

14. $\begin{bmatrix} 2x+y & 5 \\ x-2y & 7 \end{bmatrix} = \begin{bmatrix} 5 & 5 \\ 0 & 7 \end{bmatrix}$

15. $\begin{bmatrix} 3x+2y & 4 \\ 2x-3y & 0 \end{bmatrix} = \begin{bmatrix} 8 & 4 \\ 1 & 0 \end{bmatrix}$

16. $\begin{bmatrix} 4x+3y & -5 \\ 2x-4y & 2 \end{bmatrix} = \begin{bmatrix} 1 & -5 \\ 6 & 2 \end{bmatrix}$

13.2 Matrix Addition

Two matrices can be added if and only if they are of the same order. If they are of the same order they are said to be **conformable for addition**.

Examples (a) $\begin{bmatrix} 2 & 3 \\ 4 & 5 \end{bmatrix}$ and $\begin{bmatrix} 1 & 2 \\ 4 & 2 \end{bmatrix}$ are conformable for addition.

(b) $\begin{bmatrix} 1 & 2 & 5 \\ 3 & 4 & 4 \end{bmatrix}$ and $\begin{bmatrix} 2 & 1 \\ 3 & -1 \\ 4 & 6 \end{bmatrix}$ are not conformable for addition.

Sums

Matrix addition is defined as follows.

DEFINITION 13.2-1 The sum of two matrices of the same order is a matrix whose entries are each the sum of the corresponding entries of the two given matrices.

Example

$$\begin{bmatrix} 1 & 2 \\ -2 & 4 \end{bmatrix} + \begin{bmatrix} 2 & -3 \\ 4 & 5 \end{bmatrix} = \begin{bmatrix} 1+2 & 2-3 \\ -2+4 & 4+5 \end{bmatrix} = \begin{bmatrix} 3 & -1 \\ 2 & 9 \end{bmatrix}$$

Negative of a Matrix

DEFINITION 13.2-2 The negative of a matrix is a matrix whose entries are each the negative of the corresponding entry of the given matrix.

Example If $A = \begin{bmatrix} 2 & 4 \\ 3 & -1 \end{bmatrix}$, then $-A = \begin{bmatrix} -2 & -4 \\ -3 & 1 \end{bmatrix}$.

Differences

If two matrices are conformable for addition, the difference of the two matrices can be found by adding the first and the negative of the second.

Example $\begin{bmatrix} 3 & -2 \\ 1 & 4 \end{bmatrix} - \begin{bmatrix} 2 & -1 \\ 3 & 2 \end{bmatrix}$

Solution $\begin{bmatrix} 3 & -2 \\ 1 & 4 \end{bmatrix} - \begin{bmatrix} 2 & -1 \\ 3 & 2 \end{bmatrix} = \begin{bmatrix} 3 & -2 \\ 1 & 4 \end{bmatrix} + \begin{bmatrix} -2 & 1 \\ -3 & -2 \end{bmatrix} = \begin{bmatrix} 1 & -1 \\ -2 & 2 \end{bmatrix}$

EXERCISE 13.2

WRITTEN

Write each sum or difference in Problems 1–14 as a matrix if possible.

Examples (a) $\begin{bmatrix} 4 & 1 \\ 3 & -2 \end{bmatrix} + \begin{bmatrix} 2 & -7 \\ 6 & 4 \end{bmatrix}$ (b) $\begin{bmatrix} 3 & 2 & 0 \\ 4 & 1 & 0 \end{bmatrix} + \begin{bmatrix} 6 & -3 \\ 5 & 7 \end{bmatrix}$

Solutions (a) $\begin{bmatrix} 4 & 1 \\ 3 & -2 \end{bmatrix} + \begin{bmatrix} 2 & -7 \\ 6 & 4 \end{bmatrix} = \begin{bmatrix} 6 & -6 \\ 9 & 2 \end{bmatrix}$

(b) Not possible, because the matrices are not conformable for addition.

1. $\begin{bmatrix} 7 & 3 \\ -1 & 8 \end{bmatrix} + \begin{bmatrix} 5 & 10 \\ 9 & 4 \end{bmatrix}$ 2. $\begin{bmatrix} 3 & -2 \\ 1 & 8 \end{bmatrix} + \begin{bmatrix} -3 & 4 \\ -4 & -6 \end{bmatrix}$

3. $\begin{bmatrix} 2 & 5 \\ -1 & 7 \end{bmatrix} - \begin{bmatrix} 3 & 2 \\ 4 & 6 \end{bmatrix}$ 4. $\begin{bmatrix} 5 & -1 \\ -4 & 3 \end{bmatrix} - \begin{bmatrix} 2 & -1 \\ 4 & 3 \end{bmatrix}$

5. $\begin{bmatrix} 3 & 1 & 0 \\ 5 & 2 & 0 \end{bmatrix} - \begin{bmatrix} 3 & 1 \\ 5 & 2 \end{bmatrix}$

6. $\begin{bmatrix} 4 & 6 & -2 \\ 3 & 5 & -9 \end{bmatrix} + \begin{bmatrix} 2 & -5 & 1 \\ -2 & -4 & 8 \end{bmatrix}$

7. $\begin{bmatrix} 5 & 1 \\ 9 & -3 \\ -6 & 4 \end{bmatrix} + \begin{bmatrix} 2 & -3 \\ -6 & 4 \\ 5 & -2 \end{bmatrix}$

8. $\begin{bmatrix} 3 & -2 \\ 4 & 7 \\ 6 & 1 \end{bmatrix} + \begin{bmatrix} 2 & 4 & 6 \\ 3 & 5 & 7 \end{bmatrix}$

9. $\begin{bmatrix} 6 & -2 & 0 \\ 1 & 0 & 3 \\ 0 & -1 & 5 \end{bmatrix} + \begin{bmatrix} 3 & 0 & 2 \\ 0 & -1 & 0 \\ 5 & 0 & 2 \end{bmatrix}$

10. $\begin{bmatrix} 1 & 2 & -3 \\ -2 & 1 & 2 \\ 3 & -1 & 1 \end{bmatrix} + \begin{bmatrix} 0 & -2 & 3 \\ 2 & 0 & -2 \\ -3 & 1 & 0 \end{bmatrix}$

11. $\begin{bmatrix} 4 & -2 & 1 \\ 3 & 5 & 7 \\ 1 & 0 & -3 \end{bmatrix} - \begin{bmatrix} 2 & -2 & 3 \\ 4 & -1 & 6 \\ 1 & -1 & 2 \end{bmatrix}$

12. $\begin{bmatrix} -5 & 2 & -1 \\ 0 & 3 & -4 \\ 6 & 8 & -7 \end{bmatrix} - \begin{bmatrix} 1 & 0 & 3 \\ 5 & -1 & -4 \\ 4 & 5 & -5 \end{bmatrix}$

13. $\begin{bmatrix} 2 & 1 & 0 & 4 \\ 3 & -2 & 2 & -1 \\ 6 & 2 & -5 & 3 \\ 1 & 0 & 2 & 0 \end{bmatrix} + \begin{bmatrix} 4 & 2 & 3 & 0 \\ 0 & 1 & -1 & 2 \\ -4 & 1 & 5 & -3 \\ 2 & 1 & 1 & 2 \end{bmatrix}$

14. $\begin{bmatrix} -6 & 2 & 0 & 3 \\ 4 & -1 & 5 & -3 \\ 7 & 5 & 1 & 0 \\ 0 & 2 & 0 & 5 \end{bmatrix} - \begin{bmatrix} -5 & 2 & 3 & 3 \\ -2 & 3 & 4 & -3 \\ 6 & -1 & -4 & 4 \\ -2 & 2 & -3 & -1 \end{bmatrix}$

In Problems 15–20, find a matrix X such that each statement is true.

15. $X - \begin{bmatrix} 2 & 4 \\ 3 & 1 \end{bmatrix} = \begin{bmatrix} 5 & 1 \\ 4 & 0 \end{bmatrix}$

16. $X + \begin{bmatrix} 6 & 4 \\ 1 & 2 \end{bmatrix} = \begin{bmatrix} 3 & 1 \\ -2 & 4 \end{bmatrix}$

17. $\begin{bmatrix} 4 & 6 \\ -5 & 1 \end{bmatrix} + X = \begin{bmatrix} 3 & 7 \\ 4 & 5 \end{bmatrix}$

18. $\begin{bmatrix} 5 & 9 \\ 8 & -7 \end{bmatrix} - X = \begin{bmatrix} 6 & 8 \\ 6 & 1 \end{bmatrix}$

19. $X - \begin{bmatrix} 4 & 0 & 2 & 7 \\ 3 & -1 & 5 & 4 \\ 2 & 6 & 0 & 3 \end{bmatrix} = \begin{bmatrix} 1 & 1 & 1 & 1 \\ 2 & 2 & 2 & 2 \\ 3 & 3 & 3 & 3 \end{bmatrix}$

20. $\begin{bmatrix} 1 & 2 & 3 & 4 \\ 0 & -1 & -2 & -3 \\ 2 & -4 & 6 & 1 \\ 3 & -5 & 7 & 2 \end{bmatrix} - X = \begin{bmatrix} 1 & 0 & 1 & 0 \\ 0 & 1 & 0 & 1 \\ 1 & 0 & 1 & 0 \\ 0 & 1 & 0 & 1 \end{bmatrix}$

Find the missing entries in Problems 21 and 22.

21. $\begin{bmatrix} 2 & 3 & \\ & 1 & \\ 4 & 7 & \end{bmatrix} + \begin{bmatrix} & 5 & \\ 4 & 8 & \\ & 3 & \end{bmatrix} = \begin{bmatrix} 3 & -4 & -2 \\ 6 & 5 & -1 \\ 3 & 2 & -1 \end{bmatrix}$

22. $\begin{bmatrix} & 2 & 3 \\ 4 & & -2 \\ 6 & 1 & \end{bmatrix} - \begin{bmatrix} 1 & & \\ & 2 & \\ & 3 & \end{bmatrix} = \begin{bmatrix} 5 & 1 & -1 \\ 2 & 3 & 0 \\ 6 & 2 & 3 \end{bmatrix}$

13.3 Matrix Multiplication

Matrix notation can be used as a means of expressing a system of linear equations in abbreviated form. For example, the linear equation $4x + 5y = 3$ can be represented in matrix notation by

$$4x + 5y = 3 \longleftrightarrow [4 \ \ 5] \cdot \begin{bmatrix} x \\ y \end{bmatrix} = [3].$$

If notation such as this is to be useful, it is essential that the product of two matrices be defined. Examination of the statement above indicates that

$$[4 \ \ 5] \cdot \begin{bmatrix} x \\ y \end{bmatrix} = [4x + 5y],$$

and that, in this case, the product of a 1×2 matrix and a 2×1 matrix is a 1×1 matrix. The single entry in the 1×1 matrix is the sum of the products of the corresponding entries of the row of the first matrix and the column of the second matrix.

In a similar manner, the system

$$4x + 5y = 3$$
$$2x - 3y = -1$$

can be represented in matrix notation as

$$\begin{bmatrix} 4 & 5 \\ 2 & -3 \end{bmatrix} \cdot \begin{bmatrix} x \\ y \end{bmatrix} = \begin{bmatrix} 3 \\ -1 \end{bmatrix}.$$

By comparing these two forms, you should note that

$$\begin{bmatrix} 4 & 5 \\ 2 & -3 \end{bmatrix} \cdot \begin{bmatrix} x \\ y \end{bmatrix} = \begin{bmatrix} 4x + 5y \\ 2x - 3y \end{bmatrix},$$

and that the product of a 2×2 matrix and a 2×1 matrix is a 2×1 matrix, such that the first entry is the sum of the products of the entries in the first row of the first matrix and the corresponding entries in the column of the second matrix.

13.3 Matrix Multiplication

If the number of columns in the matrix on the left is equal to the number of rows in the matrix on the right, then the two matrices are said to be **conformable for multiplication** in that order. It should be noted that if the order in which the matrices are written is changed, the matrices may not be conformable for multiplication in the new order.

We extend the concept of matrix multiplication rather arbitrarily by defining the product of two conformable matrices as follows.

DEFINITION 13.3-1 If A is an $m \times n$ matrix and B is an $n \times p$ matrix, the product $C = A \cdot B$ is an $m \times p$ matrix, where each entry c_{ij} of C is obtained by multiplying the corresponding entries of the ith row of A by those of the jth column of B and then adding the results.

Example Write the product
$$\begin{bmatrix} 2 & 1 & 0 \\ 3 & -1 & 4 \end{bmatrix} \cdot \begin{bmatrix} 3 & -2 \\ 1 & -4 \\ -2 & 0 \end{bmatrix}$$
as a matrix.

Solution The first matrix is 2×3 and the second is 3×2, so the product will be a 2×2 matrix.

$$\begin{bmatrix} 2 & 1 & 0 \\ 3 & -1 & 4 \end{bmatrix} \cdot \begin{bmatrix} 3 & -2 \\ 1 & -4 \\ -2 & 0 \end{bmatrix}$$

$$= \begin{bmatrix} 2 \cdot 3 + 1 \cdot 1 + 0 \cdot (-2) & 2(-2) + 1(-4) + 0 \cdot 0 \\ 3 \cdot 3 - 1 \cdot 1 + 4 \cdot (-2) & 3(-2) - 1(-4) + 4 \cdot 0 \end{bmatrix}$$

$$= \begin{bmatrix} 7 & -8 \\ 0 & -2 \end{bmatrix}$$

It was stated in Section 13.2 that for two matrices to be conformable for addition they must have the same dimension, that is, both must be $m \times n$ matrices. Two $m \times n$ matrices are conformable for multiplication only when $m = n$.

EXERCISE 13.3

ORAL

In Problems 1–9, state the order of the matrix represented by the given product. If no such matrix exists, state this as your answer.

Examples (a) $A_{2 \times 4} \cdot B_{4 \times 1}$ (b) $A_{2 \times 3} \cdot B_{2 \times 3}$

Solutions (a) 2×1

(b) Does not exist. The two matrices are not conformable for multiplication.

1. $A_{4\times3} \cdot B_{3\times2}$
2. $A_{2\times5} \cdot B_{5\times6}$
3. $A_{3\times3} \cdot B_{3\times3}$
4. $A_{1\times4} \cdot B_{4\times1}$
5. $A_{4\times2} \cdot B_{3\times4}$
6. $A_{4\times1} \cdot B_{1\times4}$
7. $A_{3\times2} \cdot B_{2\times3}$
8. $A_{1\times5} \cdot B_{2\times5}$
9. $A_{5\times3} \cdot B_{3\times2}$

WRITTEN

Simplify each product in Problems 1–12.

Example
$$\begin{bmatrix} 4 & 2 & 0 \\ 3 & 5 & 1 \\ -1 & -2 & 2 \end{bmatrix} \cdot \begin{bmatrix} 3 & -1 & 2 \\ -4 & -2 & 3 \\ 0 & 1 & 1 \end{bmatrix}$$

Solution
$$\begin{bmatrix} 4\cdot3 + 2(-4) + 0\cdot0 & 4(-1) + 2(-2) + 0\cdot1 & 4\cdot2 + 2\cdot3 + 0\cdot1 \\ 3\cdot3 + 5(-4) + 1\cdot0 & 3(-1) + 5(-2) + 1\cdot1 & 3\cdot2 + 5\cdot3 + 1\cdot1 \\ -1\cdot3 - 2(-4) + 2\cdot0 & -1(-1) - 2(-2) + 2\cdot1 & -1\cdot2 - 2\cdot3 + 2\cdot1 \end{bmatrix}$$

$$= \begin{bmatrix} 4 & -8 & 14 \\ -11 & -12 & 22 \\ 5 & 7 & -6 \end{bmatrix}$$

1. $\begin{bmatrix} 1 & -2 \end{bmatrix} \cdot \begin{bmatrix} 3 \\ 2 \end{bmatrix}$

2. $\begin{bmatrix} 3 & -4 \\ 2 & -1 \end{bmatrix} \cdot \begin{bmatrix} 5 \\ 4 \end{bmatrix}$

3. $\begin{bmatrix} 2 & -3 \\ 4 & 1 \end{bmatrix} \cdot \begin{bmatrix} 5 & 2 \\ 3 & -2 \end{bmatrix}$

4. $\begin{bmatrix} 1 & 2 & 3 \end{bmatrix} \cdot \begin{bmatrix} 1 \\ 2 \\ 3 \end{bmatrix}$

5. $\begin{bmatrix} 2 & 2 & 1 \\ 1 & 1 & 2 \end{bmatrix} \cdot \begin{bmatrix} 1 & 2 \\ 1 & 1 \\ 1 & 2 \end{bmatrix}$

6. $\begin{bmatrix} 3 & 2 & 2 \\ 1 & 2 & 1 \end{bmatrix} \cdot \begin{bmatrix} 4 \\ 0 \\ 1 \end{bmatrix}$

7. $\begin{bmatrix} -1 & -2 & 5 \\ 1 & -1 & 3 \\ -1 & 2 & 4 \end{bmatrix} \cdot \begin{bmatrix} 2 & 2 & -1 \\ 1 & 1 & 2 \\ 1 & 0 & -1 \end{bmatrix}$

8. $\begin{bmatrix} 2 & -3 & 1 \\ 0 & 1 & -1 \\ 2 & 0 & 0 \end{bmatrix} \cdot \begin{bmatrix} 2 & -3 & 0 \\ -1 & 0 & 2 \\ -4 & -1 & 1 \end{bmatrix}$

9. $\begin{bmatrix} 2 & 2 & -1 \\ 1 & 1 & 2 \\ 1 & 0 & -1 \end{bmatrix} \cdot \begin{bmatrix} -1 & -2 & 5 \\ 1 & -1 & 3 \\ -1 & 2 & 4 \end{bmatrix}$

10. $\begin{bmatrix} 2 & -3 & 0 \\ -1 & 0 & 2 \\ -4 & -1 & 1 \end{bmatrix} \cdot \begin{bmatrix} 2 & -3 & 1 \\ 0 & 1 & -1 \\ -2 & 0 & 0 \end{bmatrix}$

11. $\begin{bmatrix} 3 & -1 & 0 & 2 \\ 4 & -5 & 2 & 1 \end{bmatrix} \cdot \begin{bmatrix} 2 & 2 \\ 1 & 0 \\ 0 & -1 \\ -2 & 1 \end{bmatrix}$

12. $\begin{bmatrix} 1 & 0 & 2 & 1 \\ 0 & 3 & 1 & 2 \end{bmatrix} \cdot \begin{bmatrix} 1 & 0 \\ 1 & 2 \\ 2 & 1 \\ 0 & 1 \end{bmatrix}$

13. Compare your answers for Problems 7 and 9. Is the product of these two matrices commutative?
14. Compare your answers for Problems 8 and 10. Is the product of these two matrices commutative?

15. Show that

$$\left(\begin{bmatrix}2\\3\\4\end{bmatrix} \cdot [2 \ 1 \ -5]\right) \cdot \begin{bmatrix}-2 & 3 & 3\\1 & -4 & -2\\2 & 1 & 0\end{bmatrix} = \begin{bmatrix}2\\3\\4\end{bmatrix} \cdot \left([2 \ 1 \ -5] \cdot \begin{bmatrix}-2 & 3 & 3\\1 & -4 & -2\\2 & 1 & 0\end{bmatrix}\right)$$

16. Write the product AB as a single matrix if

$$A = \begin{bmatrix}a_{11} & a_{12} & a_{13}\\a_{21} & a_{22} & a_{23}\\a_{31} & a_{32} & a_{33}\end{bmatrix} \quad \text{and} \quad B = \begin{bmatrix}1 & 0 & 0\\0 & 1 & 0\\0 & 0 & 1\end{bmatrix}.$$

What name can be given to B in the set of 3×3 matrices?

17. Write the products AB and BA as single matrices if

$$A = \begin{bmatrix}1 & 0 & 2\\1 & 3 & 4\\-1 & 4 & 1\end{bmatrix} \quad \text{and} \quad B = \begin{bmatrix}-13 & 8 & -6\\-5 & 3 & -2\\7 & -4 & 3\end{bmatrix}.$$

Is this product commutative? Give a name to the relationship of A to B.

18. Write the product AB as a single matrix if

$$A = \begin{bmatrix}-1 & 2\\-2 & 4\end{bmatrix} \quad \text{and} \quad B = \begin{bmatrix}2 & 2\\1 & 1\end{bmatrix}.$$

Can you note a feature of this product that is unlike the product of two real numbers?

13.4 Determinants

In Chapter 12 we saw that certain quantities cannot be represented by a single number but must be represented by an ordered pair of numbers that we call a vector. However, we also found that a vector can be associated with a number that we call its norm.

In this chapter we have seen that other quantities are best represented by a rectangular array of numbers called a matrix. The question now arises, can we associate a matrix with a real number as we did a vector? The answer to this question is partly yes and partly no. Every *square matrix* with real number entries can be paired with a real number, but other matrices cannot be so paired.

The Determinant Function

The relation that pairs every square matrix with a number is called the **determinant function** and is denoted by δ (delta). If A is a square matrix, the symbol $\delta(A)$, called the **determinant of** A, represents the number to be paired with A.

DEFINITION 13.4-1 If $A = \begin{bmatrix} a_{11} & a_{12} & \cdots & a_{1n} \\ a_{21} & & & \vdots \\ \vdots & & & \\ a_{n1} & & \cdots & a_{nn} \end{bmatrix}$,

then the determinant function is

$$\{(A, \delta(A))\mid \delta(A) = \begin{vmatrix} a_{11} & a_{12} & \cdots & a_{1n} \\ a_{21} & & & \vdots \\ \vdots & & & \\ a_{n1} & & \cdots & a_{nn} \end{vmatrix}\}.$$

We shall not attempt a comprehensive discussion of techniques for evaluating a determinant. We shall merely define techniques for evaluating determinants of 2×2 and 3×3 matrices. These determinants are of **second order** and **third order**, respectively.

Second-Order Determinants

DEFINITION 13.4-2 If $A = \begin{bmatrix} a_1 & a_2 \\ b_1 & b_2 \end{bmatrix}$, then

$$\delta(A) = \begin{vmatrix} a_1 & a_2 \\ b_1 & b_2 \end{vmatrix} = a_1 b_2 - a_2 b_1.$$

Example Evaluate $\begin{vmatrix} 3 & 5 \\ 1 & 2 \end{vmatrix}$.

Solution $\begin{vmatrix} 3 & 5 \\ 1 & 2 \end{vmatrix} = 3 \cdot 2 - 5 \cdot 1 = 6 - 5 = 1.$

Third-Order Determinants

DEFINITION 13.4-3 If $A = \begin{bmatrix} a_1 & a_2 & a_3 \\ b_1 & b_2 & b_3 \\ c_1 & c_2 & c_3 \end{bmatrix}$, then

$$\delta(A) = \begin{vmatrix} a_1 & a_2 & a_3 \\ b_1 & b_2 & b_3 \\ c_1 & c_2 & c_3 \end{vmatrix} = a_1 b_2 c_3 - a_1 b_3 c_2 + a_2 b_3 c_1 - a_2 b_1 c_3 + a_3 b_1 c_2 - a_3 b_2 c_1.$$

13.4 Determinants

Example Evaluate $\begin{vmatrix} 3 & -2 & 1 \\ 1 & 4 & 2 \\ 2 & 1 & 1 \end{vmatrix}$.

Solution
$$\begin{vmatrix} 3 & -2 & 1 \\ 1 & 4 & 2 \\ 2 & 1 & 1 \end{vmatrix} = 3 \cdot 4 \cdot 1 - 3 \cdot 2 \cdot 1 + (-2) \cdot 2 \cdot 2 - (-2) \cdot 1 \cdot 1$$
$$+ 1 \cdot 1 \cdot 1 - 1 \cdot 4 \cdot 2$$
$$= 12 - 6 - 8 + 2 + 1 - 8$$
$$= -7$$

EXERCISE 13.4

WRITTEN

Evaluate each determinant using the appropriate method illustrated in the preceding examples.

1. $\begin{vmatrix} 2 & 5 \\ 1 & 3 \end{vmatrix}$
2. $\begin{vmatrix} 5 & -1 \\ 2 & 4 \end{vmatrix}$
3. $\begin{vmatrix} -1 & -1 \\ 3 & 3 \end{vmatrix}$
4. $\begin{vmatrix} 2 & -4 \\ 2 & 4 \end{vmatrix}$

5. $\begin{vmatrix} -1 & -2 \\ -3 & -4 \end{vmatrix}$
6. $\begin{vmatrix} 1 & 2 \\ 3 & 4 \end{vmatrix}$
7. $\begin{vmatrix} 1 & 3 \\ 2 & 4 \end{vmatrix}$
8. $\begin{vmatrix} 5 & 0 \\ 10 & 2 \end{vmatrix}$

9. $\begin{vmatrix} 1 & 2 & -1 \\ 3 & 1 & 0 \\ 0 & 2 & 4 \end{vmatrix}$
10. $\begin{vmatrix} 2 & 1 & -1 \\ 3 & 1 & 4 \\ 1 & 0 & 2 \end{vmatrix}$
11. $\begin{vmatrix} 4 & 1 & -1 \\ 2 & 1 & 3 \\ 1 & 2 & 1 \end{vmatrix}$
12. $\begin{vmatrix} 1 & 2 & 3 \\ 1 & 0 & 1 \\ 2 & 4 & 6 \end{vmatrix}$

*13. Show that $\begin{vmatrix} x & y & 1 \\ x_1 & y_1 & 1 \\ x_2 & y_2 & 1 \end{vmatrix} = 0$ represents the equation of the line containing (x_1, y_1) and (x_2, y_2). (HINT: See Section 7.4.)

14. Use the equation of Problem 13 to find an equation for the line containing $(3, 2)$ and $(1, 4)$.

15. Use the equation of Problem 13 to find an equation for the line through $(3, -1)$ and $(-2, 3)$.

*16. Show that $\begin{vmatrix} a^2 & a & 1 \\ b^2 & b & 1 \\ c^2 & c & 1 \end{vmatrix} = (a - b)(b - c)(a - c)$.

13.5 Cramer's Rule

If we solve the system of linear equations

$$a_1 x + b_1 y = c_1$$
$$a_2 x + b_2 y = c_2$$

by forming a linear combination of b_2 times the first equation and $-b_1$ times the second,

$$b_2(a_1 x + b_1 y = c_1)$$
$$-b_1(a_2 x + b_2 y = c_2),$$

we obtain

$$a_1 b_2 x + b_1 b_2 y - a_2 b_1 x - b_1 b_2 y = b_2 c_1 - b_1 c_2$$
$$(a_1 b_2 - a_2 b_1) x = b_2 c_1 - b_1 c_2$$
$$x = \frac{b_2 c_1 - b_1 c_2}{a_1 b_2 - a_2 b_1}.$$

Now if we apply Definition 13.4-2 (p. 324) to both the numerator and the denominator of this fraction, we have

$$b_2 c_1 - b_1 c_2 = \begin{vmatrix} c_1 & b_1 \\ c_2 & b_2 \end{vmatrix} \quad \text{and} \quad a_1 b_2 - a_2 b_1 = \begin{vmatrix} a_1 & b_1 \\ a_2 & b_2 \end{vmatrix}.$$

Thus,

$$x = \frac{\begin{vmatrix} c_1 & b_1 \\ c_2 & b_2 \end{vmatrix}}{\begin{vmatrix} a_1 & b_1 \\ a_2 & b_2 \end{vmatrix}}.$$

In a similar manner, by solving for y, we obtain

$$y = \frac{\begin{vmatrix} a_1 & c_1 \\ a_2 & c_2 \end{vmatrix}}{\begin{vmatrix} a_1 & b_1 \\ a_2 & b_2 \end{vmatrix}}.$$

Note that the denominator of each of these fractions is the determinant formed by listing the coefficients of the variables in order. Also note that the numerator of each fraction is the same determinant, except that the constant terms replace the coefficients of the particular variable.

If D is used to represent the determinant of the coefficients, and D_x and D_y

13.5 Cramer's Rule

are used to represent the respective numerators, the equations for x and y become

$$x = \frac{D_x}{D} \quad \text{and} \quad y = \frac{D_y}{D}.$$

This is one form of **Cramer's rule**.

Example Solve by Cramer's rule.

$$2x + 5y = 9$$
$$3x - 2y = 4$$

Solution $D = \begin{vmatrix} 2 & 5 \\ 3 & -2 \end{vmatrix} = -4 - 15 = -19$

$$D_x = \begin{vmatrix} 9 & 5 \\ 4 & -2 \end{vmatrix} = -18 - 20 = -38$$

$$D_y = \begin{vmatrix} 2 & 9 \\ 3 & 4 \end{vmatrix} = 8 - 27 = -19$$

$$x = \frac{D_x}{D} = \frac{-38}{-19} = 2; \quad y = \frac{D_y}{D} = \frac{-19}{-19} = 1$$

Hence, the solution set is $\{(2, 1)\}$.

Cramer's rule can be used to solve any system of linear equations that has a unique solution. However, since our discussion has been limited to determinants of second and third order, we must limit the use of Cramer's rule to systems containing no more than three variables.

Example

$$x - y - z = -6$$
$$2x + y + z = 0$$
$$3x - 5y + 8z = 13$$

Solution $D = \begin{vmatrix} 1 & -1 & -1 \\ 2 & 1 & 1 \\ 3 & -5 & 8 \end{vmatrix}, \quad D_x = \begin{vmatrix} -6 & -1 & -1 \\ 0 & 1 & 1 \\ 13 & -5 & 8 \end{vmatrix},$

$$D_y = \begin{vmatrix} 1 & -6 & -1 \\ 2 & 0 & 1 \\ 3 & 13 & 8 \end{vmatrix}, \quad \text{and} \quad D_z = \begin{vmatrix} 1 & -1 & -6 \\ 2 & 1 & 0 \\ 3 & -5 & 13 \end{vmatrix}$$

$$x = \frac{D_x}{D} = \frac{-78}{39} = -2; \quad y = \frac{D_y}{D} = \frac{39}{39} = 1; \quad z = \frac{D_z}{D} = \frac{117}{39} = 3.$$

Hence, the solution set is $\{(-2, 1, 3)\}$.

EXERCISE 13.5

WRITTEN

Use Cramer's rule to find the solution set of each system of equations by following the preceding examples.

1. $2x - y = 4$
   $3x - 4y = 1$
2. $3x + y = 5$
   $2x - 3y = 7$
3. $4x + y = -1$
   $3x + 2y = 3$
4. $4x + 3y = -2$
   $x - y = -4$
5. $x + 6y = 6$
   $2x - 2y = 5$
6. $2x + y = 3$
   $6x - 2y = -1$
7. $3x - 2y - z = 1$
   $2x + 3y - z = 4$
   $x - y + 2x = 7$
8. $3x - 2y + 4z = 1$
   $4x + y - 5z = 2$
   $2x - 3y + z = -6$
9. $3x + 5y + 2z = 2$
   $4x + 2y - 3z = -1$
   $2x - y + 5z = -11$
10. $2x - 3y + z = 3$
    $3x - 3y + z = 0$
    $4x + y + 5z = 1$
11. $2x - 3y + z = 0$
    $2x + 3y - z = 2$
    $4x + 3y + 2z = 3$
12. $3x + 2y + 4z = 5$
    $6x + 3y - 2z = 2$
    $3x - 4y + 8z = 5$

Summary

1. A **matrix** is a rectangular array of elements called **entries**. The **dimension** or **order** of a matrix is specified by the number of rows and the number of columns, in that order. [13.1]
2. The **address** of an entry is the number of the row and the number of the column, in that order, in which the entry is found. [13.1]
3. A **square matrix** is one that has the same number of rows as columns. [13.1]
4. Two matrices are equal if and only if they are of the same order and each entry of one matrix is equal to the corresponding entry of the other. [13.1]
5. The sum of two matrices of the same order is a matrix whose entries are each the sum of the corresponding entries in the given matrices. Matrices of different order are not **conformable for addition**. [13.2]
6. The negative of a matrix is a matrix of the same order whose entries are each the negative of the corresponding entry in the given matrix. [13.2]
7. Two matrices are **conformable for matrix multiplication** if the number of *columns* in the left-hand matrix is equal to the number of *rows* in the right-hand matrix. The product of an $m \times n$ matrix and an $n \times p$ matrix, in that order, is an $m \times p$ matrix. [13.3]
8. If A and B are two matrices conformable for matrix multiplication, each entry c_{ij} of $C = AB$ is the sum of the products of each entry in the ith row of A and the corresponding entry in the jth column of B. [13.3]
9. The **determinant function** pairs every square matrix of real numbers with a real number. [13.4]
10. The second-order determinant

$$\begin{vmatrix} a_1 & b_1 \\ a_2 & b_2 \end{vmatrix} = a_1 b_2 - a_2 b_1.$$

The third-order determinant

$$\begin{vmatrix} a_1 & a_2 & a_3 \\ b_1 & b_2 & b_3 \\ c_1 & c_2 & c_3 \end{vmatrix} = a_1b_2c_3 - a_1b_3c_2 + a_2b_3c_1 - a_2b_1c_3 + a_3b_1c_2 - a_3b_2c_1. \qquad [13.4]$$

11. The solution set of a system of independent linear equations can be found by an application of determinants called **Cramer's rule**. [13.5]

REVIEW EXERCISES

SECTION 13.1

In Problems 1 and 2, find values for the variables so that the matrices in each pair are equal.

1. $\begin{bmatrix} 2 & x & -4 \\ y & 0 & 2 \\ 3 & -1 & z \end{bmatrix} = \begin{bmatrix} 2 & 5 & -4 \\ 3 & 0 & 2 \\ 3 & -1 & 0 \end{bmatrix}$

2. $\begin{bmatrix} x+y & 4 \\ x-3y & 2 \\ -1 & 5 \end{bmatrix} = \begin{bmatrix} -4 & 4 \\ 2 & 2 \\ -1 & 5 \end{bmatrix}$

SECTION 13.2

Simplify each sum in Problems 3–5 if possible.

3. $\begin{bmatrix} 2 & 3 \\ 1 & 0 \\ -1 & 2 \end{bmatrix} + \begin{bmatrix} -2 & 0 \\ -3 & 0 \\ 4 & -1 \end{bmatrix}$

4. $\begin{bmatrix} 2 & 1 & 4 \\ 3 & -1 & 0 \end{bmatrix} + \begin{bmatrix} 6 & -2 \\ 3 & 1 \\ 0 & 0 \end{bmatrix}$

5. $\begin{bmatrix} 2 & -2 & 1 \\ 1 & 1 & -2 \\ 1 & 0 & -1 \end{bmatrix} - \begin{bmatrix} -1 & -1 & 5 \\ 1 & 1 & 2 \\ 2 & 1 & -2 \end{bmatrix}$

SECTION 13.3

In Problems 6–8, find the product matrix if possible.

6. $\begin{bmatrix} 3 & 1 & -1 \\ 0 & -1 & 2 \end{bmatrix} \cdot \begin{bmatrix} 1 & -1 \\ 0 & 0 \\ 1 & 2 \end{bmatrix}$

7. $\begin{bmatrix} 1 & -1 \\ 0 & 0 \\ 1 & 2 \end{bmatrix} \cdot \begin{bmatrix} 3 & 1 & -1 \\ 0 & -1 & 2 \end{bmatrix}$

8. $\begin{bmatrix} 1 & 2 & 3 \\ 2 & 0 & -1 \\ 1 & -3 & -2 \end{bmatrix} \cdot \begin{bmatrix} x \\ y \\ z \end{bmatrix}$

SECTION 13.4

Evaluate the determinants in Problems 9–12.

9. $\begin{vmatrix} 2 & -5 \\ 3 & 1 \end{vmatrix}$

10. $\begin{vmatrix} 2 & 0 \\ -5 & -2 \end{vmatrix}$

11. $\begin{vmatrix} 1 & 2 & 3 \\ -2 & 1 & 0 \\ 3 & 0 & 2 \end{vmatrix}$

12. $\begin{vmatrix} -1 & 2 & 0 \\ 2 & -4 & 0 \\ 3 & 1 & 5 \end{vmatrix}$

SECTION **13.5**

Use Cramer's rule to find the solution set of each system of equations in Problems 13 and 14.

13. $2x - 4y = 0$
    $5x + 2y = 7$

14. $2x + y + z = 1$
    $x - 2y - 3z = 1$
    $3x + 2y + 4z = 5$

Logarithms

Recall that an expression of the form b^n is a power, and, as was shown in Chapter 3, that if b is a non-negative number, then b^n is a real number for every rational number replacement of n. In fact, b^n is a non-negative number. In more advanced work, it is shown that if $b \geq 0$, then b^n is a non-negative real number for every *real number* replacement of n. Furthermore, it can be shown that, $b > 0$ and $b \neq 1$, every non-negative real number can be represented as a unique power of b.

Now consider a few powers of 2.

$$1 = 2^0$$
$$2 = 2^1$$
$$4 = 2^2$$
$$8 = 2^3$$

Note that we can form the ordered pairs $(1, 0)$, $(2, 1)$, $(4, 2)$, $(8, 3)$ by pairing each number with the corresponding exponent of 2. The property of uniqueness mentioned above indicates that such a set of ordered pairs is a function. A function of this kind is called a **logarithmic function**. The domain of a logarithmic function is the set of positive real numbers, and the range is the set of real numbers. Each element in the range is called the **logarithm** of the corresponding element in the domain with respect to the particular base. Thus 0 is the logarithm of 1 to the base 2, 1 is the logarithm of 2 to the base 2, etc. The usual notation is

$$\log_2 1 = 0, \quad \log_2 2 = 1, \quad \log_2 4 = 2, \quad \text{etc.},$$

where the base is written as a subscript. This notation can be read in several ways. One way is "log base 2 of 1 is 0," etc.

Since $0^n = 0$ $(n \neq 0)$ and $1^n = 1$ for *any* real number replacement of n, neither 0 nor 1 can be used as the base of a logarithmic function, because every element in the domain would be paired with the same element in the range.

14.1 Properties of Logarithms

The preceding discussion indicates that the logarithm of a number with respect to some base is the exponent of the base. We now define a logarithm more formally.

DEFINITION 14.1-1 If b is a positive real number other than 1, $y \in \mathbf{R}^+$, and $x \in \mathbf{R}$, then $\log_b y = x$ if and only if $y = b^x$.

Examples (a) If $y = 3^4$, then the logarithm of y to the base 3 is 4. That is, $\log_3 y = 4$.
(b) Since $32 = 2^5$, $\log_2 32 = 5$.
(c) The logarithm of 8 to the base 2 is 3, because $8 = 2^3$.
(d) $\log_3 81 = 4$, because $81 = 3^4$.
(e) Since $\sqrt{10} = 10^{1/2}$, $\log_{10} \sqrt{10} = \frac{1}{2}$.
(f) $\log_3 \frac{1}{27} = -3$, because $\frac{1}{27} = 3^{-3}$.
(g) If $\log_5 N = 4$, then $N = 5^4 = 625$.

Properties of Logarithms to the Same Base

Since logarithms are exponents, they must obey the laws of exponents. We restate some of these laws, and by using the definition we derive some properties of logarithms to the same base.

If $M = b^x$ and $N = b^y$, then $MN = b^x b^y = b^{x+y}$. By applying the definition of a logarithm to each of these, we have $\log_b M = x$, $\log_b N = y$, and $\log_b MN = x + y$. This leads to

PROPERTY 1 $\log_b MN = \log_b M + \log_b N$

Example $\log_3 (81 \times 27) = \log_3 81 + \log_3 27$. (You can check this by converting 81 and 27 to powers of 3.)

If $M = b^x$ and $N = b^y$, then $\dfrac{M}{N} = \dfrac{b^x}{b^y} = b^{x-y}$. In the same manner as we derived Property 1, we have

PROPERTY 2 $\log_b \dfrac{M}{N} = \log_b M - \log_b N$

Example $\log_5 \dfrac{625}{25} = \log_5 625 - \log_5 25$

$\log_5 625 = 4$, $\log_5 25 = 2$, and $4 - 2 = 2$.

Check: $625 \div 25 = 25 = 5^2$.

If $N = b^x$, then $N^y = (b^x)^y = b^{xy}$. By the definition, we have $\log_b N = x$ and $\log_b N^y = xy$. Thus we have

PROPERTY 3 $\log_b N^y = y(\log_b N)$

Examples   (a) $\log 2\,(4)^3 = 3(\log_2 4)$

(b) $\log_2 \sqrt[3]{64} = \log_2 (64)^{1/3} = \tfrac{1}{3}(\log_2 64)$.

EXERCISE 14.1

ORAL

Read each statement in Problems 1–20 as a logarithm.

Example $N = 2^6$

Solution $\log_2 N = 6$

1. $N = 5^3$
2. $N = 10^4$
3. $N = 10^{-2}$
4. $N = 4^{1/3}$
5. $N = 5^{1/2}$
6. $N = 10^{5/3}$
7. $N = 10^{0.15}$
8. $N = 10^{1.2}$
9. $36 = 6^2$
10. $32 = 2^5$
11. $144 = 12^2$
12. $81 = 3^4$
13. $\dfrac{1}{49} = 7^{-2}$
14. $\dfrac{1}{27} = 3^{-3}$
15. $\dfrac{1}{256} = 4^{-4}$
16. $\dfrac{1}{625} = 25^{-2}$
17. $2 = 10^{0.301}$
18. $49 = 10^{1.69}$
19. $8 = 10^{0.903}$
20. $1 = 10^0$

Read each statement in Problems 21–38 as a power.

Example $\log_2 N = 6$

Solution $N = 2^6$

21. $\log_2 N = 3$
22. $\log_{10} N = 4$
23. $\log_7 N = 2$
24. $\log_5 N = 3$
25. $\log_{10} N = 0$
26. $\log_8 N = 1$
27. $\log_3 N = -1$
28. $\log_8 N = -2$
29. $\log_5 25 = 2$
30. $\log_3 27 = 3$
31. $\log_{25} 5 = \tfrac{1}{2}$
32. $\log_{64} 4 = \tfrac{1}{3}$
33. $\log_4 8 = \tfrac{3}{2}$
34. $\log_9 27 = \tfrac{3}{2}$
35. $\log_{10} 0.01 = -2$
36. $\log_{10} 0.001 = -3$
37. $\log_{10} 3 = 0.48$
38. $\log_{10} 10 = 1$

WRITTEN

Find the logarithm in Problems 1–8.

Example $\log_9 81$

Solution   Since $81 = 9^2$, $\log_9 81 = 2$.

1. $\log_5 25$
2. $\log_{10} 1000$
3. $\log_3 81$
4. $\log_3 243$
5. $\log_9 3$
6. $\log_{16} 2$
7. $\log_3 \tfrac{1}{27}$
8. $\log_2 \tfrac{1}{16}$

In Problems 9–20, find b, N, or x.

Examples   (a) $\log_b 4 = 2$ (b) $\log_3 N = 3$ (c) $\log_{16} 4 = x$

Solutions (a) If $\log_b 4 = 2$, then $4 = b^2$ and $b = 2$.
(b) If $\log_3 N = 3$, then $N = 3^3 = 27$.
(c) If $\log_{16} 4 = x$, then $4 = 16^x$ and $x = \frac{1}{2}$. ($4 = \sqrt{16}$)

9. $\log_b 25 = 2$
10. $\log_b 64 = 3$
11. $\log_b 2 = \frac{1}{2}$
12. $\log_b 2 = \frac{1}{3}$
13. $\log_7 N = 2$
14. $\log_3 N = -2$
15. $\log_4 N = \frac{5}{2}$
16. $\log_{49} N = \frac{3}{2}$
17. $\log_5 \frac{1}{25} = x$
18. $\log_5 125 = x$
19. $\log_{10} 10{,}000 = x$
20. $\log_{10} 0.1 = x$

Write each expression in Problems 21–32 as a sum or difference of simpler logarithmic expressions. All variables represent positive numbers.

Examples (a) $\log_b (xy)$ (b) $\log_b \left(\dfrac{x}{y}\right)$ (c) $\log_b \sqrt{xy}$

Solutions (a) $\log_b (xy) = \log_b x + \log_b y$ (Property 1)

(b) $\log_b \left(\dfrac{x}{y}\right) = \log_b x - \log_b y$ (Property 2)

(c) $\log_b \sqrt{xy} = \frac{1}{2} \log_b (xy)$ (Property 3)
$= \frac{1}{2}(\log_b x + \log_b y)$ (Property 1)

21. $\log_b (3x)$
22. $\log_b 5y$
23. $\log_b \left(\dfrac{2}{x}\right)$
24. $\log_b \left(\dfrac{y}{4}\right)$
25. $\log_b \left(\dfrac{xy}{z}\right)$
26. $\log_b \left(\dfrac{x}{yz}\right)$
27. $\log_b x^3$
28. $\log_b (xy)^2$
29. $\log_b (x\sqrt{y})$
30. $\log_b \dfrac{\sqrt{xy}}{z}$
31. $\log_b x^2 y^3$
32. $\log_b \dfrac{x^2}{y^3 z}$

Write each expression in Problems 33–44 as a single logarithmic expression with coefficient 1. All variables represent positive numbers.

Examples (a) $2 \log_b x$ (b) $\frac{1}{2}(\log_b x + \log_b y)$ (c) $\log_b x - 2 \log_b y$

Solutions (a) $2 \log_b x = \log_b x^2$
(b) $\frac{1}{2}(\log_b x + \log_b y) = \frac{1}{2} \log_b (xy) = \log_b (xy)^{1/2}$ or $\log_b \sqrt{xy}$
(c) $3 \log_b x - 2 \log_b y = \log_b x^3 - \log_b y^2 = \log_b \left(\dfrac{x^3}{y^2}\right)$

33. $\log_b x - \log_b y$
34. $\log_b 3 + \log_b x$
35. $\log_b x + \log_b y - \log_b z$
36. $\frac{1}{4}(\log_b x - \log_b y)$
37. $2 \log_b x + 5 \log_b y$
38. $\frac{1}{2} \log_b x + \frac{3}{2} \log_b z$
39. $3 \log_b x + 2 \log_b y - 2 \log_b z$
40. $\frac{1}{2}(\log_b x + 2 \log_b y - \log_b z)$
41. $\frac{2}{3} \log_b x + \frac{1}{3} \log_b y + \frac{4}{3} \log_b z$
42. $\log_b (x + y) + \log_b (x - y)$
43. $\log_b (x^2 + 2x + 1) - \log_b (x + 1)$
44. $\log_b (x^2 - x - 2) + \log_b (x - 1) - \log_b (x^2 - 3x + 2)$

45. Show that $\frac{2}{3} \log_3 27 + \frac{1}{2} \log_5 25 = \log_2 8$.

46. Show that $\frac{3}{4}\log_2 16 - \log_7 49 = \log_{10} 10$.
47. Show that $b^{2\log_b 4} - b^{3\log_b 2} = 8$.
48. Show that $\log_4 2 = \dfrac{1}{\log_2 4}$.

14.2 Logarithms to the Base 10

The decimal system of numeration is so named because the base of the system is the number 10. Also, any digit in a decimal numeral represents the coefficient of a power of 10 determined by the position or place of the digit with respect to the decimal point. For example, the decimal numeral 3472.156 can be written as

$$3 \times 10^3 + 4 \times 10^2 + 7 \times 10^1 + 2 \times 10^0 + 1 \times 10^{-1} + 5 \times 10^{-2} + 6 \times 10^{-3}.$$

Scientific Notation

The concept of place value also permits us to represent decimal numerals as the product of some number between 1 and 10 and a power of 10. This is known as **scientific notation**.

DEFINITION 14.2-1 Any positive real number r can be represented in the form $N \cdot 10^c$, where $1 \leq N < 10$ and $c \in J$.

In actual practice, the decimal numeral is rounded off to three or four significant digits, so that scientific notation is usually an approximation rather than an exact numerical representation. The numeral in the example above would be expressed in scientific notation as

$$3472.156 \approx 3.47 \times 10^3.$$

The symbol $\approx$ is read, "is approximately equal to." Numbers that have been rounded off are usually approximations, as are decimal representations of irrational numbers. However, from this point on, we use $=$ instead of $\approx$, even though the numbers are not exact.

Also, from here on, the subscript is omitted if the base of a logarithm is 10. For all other bases, the proper subscript is used.

Examples (a) Since $475 = 10^2 \times 4.75$,

$$\log 475 = \log 10^2 + \log 4.75$$
$$= 2 + \log 4.75.$$

(b) Since $0.00475 = 10^{-3} \times 4.75$,

$$\log 0.00475 = \log 10^{-3} + \log 4.75$$
$$= -3 + \log 4.75.$$

Logarithms to the Base 10

It is shown in more advanced mathematics courses that if $r > 0$ and $b > 0$, there exists one and only one real number x such that $r = b^x$. That is, every positive number r has a unique logarithm to the base b. Then, since $10^0 = 1$ and $10^1 = 10$, we can assume that for $1 \leq N < 10$, $0 \leq \log_{10} N < 1$.

By Definition 14.2-1, $r = N \cdot 10^c$, and by Property 1 [$\log_b (MN) = \log_b M + \log_b N$], $\log_{10} r = \log_{10} N + c$. Thus, the logarithm to the base 10 of any positive number can be expressed as the sum of the logarithm of an integral power of 10 and the logarithm of a number between 1 and 10. In this context, the logarithm of the power of 10 (an integer) is called the **characteristic**, and the other number (usually given as a decimal fraction) is called the **mantissa** of the logarithm of the number. Logarithms to the base 10 are called **common logarithms**.

Examples (a) and (b) above demonstrate that the characteristic of the logarithm of a number depends upon the location of the decimal point in the numeral and may be either positive, negative, or 0, and that the mantissa is independent of the position of the decimal point and is always positive.

Using a Table of Logarithms

The logarithm of a number N, $1 \leq N < 10$ is, in general, an irrational number. Since the computation of such logarithms requires advanced techniques, it is customary to use a table such as the one inside the back cover. A portion of that table is reproduced here as Table 14.2-1.

To find the logarithm of a number, locate the first two digits of the number in the column headed by N. Then locate the third digit in the top line. The logarithm of the number appears on the same line as the first two digits and in the same column as the third digit. For example, the logarithm of 2.83 is directly to the right of 2.8 and directly below 3, as indicated by the darker shading in Table 14.2-1. Thus, $\log 2.83 = 0.4518$.

The table may be used for finding logarithms of numbers greater than 10 by using Definition 14.2-1 as indicated on page 335. This technique is illustrated in

Table 14.2-1

N	0	1	2	3	4	5	6	7	8	9
2.5	.3979	.3997	.4014	.4031	.4048	.4065	.4082	.4099	.4116	.4133
2.6	.4150	.4166	.4183	.4200	.4216	.4232	.4249	.4265	.4281	.4298
2.7	.4314	.4330	.4346	.4362	.4378	.4393	.4409	.4425	.4440	.4456
2.8	.4472	.4487	.4502	.4518	.4533	.4548	.4564	.4579	.4594	.4609
2.9	.4624	.4639	.4654	.4669	.4683	.4698	.4713	.4728	.4742	.4757
3.0	.4771	.4786	.4800	.4814	.4829	.4843	.4857	.4871	.4886	.4900

the following examples. For such a number, the table gives the mantissa of the logarithm. The appropriate characteristic should be appended.

Examples Use Table 14.2-1 to find the logarithms of
(a) 2580   (b) 27.5

Solutions Think of, or write, each number in the form $10^c \cdot N$. Then,

(a) $2580 = 10^3 \times 2.58$.

$$\begin{aligned} \log 2580 &= \log 10^3 + \log 2.58 \\ &= 3 + 0.4116 \\ &= 3.4116 \end{aligned}$$

Log 2.58 was found in the same manner as log 2.83.

(b) $27.5 = 10^1 \times 2.75$ and

$$\begin{aligned} \log 27.5 &= \log 10^1 + \log 2.75 \\ &= 1 + 0.4393 \\ &= 1.4393 \end{aligned}$$

The characteristic can usually be determined by inspection rather than by the detailed method indicated in the examples above.

A slight difficulty arises when the characteristic is negative, because the mantissa is always a positive number. For example,

$$\log 2 = 0.3010;$$

then, since $0.02 = 10^{-2} \times 2.0$,

$$\log 0.02 = -2 + 0.3010.$$

If this sum is written as -1.6990, the characteristic -1 indicates that the decimal point in the number is placed just before the first nonzero digit, but the mantissa is now negative and is not the entry in the table for log 2.00.

There are two ways in which this difficulty may be overcome. One way is to place a minus sign over the characteristic and keep a positive mantissa. That is, write $-2 + 0.3010$ in the form

$$\log 0.02 = \bar{2}.3010.$$

A second technique is to add some $a - a$, $a \in N$, to the characteristic in the following manner.

$$\begin{aligned} \log 0.02 &= 0.3010 - 2 + (10 - 10) \\ &= 0.3010 + 8 - 10 \\ &= 8.3010 - 10 \end{aligned}$$

This latter form is the one ordinarily used in actual computation.

338 LOGARITHMS

Antilogarithms

When logarithms are used for computational purposes, it is also necessary to find the number that corresponds to a given logarithm. This number is called the **antilogarithm** of the given logarithm.

Examples Find antilogarithms for the given logarithms.
(a) 4.4742 (b) 0.4082 (c) 7.4698 − 10

Solutions (a) $\log N = 4.4742 = 4 + .4742$

$N = 10^4 \times 10^{.4742}$

The mantissa .4742 is located in Table 14.2-1 under 8 and to the right of 2.9. Hence, $10^{.4742} = 2.98$, and

$N = 10^4 \times 2.98 = 29{,}800$

(b) $\log N = 0.4082 = 0 + .4082$

$N = 10^0 \times 10^{.4082}$

$= 2.56$

(c) $\log N = 7.4698 - 10 = -3 + .4698$

$N = 10^{-3} \times 10^{.4698}$

$= 0.00295$

Linear Interpolation

The table of logarithms you are using is a four-place table and can be used to find logarithms of numbers with three significant digits. This table can be

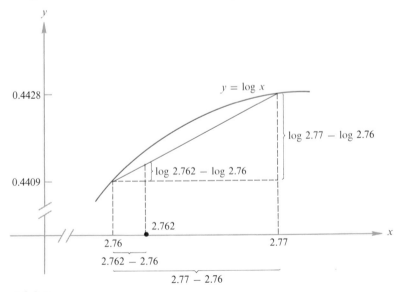

Figure 14.2-1

extended to find logarithms of numbers with four significant digits by using the process called **linear interpolation**. This process is based on the assumption that between any two points that are close together the graph of a function is very nearly a straight line.

Consider a portion of the graph of $y = \log x$ between $x = 2.76$ and $x = 2.77$ as shown in Figure 14.2-1. (NOTE: The curvature of the graph of $y = \log x$ is greatly exaggerated for purposes of our discussion.)

The figure illustrates how you can find $\log 2.762$ by linear interpolation. $\log 2.76$ and $\log 2.77$ can be read directly from the table. Since $2.76 < 2.762 < 2.77$, $\log 2.76 < \log 2.762 < \log 2.77$. Then if we assume that the portion of the graph of $y = \log x$ between $x = 2.76$ and $x = 2.77$ is a straight line segment, we have by similar triangles

$$\frac{\log 2.762 - \log 2.76}{\log 2.77 - \log 2.76} = \frac{2.762 - 2.76}{2.77 - 2.76}$$

$$\log 2.762 - \log 2.76 = (\log 2.77 - \log 2.76) \cdot \frac{2.762 - 2.76}{2.77 - 2.76}$$

$$= (0.4425 - 0.4409) \cdot \frac{0.002}{0.01}$$

$$= 0.0016 \cdot \frac{2}{10}.$$

$$\log 2.762 = \log 2.76 + 0.0016 \cdot \frac{2}{10}$$

$$= 0.4409 + 0.0003$$

$$= 0.4412.$$

The difference $\log 2.77 - \log 2.76 = 0.0016$ is called the **tabular difference**.

Antilogarithms to four significant digits can be determined in a similar manner.

Example Find antilog 0.4586.

Solution The table shows that antilog $0.4579 = 2.87$ and antilog $0.4594 = 2.88$. Then

$$\frac{0.4586 - 0.4579}{0.4594 - 0.4579} = \frac{0.0007}{0.0015} = 0.5.$$

Thus, antilog 0.4586 is 2.87 plus 0.5 times the difference between 2.87 and 2.88.

$$\text{Antilog } 0.4586 = 2.87 + 0.5(0.01) = 2.875$$

EXERCISE 14.2

WRITTEN

Write each number in Problems 1–10 in scientific notation.

1. 593
2. 3740
3. 2.51
4. 10.2
5. 0.125
6. 0.0234
7. 0.0013
8. 0.0047
9. 62,500
10. 71,400

Write each number in Problems 11–18 in decimal notation.

11. $10^2 \times 2.17$
12. $10^{-2} \times 1.18$
13. $10^{-4} \times 7.35$
14. $10^4 \times 1.07$
15. $10^6 \times 2.22$
16. $10^{-4} \times 6.43$
17. $10^0 \times 4.75$
18. $10^{-1} \times 3.46$

Write the characteristic of the logarithm of each number in Problems 19–30.

Examples (a) 2375.2 (b) 0.00107

Solutions (a) 3 (b) −3

19. 325.75
20. 471.59
21. 2.145
22. 42.731
23. 52.025
24. 1.495
25. 3475
26. 10.001
27. 0.03475
28. 0.001
29. 0.00024
30. 0.010101

Find the logarithm of each number in Problems 31–46. Round numbers off to four significant figures and use interpolation where necessary.

31. 275
32. 3.14
33. 42.5
34. 18.1
35. 0.182
36. 0.974
37. 0.0482
38. 0.0952
39. 3476
40. 25.47
41. 0.03572
42. 0.004285
43. 42.763
44. 927.46
45. 0.042517
46. 0.985425

Find the antilogarithm of each logarithm in Problems 47–62 to four significant digits.

47. 1.3075
48. 2.3139
49. 3.4533
50. 5.7782
51. $\bar{2}.5302$
52. $\bar{3}.9499$
53. 8.8993 − 10
54. 9.7796 − 10
55. 4.8085
56. 3.3951
57. 2.5960
58. 1.6900
59. 8.5725 − 10
60. 7.9000 − 10
61. 3.8350 − 5
62. 1.3860 − 3

14.3 Computations

Some arithmetic computations that are ordinarily quite tedious can be simplified by the use of logarithms. Such a use of logarithms must be an application of one or more of the properties discussed in Section 14.1. The techniques for applying the properties are illustrated in the following examples.

Examples Use logarithms to compute each of the following.

(a) $(27.4)(352)$ (b) $3.91 \div 49.3$
(c) $(2.75)^4$ (d) $\sqrt[5]{2.1}$

Solutions (a) Applying Property 1, we have

$$\log [(27.4)(352)] = \log 27.4 + \log 352.$$

In the table of logarithms we find

$$\log 27.4 + \log 352 = 1.4378 + 2.5465$$
$$= 3.9743.$$

Then,

$(27.4)(352) =$ antilog 3.9743
$= 9425.$

(The fourth digit was obtained by using the interpolation process.)

(b) Applying Property 2, we have

$\log (3.91 \div 49.3) = \log 3.91 - \log 49.3.$

From the table of logarithms, we obtain

$$\begin{aligned}\log 3.91 - \log 49.3 &= 0.5922 - 1.6928 \\ &= (10.5922 - 10) - 1.6928 \\ &= 8.8994 - 10 \\ &= \bar{2}.8994.\end{aligned}$$

Therefore,

$3.91 \div 49.3 =$ antilog $\bar{2}.8994$
$= 0.0793.$

(This result is to the nearest three significant digits. Interpolation was not used.)

(c) Applying Property 3, we have

$$\begin{aligned}\log (2.75)^4 &= 4(\log 2.75) \\ &= 1.7572.\end{aligned}$$

Then,

$(2.75)^4 =$ antilog 1.7572
$= 57.2$ to three significant digits.

(d) Since $\sqrt[5]{2.1} = (2.1)^{1/5}$, by Property 3, we have

$\log (2.1)^{1/5} = \tfrac{1}{5}(\log 2.10).$

We can use the table of logarithms to find

$$\begin{aligned}\tfrac{1}{5}(\log 2.10) &= \tfrac{1}{5}(0.3222) \\ &= 0.0644. \text{ (Rounded off)}\end{aligned}$$

Then,

$\sqrt[5]{2.1} =$ antilog 0.0644
$= 1.16$ to three significant digits.

EXERCISE 14.3

WRITTEN

Compute by means of logarithms. In Problems 1–20, find antilogs to three significant digits.

1. $(29.1)(3.06)$
2. $(9.88)(0.426)$

3. $\dfrac{69.3}{10.1}$

4. $\dfrac{5.18}{0.992}$

5. $(0.693)(0.045)(0.829)$

6. $(88.3)(0.113)(10.7)$

7. $\dfrac{(69.3)(1.15)}{10.1}$

8. $\dfrac{(701)(0.223)}{8.05}$

9. $\dfrac{0.665}{(20.9)(3.11)}$

10. $\dfrac{80.5}{392(0.616)}$

11. $(37.5)^3$
12. $(0.125)^4$
13. $(3.52)^{-3}$
14. $(4.75)^{-2}$
15. $(2)^{\sqrt{2}}$
16. $(3)^{\sqrt{3}}$
17. $\sqrt[3]{32.7}$
18. $\sqrt[3]{9.21}$
19. $\sqrt[5]{0.032}$ (HINT: See footnote.*)
20. $\sqrt[3]{0.0216}$

In Problems 21–30, interpolate as necessary to obtain antilogarithms to four significant digits.

21. $(681.4)^{2/3}(469.1)^{3/4}$

22. $(375.6)^{1/2}(64.5)^{2/3}$

23. $\dfrac{(21.71)(28.6)}{(396.4)(1.40)}$

24. $\dfrac{(791.5)(9.65)}{(6.173)(32.4)}$

25. $\left(\dfrac{(11.4)(2.86)}{\sqrt{2.41}}\right)^3$

26. $\left(\dfrac{(0.677)(81.4)}{\sqrt[3]{9010}}\right)^2$

27. $\left(\dfrac{(7.11)(0.695)}{3.98}\right)^{3/2}$

28. $\left(\dfrac{(4.17)^3}{(2.45)(1.72)}\right)^{2/3}$

29. $\sqrt[3]{\dfrac{(4.253)^3(1.745)^2}{(5.172)}}$

30. $\sqrt[4]{\dfrac{(57.26)^{1/2}(49.75)^{2/3}}{(5.723)^2}}$

Summary

1. A **logarithmic function** is a function that pairs every positive real number with the exponent of a power whose base is a positive number other than 1. [14.1]
2. The **logarithm** of a positive number to a base b is the exponent of the power of b to which b must be raised to obtain the number. [14.1]
3. Logarithms of positive numbers to the same base have the following properties:

   1. $\log_b (MN) = \log_b M + \log_b N$
   2. $\log_b \left(\dfrac{M}{N}\right) = \log_b M - \log_b N$
   3. $\log_b (M)^r = r \cdot \log_b M \qquad (r \in R)$ [14.1]

4. **Scientific notation** is the representation of any positive number in the form $N \cdot 10^c$, where N is a positive real number such that $1 \leq N < 10$ and $c \in J$. [14.2]

---

*Change the form of the characteristic by adding $a - a$ such that a is divisible by 5.

5. The logarithm of a positive number to the base 10 (called the **common logarithm**) is the sum of an integer and a positive decimal fraction less than 1. The integer is called the **characteristic**, and the decimal fraction is called the **mantissa** of the logarithm. [14.2]
6. An **antilogarithm** is the number corresponding to a given logarithm. [14.2]
7. Since the common logarithms of most numbers are irrational numbers, they are usually read from a table. The use of a table of logarithms can be extended to numbers with more significant digits than those in the table by **linear interpolation**. [14.2]
8. Computations involving products, quotients, powers, and roots can be performed by using logarithms accoring to the three properties of logarithms. [14.3]

REVIEW EXERCISES

SECTION 14.1

Write each statement in Problems 1–3 in logarithmic form.
1. $N = 3^5$ 2. $N = 6^8$ 3. $N = 5^{-3}$

In Problems 4–6, write each statement as an equation involving a power of the base.
4. $\log_2 N = 6$ 5. $\log_5 N = 4$ 6. $\log_3 N = -5$

In Problems 7–9, find the logarithm of each number to the indicated base.
7. $\log_6 36$ 8. $\log_3 \frac{1}{9}$ 9. $\log_{16} 2$

Write each of the expressions in Problems 10–12 as a sum or difference of simpler logarithmic expressions.
10. $\log_b \left(\dfrac{x}{zy}\right)$ 11. $\log_b x^4 y$ 12. $\log_b \sqrt[n]{\dfrac{x}{y}}$

Write each of the expressions in Problems 13 and 14 as a single logarithm with coefficient 1.
13. $\log_b 2x + \log_b y$ 14. $\frac{3}{2} \log_b x - \frac{1}{2} \log_b y$

SECTION 14.2

In Problems 15–17, write each number in scientific notation.
15. 3241 16. 63,000,000 17. 0.00256

Use the table to find the logarithm of each number in Problems 18–21.
18. 24.5 19. 327.8 20. 0.125 21. 0.01029

Find the antilogarithm of each logarithm in Problems 22–25.
22. 3.8142 23. 2.5508 24. $\bar{2}.6444$ 25. 7.8817 − 10

SECTION 14.3

In Problems 26–30, compute to four significant digits by using logarithms.

26. $(45.4)(18.7)$
27. $74.1 \div 18.3$
28. $\dfrac{11.92 \times 0.02236}{0.912}$
29. $(255)^3$
30. $\sqrt[4]{0.125}$

Sequences

Before we begin a discussion of the sets of numbers called sequences, it will be helpful to establish a basis for the discussion. To do this we need to consider certain subsets of the set of integers

$$J = \{\ldots, -3, -2, -1, 0, 1, 2, 3, \ldots\}.$$

Let J^* be a set of two or more integers with the following properties:

(1) The set has a least member; that is, there is one integer that is less than any other member of the set.

(2) If $a, b \in J^*$, then every integer x such that $a < x < b$ is a member of the set.

Examples Which of the following sets satisfy the conditions stated above?

(a) $\{1, 2, 3, 4, 5\}$
(b) $\{0, 1, 2, \ldots\}$
(c) $\{x \mid 5 < x \leq 100,\ x \in J\}$
(d) $\{4\}$
(e) $\{1, 3, 5, \ldots\}$
(f) $\{x \mid x < 0,\ x \in J\}$

Solutions The sets in (a), (b), and (c) satisfy the conditions in that they have at least two elements, a least element, and all integers within the stated interval. The other three do not satisfy the conditions. $\{4\}$ has only one element. $\{1, 3, 5, \ldots\}$ does not contain every integer within the interval. $\{x \mid x < 0,\ x \in J\}$ has no least element.

15.1 Sequence Functions

A function whose domain is a set of integers such as J^* described above is called a **sequence function**. The word **sequence** is usually applied to the ordered set of second components of the ordered pairs in a sequence function.

Finding the Terms of a Sequence

As with other functions, sequence functions are usually defined by an equation. If this is the case, the elements of the sequence are found in the same manner as elements in the range of any function.

Example If $n \in \{1, 2, 3, 4\}$, write the terms of the sequence defined by $f(n) = n^2 + n$.

Solution
$f(1) = (1)^2 + 1 = 2$
$f(2) = (2)^2 + 2 = 6$
$f(3) = (3)^2 + 3 = 12$
$f(4) = (4)^2 + 4 = 20$
Thus, the sequence is $\{2, 6, 12, 20\}$.

Equation for a Sequence

A sequence defined by an equation is unique; that is, for each n there is one and only one $f(n)$. However, the equation used to define a given sequence is not necessarily unique. For example, the terms of the sequence $\{1, 3, 5, 7\}$ can be made to correspond with several sets of integers by the selection of suitable equations. The set $\{1, 2, 3, 4\}$ can be used as the domain if $f(n) = 2n - 1$, the set $\{2, 3, 4, 5\}$ can be used if $f(n) = 2n - 3$, etc. In general, the determination of an equation defining a particular sequence is a matter of selecting a suitable set of integers as a domain and then using a process of trial and error until an equation is formed.

EXERCISE 15.1

WRITTEN

In Problems 1–18, write the terms of the sequence defined by each equation if $n \in \{1, 2, 3, 4\}$.

Example $f(n) = 2n + 1$

Solution $f(1) = 2 + 1 = 3$, $f(2) = 2 \cdot 2 + 1 = 5$, $f(3) = 2 \cdot 3 + 1 = 7$, and $f(4) = 2 \cdot 4 + 1 = 9$. Hence, the sequence is $\{3, 5, 7, 9\}$.

1. $f(n) = n + 5$
2. $f(n) = 3n + 1$
3. $f(n) = \dfrac{1}{n}$
4. $f(n) = \dfrac{n}{n + 1}$
5. $f(n) = \dfrac{n^2 + 2}{2}$
6. $f(n) = \dfrac{n(n + 1)}{2}$
7. $f(n) = \dfrac{n}{2n - 1}$
8. $f(n) = \dfrac{1}{2n + 1}$
9. $f(n) = \dfrac{n(n - 1)}{n + 1}$
10. $f(n) = \dfrac{n + 1}{n^2 + 1}$
11. $f(n) = 2^n$
12. $f(n) = 3^n$

13. $f(n) = (\frac{1}{2})^n$
14. $f(n) = (\frac{1}{3})^n$
15. $f(n) = (-1)^n$
16. $f(n) = (-1)^{n+1}$
17. $f(n) = \dfrac{(-1)^n(n-2)}{n}$
18. $f(n) = \dfrac{(-1)^n}{n(n+1)}$

Write a suitable equation to define each sequence in Problems 19–30 if $n \in \{1, 2, 3, 4\}$.

Example $\{5, 7, 9, 11\}$.

Solution Since 5 corresponds to 1, 7 to 2, etc., we determine by inspection that $f(n) = 2n + 3$.

19. $\{2, 4, 6, 8\}$
20. $\{3, 6, 9, 12\}$
21. $\{1, 4, 9, 16\}$
22. $\{1, 8, 27, 64\}$
23. $\{2, 4, 8, 16\}$
24. $\{3, 9, 27, 81\}$
25. $\{1, 4, 16, 64\}$
26. $\{\frac{1}{5}, 1, 5, 25\}$
27. $\{-1, 2, -3, 4\}$
28. $\{1, -2, 3, -4\}$
29. $\{-2, \frac{3}{2}, -\frac{4}{3}, \frac{5}{4}\}$
30. $\{\frac{1}{2}, -\frac{2}{3}, \frac{3}{4}, -\frac{4}{5}\}$

15.2 Arithmetic Sequences

If any term of a sequence can be found by adding some constant to the preceding term, the sequence is called an **arithmetic sequence** or an **arithmetic progression**. The constant is called the **common difference**.

Example If 2 is the first term and the common difference is 3, then $\{2, 5, 8, 11, 14, \ldots\}$ is an arithmetic sequence.

The nth Term of an Arithmetic Sequence

In the general case, a represents the first term and d the common difference, and if t_i represents any term ($i \in \{1, 2, \ldots, n\}$), then

$$t_1 = a, \quad t_2 = a + d, \quad t_3 = a + 2d, \quad t_4 = a + 3d, \quad \text{etc.}$$

Observe that the coefficient of d is 1 less than the number of the term. This indicates that the nth term can be represented by

$$t_n = a + (n - 1)d. \tag{1}$$

Example Find the tenth term of the arithmetic sequence whose first term is -3 and whose common difference is 2.

Solution By replacing n with 10, a with -3, and d with 2 in Equation (1), you have

$$t_{10} = -3 + (10 - 1)2$$
$$= -3 + 18$$
$$= 15.$$

First Term and Common Difference

If any two terms of an arithmetic sequence are known, the first term, the common difference, and any other terms can be found as required.

Example The twelfth and seventeenth terms of an arithmetic sequence are 37 and 52, respectively. Find the sixth term.

Solution By using Equation (1), you have

$$t_{12} = a + 11d = 37$$
$$t_{17} = a + 16d = 52.$$

If you subtract the first equation from the second, term by term, you have

$$5d = 15$$
$$d = 3.$$

Then by substituting in the first equation,

$$a + 33 = 37$$
$$a = 4.$$

Finally,

$$t_6 = 4 + (5)3 = 19.$$

Arithmetic Means

Numbers inserted between two given numbers to form an arithmetic sequence are called **arithmetic means**. We can use a technique similar to that of the preceding example to find arithmetic means.

Example Find three arithmetic means between 2 and 14.

Solution If we insert three arithmetic means between 2 and 14, the resulting sequence consists of five terms. Thus,

$$a = 2 \quad \text{and} \quad t_5 = a + (4)d.$$

But $t_5 = 14$, so we have

$$2 + 4d = 14$$
$$4d = 12$$
$$d = 3$$

Then,

$$t_2 = 5, \quad t_3 = 8, \quad \text{and} \quad t_4 = 11.$$

Sum of a Number of Terms

It is sometimes necessary to find the sum of a number of terms of an arithmetic sequence. This can be done by computing the required terms and then adding them together, but that is apt to be a very tedious process. There is a simpler technique.

The sum of the first n terms of an arithmetic sequence is given by

$$S_n = a + (a + d) + (a + 2d) + \cdots + a + (n - 1)d. \qquad (2)$$

The same arithmetic sequence can be expressed in reverse order by starting with the last, or nth, term and subtracting the common difference (d) successively. The sum can then be written as

$$S_n = a + (n - 1)d + a + (n - 2)d + a + (n - 3)d + \cdots + a. \qquad (3)$$

If Equations (2) and (3) are added together term by term, the result is

$$2S_n = [2a + (n - 1)d] + [2a + (n - 1)d] + [2a + (n - 1)d] + \cdots \\ + [2a + (n - 1)d],$$

in which the right-hand member contains n like terms. Thus,

$$2S_n = n[2a + (n - 1)d],$$

or

$$S_n = \frac{n}{2}[2a + (n - 1)d]. \qquad (4)$$

If we use the symbol l to represent the last term in the sequence instead of $a + (n - 1)d$, Equation (4) becomes

$$S_n = \frac{n(a + l)}{2}. \qquad (5)$$

Example Find the sum of the first ten terms of $1, 4, 7, 10, \ldots$.

Solution By inspection, we see that $a = 1$ and $d = 3$. Hence, by Equation (4),

$$S_{10} = \frac{10}{2}[2 + (9)3]$$
$$= 5(2 + 27)$$
$$= 145.$$

Example Find the sum of the first 500 natural numbers.

Solution $a = 1$ and $l = 500$, then by Equation (5),

$$S_{500} = \frac{500(1 + 500)}{2} = 125{,}250.$$

EXERCISE 15.2

WRITTEN

Find the first five terms of each arithmetic sequence in Problems 1–12 with a and d as given.

1. $a = 4, d = 3$
2. $a = 5, d = 1$
3. $a = 1, d = -5$
4. $a = 3, d = -6$

5. $a = -3, d = 3$
6. $a = -5, d = 10$
7. $a = 0, d = \frac{1}{2}$
8. $a = \frac{1}{3}, d = \frac{1}{3}$
9. $a = 3, d = \sqrt{2}$
10. $a = \sqrt{3}, d = \sqrt{3}$
11. $a = 5, d = 1 + \sqrt{2}$
12. $a = 1 - \sqrt{3}, d = 1 + \sqrt{3}$

Find the specified term for each arithmetic sequence in Problems 13–22.

Example Find t_{10} if $t_3 = 10$ and $t_{15} = 34$.

Solution
$t_3 = a + 2d = 10$
$t_{15} = a + 14d = 34$

Solving this system gives

$12d = 24$
$d = 2$

and

$a = 6.$

Then

$t_{10} = 6 + 9(2) = 24.$

13. Find t_{12} if $t_1 = 2$ and $t_5 = 14$.
14. Find t_{18} if $t_2 = 8$ and $t_4 = 2$.
15. Find t_2 if $t_{10} = 13$ and $t_{12} = 3$.
16. Find t_{20} if $t_{11} = 12$ and $t_3 = 30$.
17. Find t_{15} if $t_1 = 12$ and $t_3 = 15$.
18. Find t_{42} if $t_1 = 1.20$ and $t_3 = 0.96$.
19. Find t_5 if $t_1 = 2x$ and $t_2 = 2x - 2$.
20. Find t_1 if $t_2 = x - y$ and $t_{10} = x - 9y$.
21. Find t_{10} if $t_2 = 3a + 3b$ and $t_3 = 5a + 4b$.
22. Find t_2 if $t_1 = 3a + 4b$ and $t_3 = -3a$.
23. Find four arithmetic means between 1 and 21.
24. Find five arithmetic means between 37 and 19.
25. Find five arithmetic means between 21 and -3.
26. Find seven arithmetic means between $3/4$ and $27/4$.

Find the sum of the specified number of terms of the given arithmetic sequence in Problems 27–36.

Example 7 terms if $a = 7$ and $d = 5$.

Solution $S_n = \frac{n}{2}[2a + (n-1)d]$

Hence,

$S_7 = \frac{7}{2}[14 + 6(5)]$
$= \frac{7}{2}(44)$
$= 154.$

27. 8 terms if $a = -3$ and $d = 2$.
28. 20 terms if $a = 4$ and $d = 3$.
29. 18 terms if $a = 4$ and $d = \frac{3}{2}$.
30. 30 terms if $a = 12$ and $d = -\frac{1}{2}$.
31. 16 terms if $a = 7$ and $d = 4$.
32. 32 terms if $a = 15$ and $d = -3$.

33. 48 terms if $a = 2.6$ and $d = 0.2$.
34. 17 terms if $a = -3.4$ and $d = -1.3$.
35. 10 terms if $t_5 = 18$ and $t_{39} = 120$.
36. 7 terms if $t_3 = 11$ and $t_{11} = -5$.
37. For what value or values of k do $2k + 4$, $3k - 7$, and $k + 12$ form an arithmetic sequence?
38. For what value or values of k do $2k + 3$, $k^2 - 1$, and $2k - 5$ form an arithmetic sequence?

15.3 Geometric Sequences

If the terms of a sequence are related so that the ratio of any term, except the first, to the preceding term is constant, the sequence is called a **geometric sequence** or a **geometric progression**. If a is used to represent the first term, and r represents the common ratio, then the terms are

$$t_1 = a; \quad t_2 = ar; \quad t_3 = ar^2; \quad t_4 = ar^3; \quad \text{etc.}$$

Observe that the exponent of r in any term is one less than the number of the term. Thus, we represent the nth term as

$$t_n = ar^{n-1} \qquad (1)$$

Example Write the first four terms of the geometric sequence whose first term is 5 and whose common ratio is 2.

Solution By substituting the given values in (1), we have

$$t_1 = 5, \quad t_2 = 5(2), \quad t_3 = 5(2^2), \quad \text{and} \quad t_4 = 5(2^3).$$

Thus, the first four terms are 5, 10, 20, 40.

Finding the Common Ratio and First Term

If we are given any two terms of a geometric sequence, we can find the common ratio by extracting a root. If the root to be extracted is even, there will be two possibilities for the common ratio, and thus the two given terms can belong to two different sequences. If the root to be extracted is odd, there will be only one common ratio, and the sequence is unique. After the common ratio has been found, the first term, and therefore any other term in the sequence, can be determined. The technique to be used is illustrated in the following example.

Example If the fourth term of a geometric sequence is 54, and the seventh term is 1458, find the sixth term.

Solution In order to find the sixth term, we must know the first term and the common ratio. Since the nth term of a geometric sequence is given by $t_n = ar^{n-1}$, we have

$$t_4 = ar^3 = 54 \quad \text{and} \quad t_7 = ar^6 = 1458.$$

If we divide t_7 by t_4, we have

$$\frac{t_7}{t_4} = \frac{ar^6}{ar^3} = \frac{1458}{54},$$

from which we obtain

$r^3 = 27.$

Then, by extracting the cube root of both members,

$r = 3.$

By replacing r with 3 in either $ar^3 = 54$ or $ar^6 = 1458$, we find $27a = 54$ or $729a = 1458$, from which

$a = 2.$

Finally,

$t_6 = ar^5 = 2(3^5) = 2(243) = 486.$

Geometric Means

If one or more numbers are inserted between two given numbers so that the resulting sequence is geometric, the numbers inserted are called **geometric means**. The example above suggests a technique by which you can find geometric means.

Example Find three positive geometric means between 3 and 48.

Solution If three geometric means are inserted between 3 and 48, the resulting sequence will contain 5 terms. Thus,

$a = 3$, and $ar^4 = 48.$

If we divide both forms of the fifth term by the respective forms of the first term, we have

$$\frac{ar^4}{a} = \frac{48}{3}$$

$r^4 = 16,$

from which

$r = 2.$

(Since we want positive means, we ignore the negative root.)
We now compute the three geometric means to be

$t_2 = 3(2) = 6, \quad t_3 = 3(2^2) = 12, \quad \text{and} \quad t_4 = 3(2^3) = 24.$

Sum of n Terms

The sum of the first n terms of a geometric sequence can be expressed as

$$S_n = a + ar + ar^2 + \cdots + ar^{n-1}. \tag{2}$$

However, as with the sum of n terms of an arithmetic sequence, this equation is not always practical to use. We can multiply both members of Equation (2) by r to obtain

$$rS_n = ar + ar^2 + ar^3 + \cdots + ar^{n-1} + ar^n$$

and then subtract this equation from Equation (2), term by term, to obtain

$$S_n - rS_n = (a + ar + ar^2 + \cdots + ar^{n-1}) - (ar + ar^2 + \cdots + ar^{n-1} + ar^n)$$

$$S_n - rS_n = a - ar^n$$

$$(1 - r)S_n = a(1 - r^n)$$

$$S_n = \frac{a(1 - r^n)}{1 - r}. \qquad (3)$$

Example Find the sum of the first five terms of the geometric sequence with first term 2 and common ratio 3.

Solution Replacing a with 2, r with 3, and n with 5 in Equation (3), gives

$$S_5 = \frac{2(1 - 3^5)}{1 - 3}$$

$$= \frac{2(-242)}{-2}$$

$$= 242.$$

Infinite Geometric Sequences

If $|r| < 1$, as the number of terms in a geometric sequence approaches infinity, the value of r^n becomes very small. While r^n is never zero, the value becomes so close to zero that we can define the sum of an infinite number of terms of such a sequence.*

DEFINITION 15.3-1 If $|r| < 1$, then the sum of an infinite number of terms of a geometric sequence is

$$S_\infty = \frac{a}{1 - r}.$$

Example Find the sum of an infinite number of terms of the geometric sequence with first term 6 and common ratio $\frac{3}{4}$.

---

*It should be noted that what we give as Definition 15.3-1 is really a theorem. However, the proof of this theorem requires a discussion of limits, which is beyond the level of this book. The equation in the definition should be

$$\lim_{n \to \infty} S_n = \frac{a}{1 - r},$$

which is read, "The limit of S_n as n approaches infinity is $\frac{a}{1 - r}$."

Solution By Definition 15.3-1,

$$S_\infty = \frac{6}{1 - \frac{3}{4}}$$

$$= \frac{6}{\frac{1}{4}}$$

$$= 24.$$

EXERCISE 15.3

WRITTEN

Find the first four terms of each geometric sequence in Problems 1–12 for the given values of a and r.

Example $a = 25, r = \frac{1}{5}$.

Solution The first four terms are

$$25, 25(\tfrac{1}{5}), 25(\tfrac{1}{5})^2, 25(\tfrac{1}{5})^3$$

or

$$25, 5, 1, \tfrac{1}{5}.$$

1. $a = 3, r = 2$
2. $a = 2, r = -2$
3. $a = 81, r = \frac{1}{3}$
4. $a = 125, r = \frac{1}{5}$
5. $a = \frac{1}{3}, r = \frac{1}{2}$
6. $a = \frac{1}{5}, r = \frac{1}{5}$
7. $a = \sqrt{2}, r = \sqrt{2}$
8. $a = \sqrt{2}, r = \sqrt{3}$
9. $a = 16, r = 2^{-2}$
10. $a = 243, r = 3^{-3}$
11. $a = 1, r = \dfrac{x}{2}$
12. $a = 1, r = \dfrac{-x}{2}$

In Problems 13–22, find the common ratio, the first term, and the specified term of the geometric sequence with the given terms.

Example Find t_7 if $t_3 = 10$ and $t_5 = 0.1$.

Solution
$t_3 = ar^2 = 10$

$t_5 = ar^4 = 0.1$

$$\frac{ar^4}{ar^2} = \frac{0.1}{10}$$

$r^2 = 0.01$

$r = 0.1$ or -0.1.

If $ar^2 = 10$, then for either value of r

$a(0.01) = 10$

$a = 1000.$

The seventh term is

$t_7 = ar^6$

$= 1000(0.1)^6$

$= 0.001.$

15.3 Geometric Sequences

You should note that the two values of r will generate different sequences but in this case the odd-numbered terms are the same in both sequences.

13. Find t_6 if $t_2 = 2$ and $t_3 = 4$.
14. Find t_5 if $t_3 = 19$ and $t_4 = 57$.
15. Find t_8 if $t_2 = 5$ and $t_5 = \frac{5}{8}$.
16. Find t_{21} if $t_3 = 6$ and $t_6 = -6$.
17. Find t_9 if $t_3 = 4$ and $t_5 = 16$.
18. Find t_6 if $t_2 = 3$ and $t_4 = \frac{4}{3}$.
19. Find t_{11} if $t_3 = 2$ and $t_5 = 4$.
20. Find t_{10} if $t_2 = 1$ and $t_4 = 3$.

21. Find t_7 if $t_2 = y$ and $t_5 = y^4/x^3$ $(x, y \neq 0)$.
22. Find t_2 if $t_5 = x^2/y^3$ and $t_8 = x^5/y^6$ $(x, y \neq 0)$.

23. Find two geometric means between 3 and 24.
24. Find two geometric means between 2 and 54.
25. Find six geometric means between 1 and 128.
26. Find three positive geometric means between $\frac{1}{16}$ and 10,000.

In Problems 27–32, find the sum of the specified number of terms of the geometric sequence.

Example   First four terms if $a = 2$ and $r = 5$.

Solution   If we substitute the given values in Equation (3), page 353, we have

$$S_4 = \frac{2(1 - 5^4)}{1 - 5}$$

$$= \frac{2(-624)}{-4}$$

$$= 312.$$

27. First five terms if $a = 3$ and $r = 2$.
28. First four terms if $a = 2$ and $r = 3$.
29. First six terms if $a = 2$ and $r = -2$.
30. First five terms if $a = -1$ and $r = -3$.
31. First six terms if $a = 3$ and $r = \frac{2}{3}$.
32. First seven terms if $a = \frac{8}{125}$ and $r = \frac{5}{2}$.

In Problems 33–36, find the sum of the terms of the infinite sequence with the given values of a and r.

Example $a = 3$ and $r = \frac{1}{2}$

Solution   If we substitute the given values in $S_\infty = \dfrac{a}{1 - r}$, we obtain

$$S_\infty = \frac{3}{1 - \frac{1}{2}} = 6.$$

33. $a = 5$ and $r = \frac{1}{4}$
34. $a = 6$ and $r = \frac{1}{3}$
35. $a = 4$ and $r = -\frac{1}{2}$
36. $a = 6$ and $r = -\frac{1}{3}$

37. Find the sum of the first six terms of the geometric sequence whose second term is 2 and whose fifth term is 16.
38. Find the sum of the first eight terms of the geometric sequence whose second term is 5 and whose fifth term is $\frac{5}{8}$.

39. For what value or values of k is the sequence $\{k - 2, k - 6, 2k + 3\}$ a geometric sequence?
40. For what value or values of k is the sequence $\{3k + 4, k - 2, 5k + 1\}$ a geometric sequence?
41. Find four numbers such that the first three form an arithmetic sequence and the last three form a geometric sequence if the first number is 6 and $r = \frac{1}{2}$.
42. Write the first four terms of the sequence defined by $f(n) = 1 \cdot \left(\dfrac{x}{n+1}\right)$, $n \in \{0, 1, 2, 3, \ldots\}$. Is this sequence a geometric sequence? Why or why not?
43. A pile of logs is arranged so that there are 20 logs in the bottom layer of the pile, each successive layer contains one less log, and the top layer has only one log. How many logs are there in the pile?
44. 105 logs are piled in the same manner as in Problem 43. How many layers are there in this pile?
45. A rubber ball is dropped from a height of 6 feet and rebounds $\frac{3}{4}$ the distance it fell each time it strikes the floor. What is the approximate distance the ball will travel before coming to rest?
46. From what height should the ball in Problem 45 be dropped so that its total distance traveled is approximately 36 feet?
47. Show that the sum of the first n positive odd integers is equal to n^2.
48. Show that the sum of the first n positive even integers is equal to $n(n + 1)$.

15.4 Powers of Binomials

A special kind of sequence can be formed by writing a power of a binomial as a polynomial. Such a polynomial is called an **expansion** of the binomial. In this section we discuss the nature of binomial expansions and point out some features of the terms of the polynomials to help you develop an expression for the general term. A proof of this discussion is beyond the level of this book.

Some Specific Powers

Recall that
$$(a + b)^2 = a^2 + 2ab + b^2$$
and that
$$(a + b)^3 = a^3 + 3a^2b + 3ab^2 + b^3.$$
By computing additional powers of $a + b$, we find that
$$(a + b)^4 = a^4 + 4a^3b + 6a^2b^2 + 4ab^3 + b^4$$
$$(a + b)^5 = a^5 + 5a^4b + 10a^3b^2 + 10a^2b^3 + 5ab^4 + b^5$$
and so on.

Some Observable Features

If you examine the polynomials above, paying careful attention to the coefficients and exponents of the consecutive terms, you can observe the following for each of the four polynomials, where n is the exponent of the power of the binomial.

(1) There are $n + 1$ terms in the expansion.
(2) The sum of the exponents in any term is n.
(3) The first term of the expansion is a^n.
(4) The second term of the expansion is $na^{n-1}b$.
(5) The exponent of a is decreased by 1 and the exponent of b is increased by 1 from any term to the next. Also, the exponent of b is 1 less than the number of the term.
(6) The coefficient of any term can be found by dividing the product of the coefficient and the exponent of a in the last preceding term by the exponent of b in the new term.

If we assume that these statements are true for any value of n, then

$$(a + b)^n = a^n + \frac{1 \cdot n}{1} a^{n-1}b + \frac{1 \cdot n(n-1)}{1 \cdot 2} a^{n-2}b^2$$
$$+ \frac{1 \cdot n(n-1)(n-2)}{1 \cdot 2 \cdot 3} a^{n-3}b^3 + \cdots \quad (1)$$

Factorial Notation

Observe in Equation (1) that the denominator of the coefficient of the second term is 1, that of the third term is $1 \cdot 2$, and that of the fourth term is $1 \cdot 2 \cdot 3$. If you continue to form terms according to statement (6), above, the denominator of each successive term will contain an additional factor which is the next consecutive positive integer. Products such as these can be expressed more simply in **factorial notation**.

DEFINITION 15.4-1 For any positive integer n, the symbol $n!$ (read, "n factorial") represents the product of all the consecutive natural numbers from 1 to n, inclusive.

Examples (a) $5! = 1 \cdot 2 \cdot 3 \cdot 4 \cdot 5 = 120$
(b) $8! = 1 \cdot 2 \cdot 3 \cdot 4 \cdot 5 \cdot 6 \cdot 7 \cdot 8 = 40{,}320$

Factorial notation can also be used to represent the product of any set of consecutive natural numbers, even when the least number is not 1. For example, the product

$$10 \cdot 9 \cdot 8 \cdot 7 = \frac{10!}{6!}$$

because

$$\frac{10!}{6!} = \frac{10 \cdot 9 \cdot 8 \cdot 7 \cdot 6 \cdot 5 \cdot 4 \cdot 3 \cdot 2 \cdot 1}{6 \cdot 5 \cdot 4 \cdot 3 \cdot 2 \cdot 1} = 10 \cdot 9 \cdot 8 \cdot 7.$$

General Term of a Binomial Expansion

Let us make some more observations about the polynomial form of $(a + b)^n$. First, examine the exponents of a and b. For each term, the exponent of b is 1 less than the number of the term, and the exponent of a is n minus the exponent of b. Thus, if r is the number of the term, then the exponent of b is $r - 1$ and the exponent of a is $n - (r - 1) = n - r + 1$. That is, every term in the expansion is of the form

$$C \cdot a^{n-r+1} b^{r-1},$$

where C is the coefficient of the term.

Now examine the coefficients of the terms of the expansion. Observe that each coefficient, except the first, is a fraction. The numerator is a product of consecutive natural numbers with the first factor n and the last factor 1 more than the exponent of a. The denominator is the product of all the natural numbers from 1 up to and including the exponent of b. Thus,

$$C = \frac{1 \cdot n(n-1) \cdots (n-r+1+1)}{1 \cdot 2 \cdots (r-1)} = \frac{n(n-1) \cdots (n-r+2)}{(r-1)!}.$$

Then, substituting this expression for C, an expression for the general term of a binomial expansion is found to be

$$t_r = \frac{n(n-1) \cdots (n-r+2)}{(r-1)!} a^{n-r+1} b^{r-1}, \tag{2}$$

where a is the first term of the binomial, b is the second term of the binomial, n is the exponent of the power of the binomial, and r is the number of the term.

Example   Write the first four terms and the tenth term of $(x + y)^{15}$.

Solution   We find the first four terms by replacing a with x, b with y, and n with 15 in Equation (1) and simplifying.

$$(x + y)^{15} = x^{15} + \frac{1 \cdot 15}{1} x^{14} y + \frac{1 \cdot 15 \cdot 14}{1 \cdot 2} x^{13} y^2$$

$$+ \frac{1 \cdot 15 \cdot 14 \cdot 13}{1 \cdot 2 \cdot 3} x^{12} y^3 + \cdots$$

$$(x + y)^{15} = x^{15} + 15 x^{14} y + 105 x^{13} y^2 + 455 x^{12} y^3 + \cdots$$

To find the tenth term, we use Equation (2).

$$t_{10} = \frac{15 \cdot 14 \cdot 13 \cdot 12 \cdot 11 \cdot 10 \cdot 9 \cdot 8 \cdot 7}{1 \cdot 2 \cdot 3 \cdot 4 \cdot 5 \cdot 6 \cdot 7 \cdot 8 \cdot 9} x^6 y^9$$

$$= 5005 x^6 y^9$$

EXERCISE 15.4

WRITTEN

Write each product in Problems 1–8 in factorial notation.

Examples   (a) $6 \cdot 5 \cdot 4 \cdot 3 \cdot 2 \cdot 1$ (b) $8 \cdot 7 \cdot 6$

15.4 Powers of Binomials

Solutions (a) 6! (b) $\dfrac{8!}{5!}$

1. $8 \cdot 7 \cdot 6 \cdot 5 \cdot 4 \cdot 3 \cdot 2 \cdot 1$
2. $12 \cdot 11 \cdot 10 \cdots 1$
3. $6 \cdot 5 \cdot 4$
4. $100 \cdot 99 \cdot 98 \cdot 97$
5. $(n-2)(n-3) \cdots 1$
6. $(n+2)(n+1)(n) \cdots 1$
7. $(n)(n-1)(n-2)$
8. $(2n+2)(2n+1)(2n)(2n-1)$

Evaluate each expression in Problems 9–24.

Examples (a) 5! (b) $\dfrac{7!}{4!}$

Solutions (a) $5! = 5 \cdot 4 \cdot 3 \cdot 2 \cdot 1 = 120$

(b) $\dfrac{7!}{4!} = \dfrac{7 \cdot 6 \cdot 5 \cdot 4 \cdot 3 \cdot 2 \cdot 1}{4 \cdot 3 \cdot 2 \cdot 1} = 7 \cdot 6 \cdot 5 = 210$

9. $7!$
10. $4!$
11. $8!$
12. $10!$
13. $\dfrac{5!}{2!}$
14. $\dfrac{10!}{7!}$

15. $5! + 3!$
16. $5! - 3!$
17. $5! \cdot 3!$
18. $6! \cdot 2!$

19. $\dfrac{10!}{5! \cdot 4!}$
20. $\dfrac{7!}{4! \cdot 3!}$
21. $\dfrac{8!}{4! \cdot 3! \cdot 2!}$
22. $\dfrac{5!}{3! \cdot 2! \cdot 2!}$

23. $\dfrac{6! + 4!}{6! - 4!}$
24. $\dfrac{8! - 6!}{8! + 6!}$

Write the first four terms of the expansion of each power of a binomial in Problems 25–36.

Example $(2x - y)^5$

Solution Use Equation (1) with $a = 2x$, $b = -y$, and $n = 5$.

$$(2x - y)^5 = (2x)^5 + \dfrac{1 \cdot 5}{1}(2x)^4(-y) + \dfrac{1 \cdot 5 \cdot 4}{1 \cdot 2}(2x)^3(-y)^2$$

$$+ \dfrac{1 \cdot 5 \cdot 4 \cdot 3}{1 \cdot 2 \cdot 3}(2x)^2(-y)^3 + \cdots$$

$$= 32x^5 + 5(16x^4)(-y) + 10(8x^3)(-y)^2 + 10(4x^2)(-y)^3 + \cdots$$

$$= 32x^5 - 80x^4y + 80x^3y^2 - 40x^2y^3 + \cdots$$

25. $(x + y)^6$
26. $(x - y)^6$
27. $(x - 2)^4$
28. $(2x + 1)^5$

29. $(2x - y)^6$
30. $(x + 3y)^7$
31. $\left(\dfrac{x}{3} + 3\right)^6$
32. $\left(\dfrac{x}{2} - 2\right)^7$

33. $(x^2 - 2y)^8$
34. $\left(x^2 + \dfrac{y}{2}\right)^{10}$
35. $\left(x + \dfrac{1}{x}\right)^4$
36. $\left(x - \dfrac{1}{x}\right)^7$

In Problems 37–46, find the specified term.

Example Fifth term of $(2x - y)^{10}$

Solution Use Equation (2) with $a = 2x$, $b = -y$, $n = 10$, and $r = 5$.
Thus,

$$t_5 = \frac{10 \cdot 9 \cdots (10 - 5 + 2)}{(5 - 1)!}(2x)^{10-5+1}(-y)^{5-1}$$

$$= \frac{10 \cdot 9 \cdot 8 \cdot 7}{1 \cdot 2 \cdot 3 \cdot 4}(2x)^6(-y)^4$$

$$= 210(64x^6)(y^4)$$

$$= 13{,}440x^6y^4$$

37. Sixth term of $(x - y)^{15}$
38. Fifth term of $(x + 2)^{12}$
39. Ninth term of $(x - 2y)^{10}$
40. Third term of $(x^2 - y)^9$
41. Fourth term of $(2x - y)^7$
42. Fifth term of $\left(\dfrac{x}{2} + 2y^3\right)^{10}$
43. Term involving y^5 in the expansion of $(x - y)^{10}$
44. Term which does not contain x in the expansion of $\left(x + \dfrac{1}{x}\right)^8$
45. Middle term of $(x^2 - y^4)^8$
46. Middle term of $(y + 2x^3)^6$

Assume that the observable properties of the expansion of a binomial are true if n is either a negative integer or the reciprocal of a positive integer.

47. (a) Write the first four terms of $(1 + x)^{-1}$.

    (b) Use long division to find the first four terms of the quotient $\dfrac{1}{1 + x}$.

48. Find the first four terms of $(1 + x)^{-2}$.
49. Find $\sqrt{1.02}$ to two decimal places. [HINT: $\sqrt{1.02} = (1 + 0.02)^{1/2}$.]
50. Use $33 = 32 + 1$ to find $\sqrt[5]{33}$ to three decimal places.

Summary

1. A function whose domain is a set of two or more integers such that one of the integers is less than every other integer in the set and the set contains every integer within a stated interval, is called a **sequence function**. [15.1]
2. The word **sequence** is usually applied to the ordered set of second components of the ordered pairs in a sequence function. [15.1]
3. An **arithmetic sequence** is a sequence in which any term can be found by adding a constant, called the **common difference**, to the preceding term. [15.2]
4. If a is the first term of an arithmetic sequence and d is the common difference, then the nth term t_n is given by

$$t_n = a + (n - 1)d.$$ [15.2]

5. The sum of the first n terms of an arithmetic sequence is given by

$$S_n = \frac{n}{2}[2a + (n-1)d],$$

or by

$$S_n = \frac{n(a+l)}{2}$$

where l is the nth term. [15.2]

6. A **geometric sequence** is a sequence in which any term can be found by multiplying the preceding term by a constant called the **common ratio**. [15.3]

7. If a is the first term of a geometric sequence and r is the common ratio, then the nth term t_n is given by

$$t_n = ar^{n-1}.$$ [15.3]

8. The sum of the first n terms of a geometric sequence is given by

$$S_n = \frac{a(1-r^n)}{1-r}.$$ [15.3]

9. If $|r| < 1$, the sum of an infinite number of terms of a geometric sequence is defined as

$$S_\infty = \frac{a}{1-r}.$$ [15.3]

10. For any positive integer n, $n! = 1 \cdot 2 \cdot 3 \cdots n$. The symbol $n!$ is read, "n factorial." [15.4]

11. Some properties of the coefficients and exponents of the expanded form of a power of a binomial, such as $(a+b)^n$, were discussed. [15.4]

REVIEW EXERCISES

SECTION 15.1

Write the sequence defined by each equation in Problems 1 and 2. $n \in \{1, 2, 3, 4\}$.

1. $f(n) = \dfrac{n^2}{2n-1}$
2. $f(n) = \dfrac{(-1)^{n+1}(2n-3)}{n}$

In Problems 3 and 4, find a suitable equation for each sequence using $n \in \{1, 2, 3, 4\}$.

3. $\{1, 4, 7, 10\}$
4. $\left\{\dfrac{1}{2}, -\dfrac{1}{6}, \dfrac{1}{12}, -\dfrac{1}{20}\right\}$

SECTION 15.2

5. Find the first five terms of an arithmetic sequence if $a = -2$ and $d = 3$.
6. Find the second term and the tenth term of the arithmetic sequence whose third term is 7 and whose eighth term is 17.

7. Insert three arithmetic means between 5 and 33.
8. Find the sum of the first ten terms of the arithmetic sequence with $a = -4$ and $d = 3$.

SECTION 15.3

9. If $a = 64$ and $r = \frac{1}{2}$, find the next four terms of the geometric sequence.
10. The third term of a geometric sequence is 16 and the sixth term is 128. Find the fourth and seventh terms.
11. Insert three geometric means between 1 and 81.
12. Find the sum of the first six terms of the geometric sequence whose first term is 3 and whose common ratio is 2.
13. Find the sum of the terms of the infinite geometric sequence if $a = 3$ and $r = \frac{1}{5}$.

SECTION 15.4

Evaluate the factorials in Problems 14–16.

14. $\dfrac{10!}{7!}$ 15. $\dfrac{9!}{3!3!}$ 16. $\dfrac{5! + 3!}{5! - 3!}$

17. Find the first four terms of $(x + y)^{12}$.
18. Find the first four terms of $(x - 2y)^{10}$.
19. Find the fifth term of $(x - y)^8$.
20. Find the tenth term of $(2x - y)^{14}$.

Appendix 1 / Factoring Quadratic Trinomials over the Set of Integers

Factoring quadratic trinomials by completing the square does not involve trial and error. However, this technique is so time consuming that many people prefer to factor trinomials by trying the possible combinations of constants until two binomial factors of the trinomial are found. Unfortunately, not all trinomials with integral coefficients can be factored into two binomials with integral coefficients. The following discussion can help you to determine whether or not the trinomial can be factored over the integers and, if it can be so factored, can help you find the binomial factors with a minimum of guesswork. A set of practice problems is given at the end of the discussion.

An expression of the form $mx + n$ where m and n are constants is called a first-degree binomial in x. The product of two such binomials can be written equivalently in the form $ax^2 + bx + c$, which is called a quadratic trinomial. If the m's and n's in the binomials are integers, then a, b, and c in the trinomial are also integers.

Example Write the product $(2x + 3)(3x - 1)$ as a quadratic trinomial.

Solution $(2x + 3)(3x - 1) = (2x + 3)3x + (2x + 3)(-1)$
$= 2x(3x) + 3(3x) + 2x(-1) + 3(-1)$
$= 6x^2 + 9x - 2x - 3$
$= 6x^2 + 7x - 3$

The terms of the two binomials are named as shown in Figure A.1-1.

Observe that the term of second degree in the trinomial is the product of the first terms of the binomials, the constant term of the trinomial is the product of the last terms of the binomials, and the term of first degree (middle term) of the trinomial is the sum of the product of the outer terms and the product of the inner terms of the binomials. (See Figure A.1-2.)

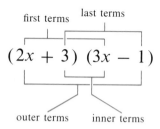

Figure A.1-1

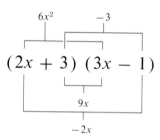

Figure A.1-2

Also notice that the product of the coefficient of the second-degree term and the constant term in the trinomial $[6(-3)]$ is equal to the product of the coefficients of the products of the outer terms and of the inner terms of the binomials $[(-2)9]$.

These observations and the fact that any integer can be written as the product of two integers lead us to state the following theorem, the proof of which is not given.

THEOREM A.1-1 If $a, b, c \in J$, a quadratic trinomial of the form $ax^2 + bx + c$ is factorable into two first-degree binomial factors with integers as coefficients if and only if the product ac can be factored into two integers whose sum is equal to b. That is,

$$ax^2 + bx + c = (mx + n)(px + q) \quad \text{(all coefficients} \in J\text{)}$$

if and only if

$$mp \cdot nq = ac \quad \text{and} \quad mq + np = b.$$

The following examples illustrate the application of the theorem.

Examples Can the following trinomials be factored into two first-degree binomials with integers as coefficients? If so, give the factored form.
(a) $4x^2 - 4x - 15$ (b) $3x^2 + 4x + 2$

Solutions (a) In this trinomial $a = 4$, $b = -4$, and $c = -15$. Hence, we need to find two factors of $4(-15)$ or -60 whose sum equals -4. Two such integers are 6 and -10, since $6(-10) = -60$ and $6 + (-10) = -4$. Therefore, the trinomial is factorable into the desired form.

Since $ax^2 + bx + c = (mx + n)(px + q)$, $mp = a = 4$, $nq = c = -15$, and $mq + np = b = -4$. However, $mq \cdot np = ac =$

−60, and thus mq can equal either 6 or −10, and np equals −10 or 6. If we select 6 as the value of mq, we begin the factoring as follows:

$$4x^2 - 4x - 15$$

$$(mx + n)(px + q)$$

with $mq = 6$

$$(2x\quad)(\quad + 3)$$

The binomials can now be completed by replacing n with −5, the other factor of −15, and p with 2, the other factor of 4, to obtain

$$(2x - 5)(2x + 3).$$

Multiplication shows that the product of these two binomials is the original trinomial.

(b) In $3x^2 + 4x + 2$, $a = 3$, $b = 4$, $c = 2$, and the product $ac = 6$. The only possible factors of 6 are 1 and 6; −1 and −6; 2 and 3; and −2 and −3. Since none of the sums of these pairs is 4, the trinomial cannot be factored over the set of integers.

This technique of factoring depends only upon the coefficients. As a result, it can be used to factor some quadratic trinomials that contain more than one variable.

Example Factor $4x^2 - 12xy + 9y^2$ into two first-degree binomials with integers as coefficients, if possible.

Solution $a = 4$, $b = -12$, $c = 9$, and $ac = 36$. Since $(-6)(-6) = 36$ and $(-6) + (-6) = -12$, the trinomial is factorable into the required form.

$$4x^2 - 12xy + 9y^2$$

$$(mx + ny)(px + qy)$$

with -6

$$(2x\quad)(\quad - 3y)$$

The binomials are now completed by replacing n with −3, the other factor of 9, and p with 2, the other factor of 4. The factored form is

$$(2x - 3y)(2x - 3y).$$

It can be verified by multiplication that the product of these two binomials is the original trinomial.

PRACTICE PROBLEMS

Follow the examples above to determine if the trinomials are factorable, and if they are, find the factored form.

1. $x^2 + 3x + 2$
2. $x^2 + 4x + 3$
3. $x^2 + 2x + 3$
4. $x^2 + 4x + 2$
5. $2x^2 - 5x - 3$
6. $5x^2 - 6x + 1$
7. $2x^2 + 9x + 10$
8. $3x^2 - 14x + 15$
9. $8x^2 + 6x - 9$
10. $6x^2 - 11x - 7$
11. $12x^2 + 16x - 5$
12. $12x^2 - 13x - 12$
13. $x^2 - 9x + 14$
14. $x^2 + 6x - 55$
15. $x^2 + 12x + 36$
16. $x^2 - 4x + 4$
17. $9x^2 + 6x + 1$
18. $x^2 - 5x + 4$
19. $x^2 + 6x - 9$
20. $x^2 + 3x + 10$
21. $2x^2 + 5x - 8$
22. $10x^2 - 23x + 12$
23. $2x^2 + 5x - 7$
24. $7x^2 + 13x + 6$
25. $2x^2 - 13x + 20$
26. $6x^2 - 23x + 15$
27. $8x^2 - 12x + 9$
28. $12x^2 + 19x - 18$
29. $8x^2 - 2x - 45$
30. $20x^2 + 9x - 18$
31. $x^2 + 8xy + 15y^2$
32. $x^2 + 5xy + 6y^2$
33. $x^2 - 8xy + 7y^2$
34. $x^2 - 10xy + 9y^2$
35. $x^2 - 9xy + 15y^2$
36. $x^2 - 3xy + 4y^2$
37. $2x^2 + xy - 3y^2$
38. $6x^2 + 7xy + 2y^2$
39. $2x^2 + 5xy - 7y^2$
40. $9x^2 - 12xy + 8y^2$
41. $20x^2 - 13xy - 21y^2$
42. $18x^2 - 45xy + 28y^2$

Appendix 2 / Table of Squares and Square Roots

N	N^2	$\sqrt{N}$	$\sqrt{10N}$	N	N^2	$\sqrt{N}$	$\sqrt{10N}$
1	1	1.000	3.162	26	676	5.099	16.125
2	4	1.414	4.472	27	729	5.196	16.432
3	9	1.732	5.477	28	784	5.292	16.733
4	16	2.000	6.325	29	841	5.385	17.029
5	25	2.236	7.071	30	900	5.477	17.321
6	36	2.449	7.746	31	961	5.568	17.607
7	49	2.646	8.367	32	1 024	5.657	17.889
8	64	2.828	8.944	33	1 089	5.745	18.166
9	81	3.000	9.487	34	1 156	5.831	18.439
10	100	3.162	10.000	35	1 225	5.916	18.708
11	121	3.317	10.488	36	1 296	6.000	18.974
12	144	3.464	10.954	37	1 369	6.083	19.235
13	169	3.606	11.402	38	1 444	6.164	19.494
14	196	3.742	11.832	39	1 521	6.245	19.748
15	225	3.873	12.247	40	1 600	6.325	20.000
16	256	4.000	12.649	41	1 681	6.403	20.248
17	289	4.123	13.038	42	1 764	6.481	20.494
18	324	4.243	13.416	43	1 849	6.557	20.736
19	361	4.359	13.784	44	1 936	6.633	20.976
20	400	4.472	14.142	45	2 025	6.708	21.213
21	441	4.583	14.491	46	2 116	6.782	21.448
22	484	4.690	14.832	47	2 209	6.856	21.679
23	529	4.796	15.166	48	2 304	6.928	21.909
24	576	4.899	15.492	49	2 401	7.000	22.136
25	625	5.000	15.811	50	2 500	7.071	22.361

(continued)

N	N^2	$\sqrt{N}$	$\sqrt{10N}$	N	N^2	$\sqrt{N}$	$\sqrt{10N}$
51	2 601	7.141	22.583	76	5 776	8.718	27.568
52	2 704	7.211	22.804	77	5 929	8.775	27.749
53	2 809	7.280	23.022	78	6 084	8.832	27.928
54	2 916	7.348	23.238	79	6 241	8.888	28.107
55	3 025	7.416	23.452	80	6 400	8.944	28.284
56	3 136	7.483	23.664	81	6 561	9.000	28.461
57	3 249	7.550	23.875	82	6 724	9.055	28.636
58	3 364	7.616	24.083	83	6 889	9.110	28.810
59	3 481	7.681	24.290	84	7 056	9.165	28.983
60	3 600	7.746	24.495	85	7 225	9.220	29.155
61	3 721	7.810	24.698	86	7 396	9.274	29.326
62	3 844	7.874	24.900	87	7 569	9.327	29.496
63	3 969	7.937	25.100	88	7 744	9.381	29.665
64	4 096	8.000	25.298	89	7 921	9.434	29.833
65	4 225	8.062	25.495	90	8 100	9.487	30.000
66	4 356	8.124	25.690	91	8 281	9.539	30.166
67	4 489	8.185	25.884	92	8 464	9.592	30.332
68	4 624	8.246	26.077	93	8 649	9.644	30.496
69	4 761	8.307	26.268	94	8 836	9.695	30.659
70	4 900	8.367	26.458	95	9 025	9.747	30.822
71	5 041	8.426	26.646	96	9 216	9.798	30.984
72	5 184	8.485	26.833	97	9 409	9.849	31.145
73	5 329	8.544	27.019	98	9 604	9.899	31.305
74	5 476	8.602	27.203	99	9 801	9.950	31.464
75	5 625	8.660	27.386	100	10 000	10.000	31.623

Answers to Odd-Numbered Exercises

Chapter 1

EXERCISE 1.1

1. $a \in \{a, b, c\}$ 3. $\frac{2}{3} \notin \{\text{natural numbers}\}$ 5. Monday $\in \{\text{days of the week}\}$
7. ▨ $\notin \{①, ⊖, ⊕, ⊠\}$ 9. $A \not\subset B$ 11. $E \not\subset D$ 13. $C \not\subset D$
15. $3 \not\subset A$ 17. $\{2, 3\} \subset A$
19. $\{a\}, \{b\}, \{a, b\}, \emptyset$ 21. $\{1\}, \{2\}, \{3\}, \{1, 2\}, \{1, 3\}, \{2, 3\}, \{1, 2, 3\}, \emptyset$
23. $\{a\}, \{b\}, \{c\}, \{d\}, \{a, b\}, \{a, c\}, \{a, d\}, \{b, c\}, \{b, d\}, \{c, d\}, \{a, b, c\}, \{a, b, d\}$, $\{a, c, d\}, \{b, c, d\}, \{a, b, c, d\}, \emptyset$
25. $\{a\}, \{b\}, \{c\}, \{d\}, \{e\}, \{a, b\}, \{a, c\}, \{a, d\}, \{a, e\}, \{b, c\}, \{b, d\}, \{b, e\}, \{c, d\}$, $\{c, e\}, \{d, e\}, \{a, b, c\}, \{a, b, d\}, \{a, b, e\}, \{a, c, d\}, \{a, c, e\}, \{a, d, e\}, \{b, c, d\}$, $\{b, c, e\}, \{b, d, e\}, \{c, d, e\}, \{a, b, c, d\}, \{a, b, c, e\}, \{a, c, d, e\}, \{a, b, d, e\}, \{b, c, d, e\}$, $\{a, b, c, d, e\}, \emptyset$
27. Yes. Every element in A must also belong to B.
29. $P = Q$. Every element of P is in Q and every element of Q is in P.
31. No. There may be elements in B that are members of A but that are not members of C.
33. $P = R$. P, Q, and R are different names for the same set.

EXERCISE 1.2

1. $\{a, b, c, d, e, f, g, h\}$ 3. $\{a, b, c, d, e, g, h\}$ 5. $\{a, c\}$ 7. $\emptyset$ 9. $\{a, c\}$
11. $\{a, b, c, d, e, f, g, h\}$ 13. $\emptyset$
15. $\{(a, e), (a, f), (a, g), (a, h), (b, e), (b, f), (b, g), (b, h), (c, e), (c, f), (c, g), (c, h), (d, e),$ $(d, f), (d, g), (d, h)\}$
17. $\{(a, b), (a, d), (a, g), (a, h), (c, b), (c, d), (c, g), (c, h), (e, b), (e, d), (e, g), (e, h), (g, b),$ $(g, d), (g, g), (g, h)\}$
19. $\{a, c, e, g\}$

369

21. 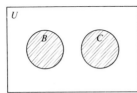   **23.**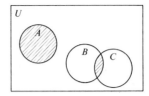

25. 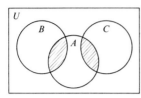   **27.**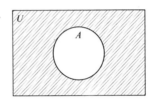

29. Yes   **31.** Yes, if $A \cap C \neq \emptyset$.

EXERCISE 1.3

1. Additive identity   **3.** Closure for addition   **5.** Multiplicative inverse
7. Additive inverse   **9.** Closure for multiplication
11. Commutative law of addition   **13.** Associative law of multiplication
15. Commutative law of multiplication   **17.** Symmetric property of equality
19. Distributive law   **21.** Definition 1.3-1   **23.** Definition 1.3-3
25. Definition 1.3-1

EXERCISE 1.4

1. Theorem 1.4-1 (Addition law)   **3.** Theorem 1.4-3 (Zero factor law)
5. Theorem 1.4-8   **7.** Theorem 1.4-14   **9.** Theorem 1.4-9
11. (a) Hypothesis
    (b) Closure for multiplication
    (c) Reflexive property of equality
    (d) Substitution
13. (a) Definition 1.3-3
    (b) Distributive law
    (c) Definition 1.3-3
    (d) Transitive property of equality
15. (a) Hypothesis
    (b) Multiplicative inverse
    (c) Multiplication law
    (d) Associative law of multiplication
    (e) Multiplicative inverse
    (f) Multiplicative identity

EXERCISE 1.5

1. $a \in \mathbf{R}^+, a > 0$ **3.** Axiom O-2   **5.** Theorem 1.5-3   **7.** Theorem 1.5-8
9. Theorem 1.5-7   **11.** Theorem 1.5-1   **13.** Theorem 1.5-10
15. Definition 1.5-1   **17.** Axiom O-1   **19.** Theorem 1.5-2
21. (a) $x < z$ and $x + y = z$ Hypothesis
    (b) $y \in \mathbf{R}^+$ Definition 1.5-1
    (c) $5y \in \mathbf{R}^+$ Axiom O-2
23. (a) $x, y > 0$ Hypothesis
    (b) $x + y > y$ Theorem 1.5-2
    (c) $\dfrac{1}{x+y} < \dfrac{1}{y}$ Theorem 1.5-9

Answers to Odd-Numbered Exercises **371**

25. (a) Hypothesis
(b) Definition 1.5-1
(c) Multiplication law
(d) Distributive law
(e) Axiom O-2
(f) Definition 1.5-1
(g) Commutative law of multiplication

EXERCISE 1.6

1. 12 **3.** 32 **5.** 5 **7.** 3 **9.** −5 **11.** −24 **13.** −15 **15.** −80
17. $\frac{5}{6}$ **19.** $\frac{47}{75}$ **21.** $\frac{-5}{4}$ **23.** $\frac{-1}{54}$ **25.** $\frac{-34}{35}$ **27.** $\frac{-151}{144}$ **29.** 7 **31.** −18
33. 60 **35.** 211 **37.** −8 **39.** 27 **41.** $\frac{1}{6}$ **43.** $\frac{11}{75}$ **45.** $\frac{7}{6}$ **47.** $\frac{25}{42}$
49. $\frac{-5}{24}$ **51.** $\frac{-1}{5}$

EXERCISE 1.7

1. 35 **3.** 135 **5.** −24 **7.** −378 **9.** −126 **11.** −2772 **13.** 56
15. 1596 **17.** $\frac{1}{6}$ **19.** $\frac{20}{17}$ **21.** $\frac{-3}{8}$ **23.** $\frac{15}{16}$ **25.** $\frac{8}{35}$ **27.** $\frac{63}{88}$ **29.** 5 **31.** 6
33. −3 **35.** −9 **37.** 5 **39.** 7 **41.** $\frac{3}{4}$ **43.** $\frac{2}{3}$ **45.** $\frac{-4}{3}$ **47.** $\frac{-35}{8}$
49. $\frac{7}{12}$ **51.** $\frac{5}{2}$

EXERCISE 1.8

1. As the quotient of two integers, as a terminating decimal, or as a repeating decimal
3. $\frac{1}{8}$ **5.** $\frac{121}{100}$ **7.** $\frac{27}{40}$ **9.** $\frac{-271}{50}$ **11.** $\frac{96}{25}$ **13.** $\frac{-7}{200}$ **15.** 1 **17.** $\frac{5}{9}$
19. $\frac{-23}{99}$ **21.** $\frac{113}{99}$ **23.** $\frac{41}{333}$ **25.** $\frac{41}{330}$ **27.** $\frac{37}{330}$
29. (a) {0} (b) {0} (c) $\{-5.75, 0, 1.25\overline{5}, \sqrt{2.25}, \frac{14}{3}\}$
(d) $\{-\sqrt{23}, -\sqrt{6.75}, -3.45\ldots, \sqrt{3}, 2.01\ldots\}$
31. (a) ∅ (b) $\left\{\frac{-8}{4}\right\}$
(c) $\left\{\frac{-32}{13}, \frac{-8}{4}, -1.22, -\sqrt{1.44}, 0.03\overline{3}, \sqrt{1.69}, 2\sqrt{0.49}, 25\sqrt{0.01}\right\}$ (d) ∅
33. 7 **35.** −1.2 **37.** $\frac{-5}{7}$ **39.** $|x|$ **41.** 2 **43.** $\frac{2}{3}$ **45.** 3.146
47. 4.851 **49.** 1.581 **51.** 3.837 **53.** 8.302

REVIEW EXERCISES, Chapter 1

1. List **2.** {a, b, c, d} **3.** Equivalent **4.** x **5.** Variable, replacement set
6. One-to-one **7.** Subset, $A \subset B$ **8.** Binary **9.** Union, (1, 2, 3, 4, 5)
10. Intersection, {2, 4} **11.** Set-builder notation
12. Cartesian product, {(1, 3), (1, 4), (2, 3), (2, 4)} **13.** Ordered **14.** Commutative

15. Natural numbers, whole numbers, integers, rational numbers, irrational numbers
16. Reflexive property   17. $b = a$ 18. $a = c$, transitive
19. Closed   20. $(a + b) + c, a \cdot b \cdot c$
21. A-1: $a + b \in R$ A-2: $(a + b) + c = a + (b + c)$ A-3: $0 + a = a + 0 = a$
    A-4: $a + (-a) = 0$ A-5: $a + b = b + a$ M-1: $ab \in R$ M-2: $(ab)c = a(bc)$
    M-3: $a \cdot 1 = 1 \cdot a = a$ M-4: $a \cdot \frac{1}{a} = \frac{1}{a} \cdot a = 1$ M-5: $ab = ba$
    D: $a(b + c) = ab + ac$ and $(a + b)c = ac + bc$
22. Theorem   23. Hypothesis, conclusion   24. Deductive   25. Origin
26. Graph   27. Coordinate   28. $a < b$ and $c > b$ 29. $a < 0, a = 0, a > 0$
30. Positive numbers   31. Positive number   32. $a < c$ 33. $<$ 34. $>$
35. Absolute value   36. $-a > 0$ 37. $-(|a| + |b|)$ 38. $|a| - |b|$
39. $-(|b| - |a|)$ 40. $(-b)$ 41. Positive   42. Negative   43. -28
44. -12 45. $\frac{63}{16}$ 46. Infinite   47. Terminating, periodic   48. Irrational
49. Rational   50. 3

Chapter 2

EXERCISE 2.1

1. $10x - 15 = 3x, x \neq 0$ 3. $2x + 4 = 5x - 5, x \neq 1, -2$
5. $4 - 28x < 3x, x > 0$ 7. $4 + 5x - 15 > 2x - 6, x > 3$
9. $3x + 6 > 4x - 28, x > 7$ 11. $x + 1 > 5x + 5 - x, x > -1$ 13. $x = a$
15. $x = \frac{-b}{a^2 - a}$ 17. $x = \frac{b^2 - a^2}{b - a}$ 19. $y > \frac{x - 3}{2}$ 21. $y > 4x - 3$
23. $y < x - 2$ 25. $x = \frac{5y + 40}{7}$ 27. $x = \frac{abc}{a + b}$

EXERCISE 2.2

1. $\left\{\frac{-5}{2}\right\}$ 3. $\emptyset$ 5. $\left\{\frac{-5}{4}\right\}$ 7. $\{0\}$
9. $\{x | x < -5\}$
11. $\{x | x > 4\}$ 13. Absolute inequality
15. $\{3\}, x \neq -2$ 17. $\{4\}, x \neq -1$ 19. $\{3\}, x \neq 2$ 21. $\{3\}, x \neq 1$
23. Identity, $x \neq 0$ 25. $\{x | x \leq \frac{13}{2}\}$
27. $\{x | x \geq -15\}$
29. $\{x | x > 8\}$

31. $\{x \mid x < \frac{-6}{7}\}$

33. $\{x \mid x \leq -4\}$

35. $\{x \mid x \leq -\frac{9}{2}\}$

37. $\{x \mid 0 < x < \frac{1}{3}\}$

39. $\{x \mid -1 < x < \frac{-3}{5}\}$

41. $\{x \mid -12 < x < -3\}$

43. $\{x \mid -2 < x < -1\}$

EXERCISE 2.3

1. $\{-6, 6\}$ **3.** $\{\frac{1}{4}, \frac{5}{4}\}$ **5.** $\{\frac{-8}{3}, -2\}$ **7.** $\{-1, \frac{-1}{3}\}$ **9.** $\{\frac{-1}{12}, \frac{1}{4}\}$

11. $\{x \mid -2 < x < 2\}$

13. $\{x \mid -7 < x < 1\}$

15. $\{x \mid x < 1\} \cup \{x \mid x > 4\}$

17. $\{x \mid -4 < x < 16\}$

19. $k = 3$

21. The absolute value of a number is non-negative.

23. $\{x \mid x < \frac{-1}{3}\} \cup \{x \mid x > 3\}$ **25.** $\emptyset$ **27.** $\{x \mid x < \frac{1}{2}\}$

EXERCISE 2.4

1. (1) Let x represent an odd integer and $x + 2$ represent the next consecutive odd integer.
   (2) $x + x + 2 = 72$
   (3) $\{35\}$
   (4) The integers are 35 and 37.

3. (1) Let x represent the width and $3x$ the length.
(2) $(3x + 5)(x - 2) = x(3x) - 21$
(3) $\{11\}$
(4) The dimensions are 11 feet and 33 feet.

5. (1) Let x, $x + 10$, and $x + 20$ represent the amounts to be paid by John, Jerry, and Paul, respectively.
(2) $x + x + 10 + x + 20 = 90$
(3) $\{20\}$
(4) John pays $20, Jerry pays $30, and Paul pays $40.

7. (1) Let x represent the number of quarters and $x + 32$ the number of dimes.
(2) $25x + 10(x + 32) = 1160$
(3) $\{24\}$
(4) There are 24 quarters and 56 dimes in the collection.

9. (1) Let x represent the number of adult tickets and $2500 - x$ the number of children's tickets.
(2) $150x + 75(2500 - x) = 245{,}625$
(3) $\{775\}$
(4) There were 775 adult tickets and 1725 children's tickets sold.

11. (1) Let x represent the distance from the fulcrum to the 250-pound weight.
(2) $250x = 300(10)$
(3) $\{12\}$
(4) The 250-pound weight is 12 feet from the fulcrum.

13. (1) Let x represent the distance from the fulcrum to the 40-pound weight.
(2) $40x + 5(30) = 7(50)$
(3) $\{5\}$
(4) The 40-pound weight should be placed 5 feet from the fulcrum on the same side as the 30-pound weight.

15. (1) Let x represent the number of days required to fill the reservoir using both pipes.
(2) $\dfrac{1}{x} = \dfrac{1}{3} + \dfrac{1}{5}$
(3) $\{\tfrac{15}{8}\}$
(4) It would require $1\tfrac{7}{8}$ days to fill the reservoir.

17. (1) Let x represent the number of hours required if all three machines are working.
(2) $\dfrac{1}{x} = \dfrac{1}{8} + \dfrac{1}{6} + \dfrac{1}{4}$
(3) $\{\tfrac{24}{13}\}$
(4) $1\tfrac{11}{13}$ hours are needed if all three machines are used.

19. (1) Let x represent the speed of the airplane with no wind.
(2) $\dfrac{840}{x + 30} = \dfrac{660}{x - 30}$
(3) $\{250\}$
(4) The speed of the airplane with no wind is 250 mph.

21. (1) Let x represent the number of seconds it will take the loser.
(2) $\dfrac{1500}{x} = \dfrac{1430}{220}$
(3) $\left\{\dfrac{33{,}000}{143}\right\}$
(4) It would take the loser approximately 231 seconds.

23. (1) Let x represent the score on the sixth test.
  (2) $80 \leq \dfrac{75 + 82 + 61 + 86 + 78 + x}{6} \leq 90$
  (3) $\{x \mid 98 \leq x \leq 158\}$
  (4) He must have a score of 98 or better.
25. (1) Let x represent the number of centimeters in the height.
  (2) $4960 \leq 8 \cdot 10 \cdot 10 \cdot x \leq 5040$
  (3) $\{x \mid 6.2 \leq x \leq 6.3\}$
  (4) The height of the block may be between 6.2 cm and 6.3 cm, inclusive.

REVIEW EXERCISES, Chapter 2

1. Commutative law of addition **2.** Distributive law **3.** Theorem 1.4-1
4. Theorem 1.4-2 **5.** Theorem 1.5-4 **6.** Theorem 1.5-5 **7.** $x = \dfrac{y}{1-y}, y \neq 1$
8. $x \neq y$. If $x - y > 0$, then $y < \dfrac{x}{2}$. If $x - y < 0$, then $y > x$. **9.** $\{2\}$
10. $\{0\}$ **11.** Identity **12.** $\{\tfrac{7}{3}\}$ **13.** $\{x \mid x > 1\}$ **14.** $\{x \mid x < 8\}$
15. Absolute inequality **16.** $\left\{x \mid x < \dfrac{-9}{2}\right\}$ **17.** $\{3, 9\}$ **18.** $\{-10, 5\}$
19. $\left\{\dfrac{-1}{4}, \dfrac{3}{4}\right\}$ **20.** $\{x \mid x < -3\} \cup \{x \mid x > 4\}$ **21.** $\{x \mid 1 < x < 4\}$
22. Absolute inequality
23. (1) Let x represent the number.
  (2) $2x - \tfrac{2}{3}x = 32$
  (3) $\{24\}$
  (4) The number is 24.
24. (1) Let x represent the number of dimes and $x - 3$ the number of nickels.
  (2) $10x + 5(x - 3) = 180$
  (3) $\{13\}$
  (4) There are 13 dimes and 10 nickels.
25. (1) Let x represent the number of articles to be manufactured.
  (2) $250 + \dfrac{100{,}000}{x} < 300$
  (3) $\{x \mid x > 2000\}$
  (4) At least 2001 articles must be manufactured.

Chapter 3

EXERCISE 3.1

1. Theorem 3.0-5 **3.** Theorem 3.0-4 **5.** Theorem 3.0-2
7. Theorem 3.0-5 and Theorem 3.0-4 **9.** Theorem 3.0-2 and Theorem 3.0-3
11. Theorem 3.0-3 and Definition 3.1-2 **13.** $3xy^3$ **15.** $6x^7$ **17.** $\dfrac{3x}{2y^2}$ **19.** $8y$
21. x^{4-n} **23.** x^{6n+8} **25.** x^{2n} **27.** $x^4 y^{2n}$ **29.** $a^{3n+3} t^{6n+3}$ **31.** $2x^2 y^3$
33. $x^3 y^{-2}$ **35.** $3x^2 y^{-1} z^{-3}$ **37.** $4xy^4$ **39.** $x^{-3} y^6$

EXERCISE 3.2

1. 8   3. $\frac{1}{64}$ 5. 8   7. $\frac{1}{256}$ 9. $\frac{1}{25}$ 11. Theorem 3.0-5   13. Theorem 3.0-4
15. Theorem 3.0-2   17. (a) Theorem 3.0-2   (b) Theorem 3.0-3
19. (a) Theorem 3.0-5   (b) Theorem 3.0-4   (c) Theorem 3.0-3   21. x^2 23. $xy^{1/4}$
25. $x^{1/2}y^{1/2}$ 27. $x^{1/2}y$ 29. $x^{11/15}y^{5/3}$ 31. $\frac{y^{1/4}}{x}$ 33. $\frac{y^{1/2}}{x^2}$ 35. $\frac{y^{1/2}}{z^{2/3}}$
37. $\frac{2^3 y^{1/6}}{x^6}$ 39. $\frac{y^{1/2}}{3x^4}$ 41. $x^{4/5}y^{6/5}$ 43. $y^{3/2}$ 45. $\frac{x^{4/3}}{y^{2/3}}$ 47. $\frac{x^{9/4}}{y^{3/2}}$
49. $\frac{1}{x^{1/2}}$ 51. $x^{2/n}$ 53. $x^{(n-8)/n}$ 55. $\frac{x^{6/(n+1)}}{y^{10/(n+1)}}$

EXERCISE 3.3

1. $4x^2 - x + 2$ 3. $y^2 + 5y - 3$ 5. $7x^2y^2 - xy + 2$ 7. $5y^{-2} + y^{-1} + 8$
9. $7x^{3/2} - 6x^{1/2} - 4$ 11. $-x^2 - 5xy + 6y^2$ 13. $6y^2 - 8$
15. $-x^{-2} - x^{-1} - 4$ 17. $8x^{3/2} - 7x^{1/2} - 1$ 19. $-x^{1/2}y^{-2} + 4x^{1/2}y^{-1} - 5xy$
21. $6x^2 - 8x + 4$ 23. $4x^2y^2 + 3xy - 1$ 25. $3x^{3/2} + 2x^{1/2} - 11$
27. $y^{-2} + 3y^{-1} - 4$ 29. $3x^{-2}y^{3/2} - 2x^{-1}y^{1/2} - 4$

EXERCISE 3.4

1. $3x^3 + x^2$ 3. $4x + 1$ 5. $x^{5/2} + x^{-1/2}$ 7. $1 - x^{-2/3}$ 9. $y - 1$
11. $(x + 2)^{5/2} + (x + 2)^{-1/2}$ 13. $y + z - 1$ 15. $x^{2n+1} + 1$
17. $z^{3n/2} - z^{-n/2}$ 19. $(x - 3)^{2n-1} + (x - 3)^{(3n-4)/2}$ 21. $x^2 - 4x - 21$
23. $2x^3 + 3x^2 - 8x - 12$ 25. $2x^3 + x^2 - 5x + 2$ 27. $x^{-2} + 6x^{-1} + 8$
29. $3x^{-3} + 2x^{-2} - x^{-1}$ 31. $x + x^{2/3} - 1 - x^{-1/3}$ 33. $x + 1$
35. $5x^2(x^2 + 3x - 4)$ 37. $xy(3x^2 - 4xy + 7y^2)$ 39. $x^{-2}(x^{-1} + 1 - x^3)$
41. $x^{1/2}(x + 1)$ 43. $y^{1/3}(y^{2/3} - y^{1/3})$ 45. $x^{-1/2}(x^{-1} - 1)$
47. $x^{-1/5}(x + 1)$ 49. $(x + 2)(x + 3)$ 51. $(x - 1)^{-1}[(x - 1)^{-1} - 1]$
53. $(x - 2)^{-1/3}(x - 1)$

EXERCISE 3.5

1. $2\sqrt{2}$ 3. $x\sqrt[3]{x}$ 5. $7\sqrt{2}$ 7. $2\sqrt[3]{3}$ 9. $2x^2\sqrt{x}$ 11. $2x\sqrt[3]{2x}$ 13. $xy\sqrt[3]{3xy}$
15. $-2y\sqrt[3]{4y}$ 17. $\sqrt{3}$ 19. $\sqrt{xy}$ 21. $\sqrt[3]{3}$ 23. $\sqrt[3]{2xy^2}$ 25. $\sqrt{5}$
27. $\sqrt[3]{4}$ 29. $\frac{\sqrt{2}}{2}$ 31. $\sqrt{x}$ 33. $\frac{\sqrt{14}}{4}$ 35. $\frac{2\sqrt[3]{9x}}{3x}$ 37. $\sqrt[5]{2}$ 39. $\frac{\sqrt{x+1}}{x+1}$
41. $\sqrt{3}$ 43. $2\sqrt{x}$ 45. $x\sqrt{6x}$ 47. $y^2\sqrt{5y}$ 49. $\frac{3}{\sqrt{33}}$ 51. $\frac{3x}{\sqrt{6xy}}$
53. $\frac{3}{\sqrt[3]{12}}$ 55. $\frac{2x}{\sqrt[3]{6xy}}$ 57. $\frac{3x}{\sqrt[3]{15xy}}$

EXERCISE 3.6

1. $\sqrt{27}$ or $3\sqrt{3}$ 3. $\sqrt[3]{6xy}$ 5. $\sqrt[3]{6x}$ 7. $\sqrt[6]{8x^5y^5}$ 9. $\sqrt{6}$ 11. $\sqrt{2x}$
13. $\sqrt[3]{36x}$ 15. $\sqrt[10]{27x^3}$ 17. $\sqrt{30xy} + \sqrt{65xz}$ 19. $\sqrt[3]{14y} - \sqrt[3]{18z^2}$

21. $\sqrt{6} - \sqrt{10} + \sqrt{21} - \sqrt{35}$ **23.** $3\sqrt{15} + 6\sqrt{35} - 4\sqrt{6} - 8\sqrt{14}$
25. $3(\sqrt{2} + 1)$ **27.** $\dfrac{2 + \sqrt{2}}{2}$ **29.** $\dfrac{x(\sqrt{x} + 3)}{x - 9}$ **31.** $\dfrac{x + 2\sqrt{xy} + y}{x - y}$
33. $\dfrac{-1}{5 - 2\sqrt{6}}$ **35.** $\dfrac{x - y^2}{x\sqrt{x} - xy + \sqrt{xy} - y\sqrt{y}}$ **37.** $\dfrac{2x - 3y}{x\sqrt{6} + 5\sqrt{xy} + y\sqrt{6}}$
39. $\dfrac{x^4 - 2xy}{x^3 - x\sqrt{2xy} - x^2\sqrt{xy} + xy\sqrt{2}}$ **41.** $\sqrt[12]{128}$ **43.** $\sqrt[4]{20}$ **45.** $\sqrt[6]{x^5}$
47. $x\sqrt[12]{x}$ **49.** $\sqrt[4]{2}$ **51.** $\dfrac{\sqrt[12]{2048}}{2}$ **53.** $\sqrt[6]{x}$ **55.** $\dfrac{\sqrt[4]{xy^2}}{y}$

EXERCISE 3.7

1. $-\sqrt{3}$ **3.** $\sqrt{5}$ **5.** $4\sqrt{xy}$ **7.** 0 **9.** $11\sqrt[3]{2}$ **11.** $8\sqrt[3]{2y}$
13. $8\sqrt{5} - 2\sqrt{10}$ **15.** $\dfrac{5 - \sqrt{3}}{7}$ **17.** $\dfrac{4\sqrt{3} - 3\sqrt{2}}{6}$ **19.** $\dfrac{(2x - y)\sqrt{3}}{3}$

REVIEW EXERCISES, Chapter 3

1. x^5 **2.** x^3 **3.** $\dfrac{1}{y^3}$ **4.** $\dfrac{x}{y}$ **5.** x^4y^6 **6.** $\dfrac{x^4}{y^6}$ **7.** x^3y^2 **8.** y^2 **9.** $\dfrac{x}{y^2z^2}$
10. x^2 **11.** $x^{2n}y^{2n-2}$ **12.** $x^{3n+3}y^{3n}$ **13.** $x^{1/3}$ **14.** $x^{13/10}$ **15.** $\dfrac{z}{x^4y^{3/2}}$
16. $\dfrac{y^{2/5}}{x^{1/3}}$ **17.** $x^{3/4}$ **18.** $\dfrac{y^2}{x^3z}$ **19.** $2x^3 + 3x^2 - 3x + 1$ **20.** $5x^{-2} + 4x^{-1} - 8$
21. $9x^{3/2} + 3x^{1/2} + 9$ **22.** $3x^{1/2}y^{-2} + 2y^{-3/2} - y$ **23.** $x^{7/3} - x$
24. $6x - 1 - 2x^{-1}$ **25.** $x - 2$ **26.** $x^{-1}(x^2 + x + 1)$
27. $x^{1/2}(x^{1/4} + 2x^{1/2} + 1)$ **28.** $(2x - 4)^{1/3}(5 - 2x)$ **29.** $6xy\sqrt{2x}$
30. $-2xy^{-1}\sqrt[3]{2x}$ **31.** $y\sqrt{6y}$ **32.** $\dfrac{2\sqrt{5}}{5}$ **33.** $\dfrac{\sqrt[3]{4x}}{2x}$ **34.** $\dfrac{2\sqrt{6}}{3}$
35. $2x - 5\sqrt{xy} - 12y$ **36.** $\dfrac{x + 4\sqrt{x} + 4}{x - 2}$ **37.** $\sqrt{432}$ **38.** $\dfrac{\sqrt[6]{108}}{2}$
39. $5\sqrt{6x}$ **40.** $3\sqrt[3]{2x^2}$ **41.** $5\sqrt{2}$

Chapter 4

EXERCISE 4.1

1. $1; 0; 1; 16$ **3.** $5; 0; -4; 0$ **5.** $0; -1; 0$ **7.** $0; 12; 0$
9. $a^2 + 3a + 3$ **11.** $x^2 + 2hx + h^2 - 3x - 3h + 2$ **13.** $2x + h - 3$

EXERCISE 4.2

1. $13x^2 - 4x - 9$ **3.** 0 **5.** $x^2 - x + 10$ **7.** $4x^5 + x^4 - 3x^3 - 2x^2 + x + 5$
9. $9x^2 - xy - 2y^2$ **11.** $4x^2 + 4x^2y - 3xy^2 - 2xy + 2x + 5y$
13. $x^4 - x^2 - 2x - 1$ **15.** $3x^5 - x^4 - 14x^3 + 8x^2$
17. $6x^6 + 3x^5 - 22x^4 + 38x^3 - 16x^2 + 3x + 4$

19. $6x^7 - x^5 + 2x^4 + 6x^3 - 3x^2 + 24x + 6$ 21. $x^2 + 2xy + y^2 - z^2$
23. $4x^2 - 16x + 16$ 25. $4x^4 - 25y^2$
27. $4x^2 + 16y^2 + 9z^2 - 16xy + 12xz - 24yz$
29. $x^4 - 4x^3y + 6x^2y^2 - 4xy^3 + y^4$ 31. $x^3 + 9x^2y + 27xy^2 + 27y^3$
33. $x^6 + 3x^4y^2 + 3x^2y^4 + y^6$

EXERCISE 4.3

1. $x + 3$ 3. $x + 1 + \dfrac{4}{x+3}$ 5. $2x + 3 + \dfrac{2}{x+2}$ 7. $x^2 + 7 + \dfrac{33}{x^2-3}$

9. $2x^2 - 3x + 5$ 11. $2x^2 - 5x + 4 + \dfrac{26}{3x-1}$

13. $4x^2 + 4x + 23 + \dfrac{54x+81}{x^2-2x-3}$ 15. $x^2 + 2x + 4$ 17. $3x + y$

19. $2x^2 + 7y$ 21. $4x + 10 + \dfrac{33}{x-3}$ 23. $-2x - 6 - \dfrac{5}{x-2}$

25. $3x^2 - 7x + 16 - \dfrac{25}{x+2}$ 27. $-2x^2 + x - 3 + \dfrac{16}{x+3}$

29. $2x^2 + 3x - 7 + \dfrac{2}{x+1}$ 31. $3x^3 + 4x^2 + 9x + 17 + \dfrac{41}{x-2}$

33. $x^2 + 6x + 17 + \dfrac{54}{x-3}$ 35. $x^3 + 2x^2 + 6x + 12 + \dfrac{26}{x-2}$

EXERCISE 4.4

1. $\pm 8x$ 3. $\pm 14x$ 5. $\pm 18x$ 7. 1   9. 9   11. 16   13. 25
15. y^2 17. 1   19. $\frac{9}{4}$ 21. $(x^2 + 2x + 1) - 4$ 23. $(x^2 - 6x + 9) - 25$
25. $2[(x^2 + 2x + 1) - \frac{9}{2}]$ 27. $5[(x^2 + 6x + 9) - \frac{49}{5}]$ 29. $2[(x^2 + \frac{3}{2}x + \frac{9}{16}) - \frac{23}{16}]$
31. $3[(x^2 + \frac{2}{3}x + \frac{1}{9}) - \frac{1}{3}]$

EXERCISE 4.5

1. $(x + 3)(x - 3)$ 3. $(3x + 2)(3x - 2)$ 5. $(x + \sqrt{2})(x - \sqrt{2})$
7. $(2x + \sqrt{11})(2x - \sqrt{11})$ 9. $(x + 7)(x - 1)$ 11. $(3x + 5)(3x - 1)$
13. $(2x + 2)(2x - 8) = 4(x + 1)(x - 4)$ 15. $(x + y + 3)(x - y - 1)$
17. $(2x + 3y + 3)(2x - 3y - 1)$
19. $(4x + 2y + 4)(4x - 2y + 2) = 4(2x + y + 2)(2x - y + 1)$
21. $(x + \frac{7}{2})^2 - (\frac{1}{2})^2 = (x + 3)(x + 4)$ 23. $(x + 2)^2 - 1 = (x + 3)(x + 1)$
25. $(x - 3)^2 - (2)^2 = (x - 1)(x - 5)$
27. $3[(x - \frac{5}{6})^2 - (\frac{13}{6})^2] = (x - 3)(3x + 4)$
29. $6[(x + \frac{11}{12})^2 - (\frac{5}{12})^2] = 6(x + \frac{4}{3})(x + \frac{1}{2}) = (3x + 4)(2x + 1)$
31. $(x - \frac{5}{2})^2 - (\frac{3}{2})^2 = (x - 1)(x - 4)$
33. $\left(x - \dfrac{7}{2}\right)^2 - \left(\dfrac{\sqrt{41}}{2}\right)^2 = \left(x - \dfrac{7}{2} + \dfrac{\sqrt{41}}{2}\right)\left(x - \dfrac{7}{2} - \dfrac{\sqrt{41}}{2}\right)$
35. $2[(x + 2)^2 - \frac{13}{2}] = 2\left(x + 2 + \dfrac{\sqrt{26}}{2}\right)\left(x + 2 - \dfrac{\sqrt{26}}{2}\right)$

37. $2\left[\left(x + \frac{3}{4}\right)^2 - \left(\frac{\sqrt{17}}{4}\right)^2\right] = \left(2x + \frac{3 + \sqrt{17}}{2}\right)\left(x + \frac{3 - \sqrt{17}}{2}\right)$

39. $(x - y)^2 - (3y)^2 = (x - 4y)(x + 2y)$

41. $(x + 3y)^2 - (4y)^2 = (x - y)(x + 7y)$

43. $2\left[\left(x - \frac{y}{4}\right)^2 - \left(\frac{5y}{4}\right)^2\right] = (2x - 3y)(x + y)$

EXERCISE 4.6

1. $(x - y)(x^2 + xy + y^2)$ **3.** $(x + 2)(x^2 - 2x + 4)$
5. $(5 + x)(25 - 5x + x^2)$ **7.** $(1 - x)(1 + x + x^2)$
9. $(2x - 3)(4x^2 + 6x + 9)$ **11.** $(4x - 3y)(16x^2 + 12xy + 9y^2)$
13. $(x^2 + x + 3)(x^2 - x + 3)$ **15.** $(3x^2 + 2x + 1)(3x^2 - 2x + 1)$
17. $(x^2 + 2x - 2)(x^2 - 2x - 2)$ **19.** $(2x^2 + 2x + 1)(2x^2 - 2x + 1)$
21. $(3x^2 + x - 1)(3x^2 - x - 1)$ **23.** $(2x^2 + 2x - 3)(2x^2 - 2x - 3)$
25. $(2a + b)(x + y)$ **27.** $(4h + 5c)(x - y)$ **29.** $(x^2 + 1)(x - 3)$
31. $(x^2 + 2)(3x - 2)$ **33.** $(x - 3)(x + 1)(x - 1)$
35. $(x + 1)^2 - (3y)^2 = (x + 3y + 1)(x - 3y + 1)$
37. $(y + z)^2 - (2x)^2 = (y + z + 2x)(y + z - 2x)$
39. $(2x)^2 - (3y + 1)^2 = (2x + 3y + 1)(2x - 3y - 1)$
41. $(x + 3)^2 - (y + 1)^2 = (x + y + 4)(x - y + 2)$

REVIEW EXERCISES, Chapter 4

1. $0; -4; -6; 0$ **2.** $0; 0; -12; 0$ **3.** $2x + h + 5$ **4.** $8x^2 + 7x - 4$
5. $6x^2 + 3x + 2$ **6.** $2x^2 + 4xy$ **7.** $6x^2 + 5x - 6$
8. $15x^4 + 11x^3 + 12x^2 + 24x + 8$ **9.** $16x^2 - 40x + 25$ **10.** $4x^2 - 9y^2$
11. $x^4 + 2x^3 + 3x^2 + 2x + 1$ **12.** $8x^3 - 36x^2y + 54xy^2 - 27y^3$ **13.** $x - 2$
14. $x - 8 + \frac{47}{2x + 4}$ **15.** $x^2 + 4x + 5 - \frac{8}{2x + 3}$
16. $2x^2 - x + 3 - \frac{5x - 2}{2x^2 + x - 3}$ **17.** $x - 4$ **18.** $4x^2 + 2x - 7 + \frac{3}{x - 1}$
19. $x^3 - 2x^2 + 7x - 14 + \frac{20}{x + 2}$ **20.** $x^4 + x^3 + x^2 + x + 1$ **21.** $\pm 4x$
22. $\pm 10x$ **23.** 36 **24.** 1 **25.** $(x + 4)^2 - 28$ **26.** $(x + 3)^2 + 3$
27. $3[(x + 1)^2 - \frac{16}{3}]$ **28.** $2[(x + \frac{3}{2})^2 - \frac{3}{4}]$ **29.** $(2x + 3)(2x - 3)$
30. $(3x + 4)(3x - 4)$ **31.** $(2x - 1)(2x + 7)$ **32.** $(x - 1)(x + 5)$
33. $(x - 4)(x - 3)$ **34.** $(3x - 1)(x + 2)$ **35.** $(3x - 2y)(9x^2 + 6xy + 4y^2)$
36. $(5x + 4y)(25x^2 - 20xy + 16y^2)$ **37.** $(x + 1)(x - 1)(x + 3)(x - 3)$
38. $(4x + 2y - z)(4x - 2y + z)$ **39.** $(2x + 3)(y - 4)$ **40.** $(x + 3y - 4)(x - 3y)$

Chapter 5

EXERCISE 5.1

1. $\{5, -2\}$ **3.** $\{3, -4\}$ **5.** $\{0, \frac{2}{3}\}$ **7.** $\{4, -4\}$ **9.** $\{2, 4\}$
11. $\left\{\frac{-3 - \sqrt{29}}{2}, \frac{-3 + \sqrt{29}}{2}\right\}; \{-4.193, 1.193\}$ **13.** $\{-3\}$ **15.** $\{-2, -\frac{1}{3}\}$

17. $\{-1-\sqrt{3}, -1+\sqrt{3}\}; \{0.732, -2.732\}$ 19. $\{-\frac{1}{3}\}$
21. $\left\{\frac{3-\sqrt{5}}{2}, \frac{3+\sqrt{5}}{2}\right\}; \{0.382, 2.618\}$ 23. $\{-\frac{9}{4}, \frac{5}{2}\}$ 25. $\{0, -7\}$
27. $\{-3, 4\}$ 29. $\{\frac{3}{4}, \frac{3}{2}\}$

EXERCISE 5.2

1. Not real numbers   3. Real numbers   5. Real numbers   7. $\{-\frac{4}{3}, \frac{3}{2}\}$ 9. $\{\frac{2}{3}\}$
11. $\left\{\frac{1+\sqrt{3}}{3}, \frac{1-\sqrt{3}}{3}\right\}; \{0.91, 0.24\}$ 13. $\{\frac{3}{5}, \frac{1}{2}\}$ 15. $\{-\frac{4}{7}, \frac{4}{7}\}$ 17. $\{-\frac{2}{5}, 0\}$
19. $\{-4, 6\}$ 21. $\{2+\sqrt{2}, 2-\sqrt{2}\}; \{0.59, 3.41\}$
23. $\left\{\frac{5+\sqrt{7}}{2}, \frac{5-\sqrt{7}}{2}\right\}; \{3.823, 1.823\}$ 25. 4   27. 9
29. $\{-2\sqrt{3}, 2\sqrt{3}\}; \{-3.46, 3.46\}$ 31. $k < 16$ 33. $|k| > 6$ 35. $k < -2$

EXERCISE 5.3

1. $\{16\}$ 3. $\{22\}$ 5. $\{4\}$ 7. $\varnothing$ 9. $\{3\}$ 11. $\{13\}$ 13. $\{4\}$
15. $\varnothing$ 17. $\{-\frac{1}{2}, \frac{3}{2}\}$ 19. $\{\frac{1}{8}\}$ 21. $\{-\frac{8}{9}\}$ 23. $\{4\}$ 25. $\{0\}$
27. $\{\frac{1}{2}\}$ 29. $\{4\}$

EXERCISE 5.4

1. $\{\frac{1}{4}, 9\}$ 3. $\{-2, -1, 1, 2\}$ 5. $\{-\sqrt{6}, \sqrt{6}\}$ 7. $\{-8, 64\}$ 9. $\{-\frac{1}{2}, \frac{1}{3}\}$
11. $\{-1, -\frac{1}{3}, \frac{1}{3}, 1\}$ 13. $\{-64, \frac{125}{27}\}$ 15. $\{16\}$ 17. $\{-2, 6\}$
19. $\{-7, -3, -2, 2\}$ 21. $\{-\sqrt{14}, -3, 3, \sqrt{14}\}$

EXERCISE 5.5

1. $\{x \mid x < 0\} \cup \{x \mid x > 1\}$

3. $\{x \mid x < -2\} \cup \{x \mid x > 3\}$

5. $\{x \mid x < -\frac{3}{2}\} \cup \{x \mid x > 2\}$

7. $\{x \mid x < 2\} \cup \{x \mid x > 5\}$

9. $\{x \mid x < -\frac{3}{2}\} \cup \{x \mid x > -1\}$

11. $\{x \mid 0 \leq x \leq \frac{7}{2}\}$

13. $\{x \mid -5 < x < -2\}$

15. $\{x | 2 < x < 3\}$

17. $\{x | x < -4\} \cup \{x | x > 4\}$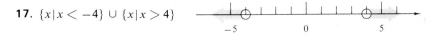

19. $\{x | -1 \leq x \leq \frac{5}{3}\}$

21. $\{x | x < -2\} \cup \{x | 2 < x < 5\}$

23. $\{x | -4 < x < -3\} \cup \{x | x > 1\}$

25. $\{x | x < -1\} \cup \{x | 3 < x < 4\}$

27. If $x > 1$, you can multiply both members by x to obtain $x^2 > x$, or $x < x^2$.

EXERCISE 5.6

1. (1) Let x represent a positive even integer and $x + 2$ represent the next consecutive even integer.
   (2) $x(x + 2) = 168$
   (3) $\{-14, 12\}$
   (4) The integers are 12 and 14.
3. (1) Let x represent one of the numbers and $x - 7$ the other.
   (2) $x(x - 7) = 198$
   (3) $\{-11, 18\}$
   (4) The two numbers are -11 and -18, or 18 and 11.
5. (1) Let x represent an even integer and $x + 2$ the next consecutive even integer.
   (2) $x(x + 2) = 288$
   (3) $\{-18, 16\}$
   (4) The two integers are -18 and -16, or 16 and 18.
7. (1) Let x represent one part and $42 - x$ the other.
   (2) $x = (42 - x)^2$
   (3) $\{36, 49\}$
   (4) The two parts are 6 and 36. (49 and -7 are not considered in this case since 42 is a positive number.)
9. (1) Let x represent the first positive integer, $3x$ the second, and $2x^2$ the third.
   (2) $x + 2x + 2x^2 = 44$
   (3) $\{-\frac{11}{2}, 4\}$
   (4) The positive integers are 4, 8, and 32.
11. (1) Let x represent the measure of the altitude and $x + 5$ the measure of the base.
    (2) $\frac{x}{2}(x + 5) = 42$
    (3) $\{-12, 7\}$
    (4) The altitude is 7 feet and the base is 12 feet.

13. (1) Let x represent the measure of the width and $x + 3$ the measure of the length.
(2) $x(x + 3) = 9(42)$
(3) $\{-21, 18\}$
(4) The dimensions are 18 feet and 21 feet.

15. (1) Let x represent the length of a side of the square.
(2) $(x + 2)^2 = x^2 + 24$
(3) $\{5\}$
(4) The side of the original square is 5 inches.

17. (1) Let x represent the length of a side of the square.
(2) $2(x - 4)^2 = 128$
(3) $\{-4, 12\}$
(4) The side of the original square is 12 inches.

19. (1) Let x represent the width and $2x$ the length.
(2) $2(x - 4)(2x - 4) = 480$
(3) $\{-8, 14\}$
(4) The dimensions are 14 inches and 28 inches.

21. (1) Let x represent the speed in miles per hour.
(2) $\frac{x}{5}(x) = \frac{32}{5}(x + 10)$
(3) $\{-8, 40\}$
(4) His average speed was 40 miles per hour.

23. (1) Let x represent the measure of one dimension and $3x$ the other.
(2) $(x + 4)(3x - 2) = 6x^2$
(3) $\{\frac{4}{3}, 2\}$
(4) The dimensions are $\frac{4}{3}$ inches and 4 inches, or 2 inches and 6 inches.

25. (1) Let x represent a positive integer and $x + 1$ the next consecutive integer.
(2) $x(x + 1) < 56$
(3) $\{x \mid -8 < x < 7\}$
(4) The smaller integer must be between 0 and 7.

27. (1) Let x represent the smaller number and $x + 2$ the larger.
(2) $x(x + 2) < 63$.
(3) $\{x \mid -9 < x < 7\}$
(4) The greatest integer less than the smaller number is -9. The least integer greater than the smaller number is 7.

29. (1) Let x represent the measure of a side of the square.
(2) $108 < 2(x - 4)^2 < 128$
(3) $\{x \mid 11 < x < 12\}$
(4) The length of a side of the square is between 11 inches and 12 inches.

REVIEW EXERCISES, Chapter 5

1. $\{2, 3\}$ **2.** $\{-\frac{3}{2}, \frac{3}{4}\}$ **3.** $\{-1, \frac{7}{3}\}$ **4.** $\{-\frac{1}{4}, \frac{3}{2}\}$ **5.** $\{0, \frac{7}{2}\}$ **6.** $\{-3, -\frac{4}{3}\}$
7. Real numbers **8.** Not real numbers **9.** Real numbers **10.** Real numbers
11. $\{-5, 3\}$ **12.** $\{-6, 2\}$ **13.** $\left\{\dfrac{-\sqrt{10}}{2}, \dfrac{\sqrt{10}}{2}\right\}$ **14.** $\{-\frac{3}{2}, \frac{2}{3}\}$
15. $\left\{\dfrac{-7 + \sqrt{109}}{6}, \dfrac{-7 - \sqrt{109}}{6}\right\}$ **16.** $\left\{\dfrac{-3 + \sqrt{73}}{4}, \dfrac{-3 - \sqrt{73}}{4}\right\}$ **17.** $\{16\}$
18. $-3 < k < 3$ **19.** $\{-7\}$ **20.** $\{24\}$ **21.** $\{-3, -2\}$ **22.** $\{4\}$

Answers to Odd-Numbered

23. $\{5\}$ 24. $\{-3\}$ 25. $\{1, \frac{9}{4}\}$ 26. $\{8, 27\}$ 27. $\{\frac{1}{2}, 3\}$
28. $\{-3, -2, 2, 3\}$ 29. $\{-3, -2, 2, 3\}$ 30. $\{-\sqrt{7}, -\sqrt{2}, \sqrt{2}, \sqrt{7}\}$
31. $\{x \mid x < -3\} \cup \{x \mid x > 2\}$ 32. $\{x \mid -\frac{3}{2} < x < \frac{2}{3}\}$ 33. $\{x \mid 2 < x < $
34. $\{x \mid x < -1\} \cup \{x \mid x > \frac{3}{2}\}$ 35. $\{x \mid -2 < x < 1\} \cup \{x \mid x > 5\}$
36. (1) Let x represent the number of trees bought the first time and $x + 10$ t. bought the second time.
    (2) $\dfrac{100}{x} - \dfrac{100}{x + 10} = 5$
    (3) $\{-20, 10\}$
    (4) The man bought 10 trees the first time and 20 trees the second time.
37. (1) Let x represent a positive even integer and $x + 2$ the next consecutive even integer.
    (2) $x(x + 2) < 48$
    (3) $\{x \mid -8 < x < 6\}$
    (4) The greatest value of the smaller integer is 4.

Chapter 6

EXERCISE 6.1

1. No 3. No 5. Yes 7. No 9. No 11. $\dfrac{x + 1}{3x + 2}$ 13. $\dfrac{-2}{x}$

15. $\dfrac{x + 2}{x - 2}$ 17. $\dfrac{2 - x}{2}$ 19. $-3x - 3$ 21. $x^2 - 4$ 23. $4x^2 - 2x - 30$

25. $(1 - x)(2 + x)$ 27. $x^3 + 2x^2 + 2x + 1$

EXERCISE 6.2

1. $\dfrac{3x + 2}{y}, y \neq 0$ 3. $\dfrac{5x + 4y}{xy}, x, y \neq 0$ 5. $\dfrac{x^2 + x + 2}{y^2}, y \neq 0$

7. $\dfrac{2x + 1}{y}, y \neq 0$ 9. $\dfrac{3x + y}{z}, z \neq 0$ 11. $\dfrac{3x + 5}{x + 1}, x \neq -1$

13. $\dfrac{3x + 1}{x + 5}, x \neq -5$ 15. $x + 5, x \neq -2$ 17. $\dfrac{3x^2 + 3x - 5}{x - 1}, x \neq 1$

19. $\dfrac{2x - 10}{x + 2}, x \neq -2, -3$ 21. $12x^2y^3$ 23. $18x^2y^3$ 25. $(x + 1)(x - 1)(x - 1)$

27. $(x - 3)(x + 2)(x + 3)$ 29. $(x + 1)(x - 1)(x^2 + x + 1)(x^2 - x + 1)$

31. $(x^2 + x + 1)(x^2 - x + 1)(x + 1)(x - 1)$ 33. $\dfrac{2x^2y + 12}{3xy^2}$ 35. $\dfrac{35 - 12x^3}{15x^2y}$

37. $\dfrac{10y^2 + 4xy - 5}{5xy^2}$ 39. $\dfrac{15x - 2}{(3x - 2)(3x + 2)}$ 41. $\dfrac{4x^2 - 8x - 8}{(2x + 3)(3x - 2)}$

43. $\dfrac{x^2 - 17x + 1}{(x + 4)(x - 3)}$ 45. $\dfrac{9 + 9x - 12x^2}{2(2x + 3)(2x + 3)}$ 47. $\dfrac{4x}{(x - 4)(x + 4)(x + 4)}$

49. $\dfrac{6x + 6y - x^2 + 5xy - 6y^2}{2(x - 3y)(x + 2y)(x + y)}$ 51. $\dfrac{7x^2 - x - 3}{2(x - 1)(x^2 + x + 1)}$

53. $\dfrac{24x^2 + 24x + 45}{(2x - 3)(4x^2 + 6x + 9)(2x - 3)}$ 55. $\dfrac{x^3 + x}{(x^2 + x + 1)(x^2 - x + 1)}$

7. $\dfrac{4x^4 - 38x^2 - 21x + 63}{2(2x+3)(2x-3)(x-3)}$ 59. $\dfrac{6x - 3xy - 3z}{(z+2x)(z-2x)(y-3)(y-3)}$

EXERCISE 6.3

1. $\dfrac{4y}{x}$ 3. $\dfrac{6vx}{7w^2y^3}$ 5. $4y$ 7. $\dfrac{xz^2}{12w^2y}$ 9. $\dfrac{x-6}{x-2}$ 11. $\dfrac{6x^2 - 17x + 5}{2x}$

13. $\dfrac{4x^2}{x+3}$ 15. $x^3 + x^2 - 2x + 12$ 17. $\dfrac{x^3 + 64}{(x+2)(x-2)}$ 19. $\dfrac{2x-4}{x+1}$

21. $\dfrac{2+x}{3-x}$ 23. $\dfrac{x^2 + 3x + 2}{x}$ 25. $\dfrac{x-1}{x-7}$ 27. $\dfrac{2-x}{1+x}$ 29. $\dfrac{x^2 - 2x + 2}{y^2 + y + 1}$

31. 1 33. $\dfrac{x-1}{x-3}$ 35. $\dfrac{1 + 3x - 18x^2}{1 + 3x}$ 37. $\dfrac{2x + 3y}{x + 3y}$ 39. $\dfrac{x-y}{x+y}$

41. $\dfrac{-x^2 + 4xy - 4y^2}{y^2}$ 43. $\dfrac{3x^2 - 3xy + 8}{2(2x+y)}$ 45. $\dfrac{y^2(x^2 - y^2)}{3x^2 - 2y^2}$ 47. $\dfrac{x^2 + y^2}{2xy}$

EXERCISE 6.4

1. $\{-\tfrac{7}{5}, 2\}$ 3. $\{-\tfrac{17}{2}, 2\}$ 5. $\{-\tfrac{9}{4}, 2\}$ 7. $\{3, 6\}$ 9. $\{-\tfrac{4}{5}, 1\}$
11. $\{-3, 3\}$ 13. $\{-\tfrac{9}{5}, 3\}$ 15. $\{-\tfrac{1}{2}, 2\}$ 17. $\{-\tfrac{59}{9}, 2\}$
19. $\{2 - \sqrt{11}, 2 + \sqrt{11}\}$ 21. $\{8\}$
23. (1) Let x represent an odd integer and $x + 2$ the next consecutive odd integer.
(2) $\dfrac{x}{x+1} + \dfrac{x+2}{x+3} = \dfrac{19}{12}$ (3) $\{-\tfrac{11}{5}, 3\}$ (4) The integers are 3 and 5.

25. (1) Let x represent the rate of the boat in still water.
(2) $\dfrac{30}{x+3} + \dfrac{18}{x-3} = 4$ (3) $\{0, 12\}$

(4) The speed of the boat is 12 miles per hour.
27. (1) Let x represent the number of boys.
(2) $\dfrac{40}{x-1} = \dfrac{250}{x} - 40$ (3) $\{\tfrac{5}{4}, 5\}$ (4) There were 5 boys in the group.

EXERCISE 6.5

1. $\tfrac{6}{7}$ 3. $\dfrac{x+1}{x-1}$ 5. $2x - 1$ 7. $\dfrac{2x-1}{x}$ 9. $x - 1$ 11. $\dfrac{x^2 + x - 2}{x^2 + 2x - 6}$

13. $\dfrac{x-2}{x-1}$ 15. $\dfrac{-2}{x+2}$ 17. $\dfrac{x}{x-2}$ 19. 1 21. $\dfrac{2x-6}{2x-3}$ 23. $\dfrac{x+2}{x-1}$

25. $\dfrac{x-2}{2x-1}$ 27. $\dfrac{2x}{x^2+1}$ 29. $\dfrac{x+2}{2(x+1)^{3/2}}$ 31. $\dfrac{x^3 - 8x}{(x^2 - 4)^{3/2}}$

33. $\dfrac{2 - 2x}{x^2(3x-2)^{2/3}}$

REVIEW EXERCISES, Chapter 6

1. Yes   2. No   3. $2x + 7$ 4. $4x^2 + 6x + 9$ 5. $x + y$ 6. $x^2 - xy + y^2$
7. $x^2 - x$ 8. $2x + 2y$ 9. $x^2 - 8x + 16$ 10. $x^3 - 2x^2y + 2xy^2 - y^3$

11. $\dfrac{3x+1}{x^2-x+2}$ 12. $\dfrac{-x^2-4xy+3y^2}{(x+y)(x-y)}$ 13. $\dfrac{2x^2-4x-4}{(x+4)(x-4)(x-1)}$
14. $\dfrac{x^2}{x^3-1}$ 15. $\dfrac{(x+2)^2}{x^2-16}$ 16. $\dfrac{x+3}{x-5}$ 17. $\dfrac{x-5}{x+5}$ 18. $\dfrac{x-1}{x-3}$ 19. $\{2\}$
20. $\{2\}$ 21. $\{-\tfrac{13}{4}\}$ 22. $\{\tfrac{3}{2}\}$
23. (1) Let x represent an even integer and $x+2$ the next consecutive even integer.
    (2) $\dfrac{x(x+2)}{x+(x+2)} = \dfrac{x}{2} + \dfrac{5}{11}$
    (3) $\{10\}$
    (4) The integers are 10 and 12.
24. $\dfrac{x^3+1}{x^2+1}$ 25. $\dfrac{-(x^2+y^2)}{4xy}$

Chapter 7

EXERCISE 7.1

1. $\{(1,1), (1,2), (1,5), (2,1), (2,3), (3,2), (3,3), (5,1)\}$
3. $\{(1,1), (2,1), (2,2), (3,1), (3,3), (5,1), (5,5), (6,1), (6,2), (6,3), (6,6)\}$
5. $\{(2,1), (3,1), (3,2), (5,1), (5,2), (5,3), (6,1), (6,2), (6,3), (6,5)\}$
7. (a) $\{-3, -2, -1, 0, 1, 2, 3\}$
   (b) $\{(-3,-5), (-2,-4), (-1,-3), (0,-2), (1,-1), (2,0), (3,1)\}$
9. (a) $\{-3, -2, -1, 0, 1, 2, 3\}$
   (b) $\{(-3,3), (-2,2), (-1,1), (0,0), (1,1), (2,2), (3,3)\}$
11. (a) $\{-3, -1, 0, 1, 3\}$ (b) $\{(-3,-2), (-1,2), (0,1), (1,2), (3,-2)\}$
13. (a) $\{-2, -1, 0, 1, 2\}$ (b) $\{(-2,0), (-1, \sqrt{3}), (0,2), (1, \sqrt{3}), (2,0)\}$
15. (a) $\{-1, 0, 1\}$ (b) $\{(-1, 2/\sqrt{3}), (0,1), (1, 2/\sqrt{3})\}$
17. Domain: $\{x \mid x \in R\}$ Range: $\{y \mid y \geq -1\}$
19. Domain: $\{x \mid -2 \leq x \leq 2\}$ Range: $\{y \mid 0 \leq y \leq 2\}$
21. Domain: $\{x \mid -3 \leq x \leq 3\}$ Range: $\{y \mid -3 \leq y \leq 3\}$
23. Domain: $\{x \mid x \in R, x \neq -1\}$ Range: $\{y \mid y \in R, y \neq 0\}$
25. Domain: $\{x \leq -4\} \cup \{x \mid x \geq 4\}$ Range: $\{y \mid y \in R\}$
27. Domain: $\{x \mid -2 < x < 2\}$ Range: $\{y \mid y \in R\}$
29. 17, 19, 23, 27

EXERCISE 7.2

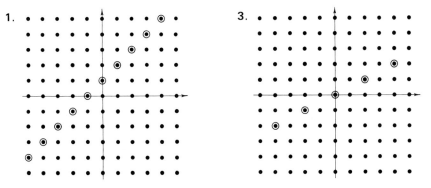

386 ANSWERS TO ODD-NUMBERED EXERCISES

5.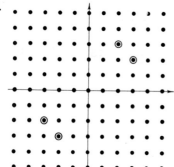

7.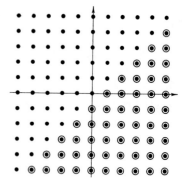

9.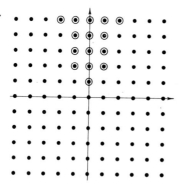

EXERCISE 7.3

1.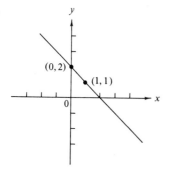

3.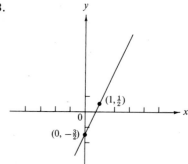

5.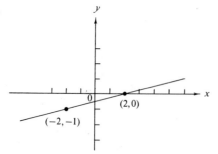

7.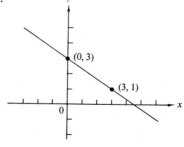

9.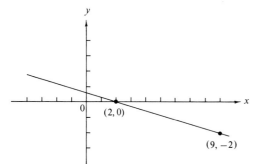

11.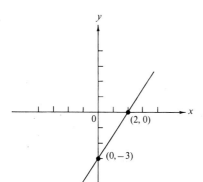

13.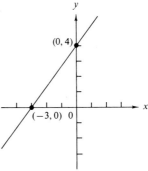

15.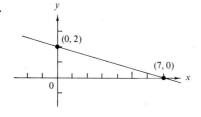

17.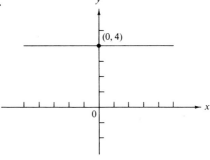

19.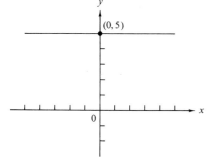

21.

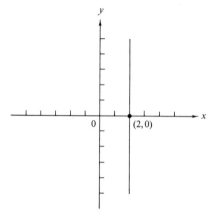

EXERCISE 7.4

1. (a) 5 (b) $\frac{3}{4}$ **3.** (a) 5 (b) $\frac{4}{3}$ **5.** (a) 13 (b) $-\frac{12}{5}$ **7.** (a) $2\sqrt{5}$ (b) 2
9. (a) $\sqrt{61}$ (b) $-\frac{5}{6}$ **11.** (a) 7 (b) 0 **13.** $5x - 2y = 7$
15. $x + y = -1$ **17.** $y = 7$ **19.** $3y - 2x = 20$ **21.** $2x + y = 5$
23. $2y - 5x = 4$ **25.** $4x + 5y = 16$ **27.** $2x - y = 4$ **29.** $y - x = 2$

EXERCISE 7.5

1. $(-4, 5), (-6, 2), (3, 4), (-3, -2)$
3. $(-4, 0), (-6, 2), (-9, 1), (-10, -1)$
5. $(-4, 3), (-6, 5), (1, -2), (4, 2)$
7. $(4, -5), (6, -2), (-3, -4), (3, 2)$
9. $(3, -9), (2, -7), (-4, 1), (-2, 5)$
11. (a) x-intercepts: $2, -2$;
  y-intercept: -4
  (b) $(0, -4)$, a minimum
  (c) $x = 0$
  (d)

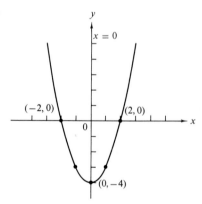

13. (a) *x*-intercepts: 4, −4;
y-intercept: 16
(b) (0, 16), a maximum
(c) $x = 0$
(d)

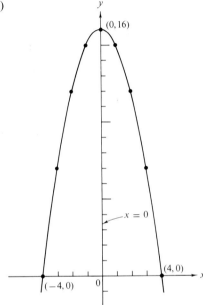

15. (a) *x*-intercepts: −3, 2
y-intercept: 6
(b) $(-\frac{1}{2}, \frac{25}{4})$, a maximum
(c) $x = -\frac{1}{2}$
(d)

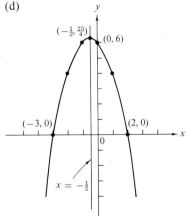

17. (a) *x*-intercepts: −3, 2;
y-intercept: −6
(b) $(-\frac{1}{2}, -\frac{25}{4})$, a minimum
(c) $x = -\frac{1}{2}$
(d)

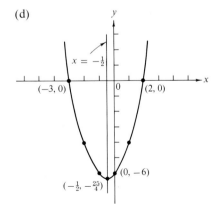

19. (a) x-intercepts: $-\frac{5}{2}, 3$;
   y-intercept: -15
   (b) $(\frac{1}{4}, -\frac{121}{8})$, a minimum
   (c) $x = \frac{1}{4}$
   (d)

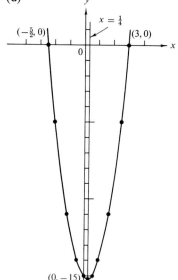

21. (a) x-intercepts: $-\frac{7}{2}, 2$;
   y-intercept: -14
   (b) $(-\frac{3}{4}, -\frac{121}{8})$, a minimum
   (c) $x = -\frac{3}{4}$
   (d)

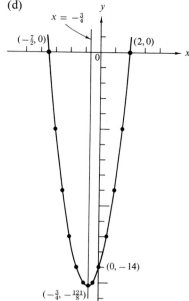

23. (a) x-intercepts: $-4, 1$;
   y-intercept: -4
   (b) $(-\frac{3}{2}, -\frac{25}{4})$, a minimum
   (c) $x = -\frac{3}{2}$
   (d)

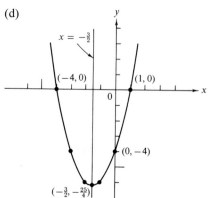

25. (a) x-intercepts: $-3, 1$;
   y-intercept: -3
   (b) $(-1, -4)$, a minimum
   (c) $x = -1$
   (d)

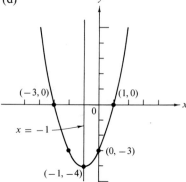

27. (a) x-intercepts: none;
  y-intercept: 4
  (b) $(0, 4)$, a minimum
  (c) $x = 0$

(d)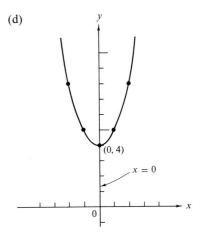

REVIEW EXERCISES, Chapter 7

1. (a) Domain: $\{1, 2, 3, 4, 5, 6\}$ (b) Range: $\{1, 2, 3, 4, 5, 6\}$
   (c) $\{(1, 1), (1, 2), (1, 3), (1, 4), (1, 5), (1, 6), (2, 2), (2, 4), (2, 6), (3, 3), (3, 6), (4, 4), (5, 5), (6, 6)\}$
2. Domain: $\{-2, -1, 0, 1, 2\}$ $\{(-2, 0), (-1, \sqrt{3}), (0, 2), (1, \sqrt{3}), (2, 0)\}$
3. (a) Function (b) Relation (c) Function (d) Function

4.

5.

6.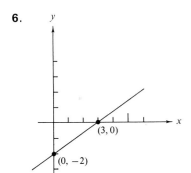

7.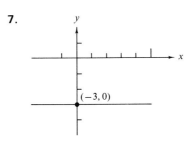

8.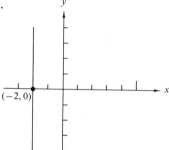

9. Distance: 5 Slope: $-\frac{3}{4}$

10. Distance: 13 Slope: $\frac{12}{5}$ **11.** $x + y = -2$ **12.** $3x + 5y = 15$
13. $3x - 4y = 8$ **14.** $2x + 3y = -18$ **15.** Slope: $\frac{5}{3}$ y-intercept: -2
16. $(-3, 2), (1, 3), (2, -1), (-4, -5)$ **17.** $(1, 0), (-3, -1), (-4, 3), (2, 7)$
18. $(-2, 1)$, a minimum   **19.** $(3, 13)$, a maximum

20. x-intercepts: $-2, 3$
y-intercept: -6
Vertex: $(\frac{1}{2}, -\frac{25}{4})$, a minimum
Axis: $x = \frac{1}{2}$

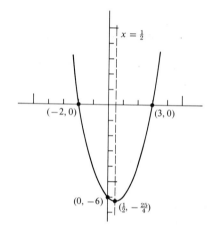

Chapter 8

EXERCISE 8.1

1. $x^2 + y^2 - 4x - 8y - 5 = 0$ **3.** $x^2 + y^2 + 2x - 4y + 1 = 0$
5. $x^2 + y^2 - 6x + 4y - 36 = 0$ **7.** $x^2 + y^2 + 4x + 4y - 41 = 0$
9. $x^2 + y^2 - 6x - 55 = 0$ **11.** $x^2 + y^2 - 8y = 0$
13. Center: $(4, 1)$ Radius: 1   **15.** Center: $(-2, 1)$ Radius: 4
17. Center: $(2, 0)$ Radius: 2   **19.** Center: $(-\frac{3}{2}, \frac{5}{2})$ Radius: 3
21. Center: $(\frac{1}{2}, 2)$ Radius: $\dfrac{3\sqrt{5}}{2}$ **23.** Center: $(-\frac{3}{4}, -\frac{5}{4})$ Radius: $\dfrac{3\sqrt{2}}{4}$

EXERCISE 8.2

1. $y^2 = 12x$ **3.** $x^2 = 8y$ **5.** $y^2 = -16x$ **7.** $x^2 = -24y$ **9.** $y^2 = 6x$
11. $x^2 = 10y$ **13.** $y^2 = -5x$ **15.** $x^2 = -8y$ **17.** $(y - 4)^2 = 12(x - 2)$
19. $(y - 5)^2 = 12(x - 3)$ **21.** $(y - 3)^2 = -40(x - 6)$

23. $(y - 3)^2 = 4(x + 2)$ **25.** $y^2 - 4y - 10x + 19 = 0$
27. $x^2 - 2x - 16y - 15 = 0$
29. Vertex: $(0, 0)$
Focus: $(2, 0)$
Directrix: $x = -2$
Axis: $y = 0$
31. Vertex: $(0, 0)$
Focus: $(0, 1)$
Directrix: $y = -1$
Axis: $x = 0$
33. Vertex: $(-1, 2)$
Focus: $(\frac{1}{2}, 2)$
Directrix: $x = -\frac{5}{2}$
Axis: $y = 2$
35. Vertex: $(1, 1)$
Focus: $(1, \frac{5}{2})$
Directrix: $y = -\frac{1}{2}$
Axis: $x = 1$
37. Vertex: $(-5, -1)$
Focus: $(-\frac{19}{4}, 1)$
Directrix: $x = -\frac{21}{4}$
Axis: $y = -1$
39. Vertex: $(-\frac{3}{8}, -\frac{9}{16})$
Focus: $(-\frac{3}{8}, -\frac{1}{2})$
Directrix: $y = -\frac{5}{8}$
Axis: $x = -\frac{3}{8}$

EXERCISE 8.3

1. $a = 5, b = 4$,
Foci: $(-3, 0), (3, 0)$
Vertices: $(-5, 0), (5, 0)$
$e = \frac{3}{5}$

3. $a = 4, b = 3$
Foci: $(\sqrt{7}, 0), (-\sqrt{7}, 0)$
Vertices: $(-4, 0), (4, 0)$
$e = \dfrac{\sqrt{7}}{4}$

5. $a = 5, b = 3$
Foci: $(0, -4), (0, 4)$
Vertices: $(0, -5), (0, 5)$
$e = \frac{4}{5}$

7. $a = 4, b = 2$
Foci: $(0, -2\sqrt{3}), (0, 2\sqrt{3})$
Vertices: $(0, -4), (0, 4)$
$e = \dfrac{\sqrt{3}}{2}$

9. $a = \sqrt{3}, b = \sqrt{2}$
Foci: $(0, -1), (0, 1)$
Vertices: $(0, -\sqrt{3}), (0, \sqrt{3})$
$e = \dfrac{\sqrt{3}}{3}$

11. $\dfrac{x^2}{25} + \dfrac{y^2}{9} = 1$

13. $\dfrac{x^2}{16} + \dfrac{y^2}{25} = 1$ **15.** $\dfrac{x^2}{100} + \dfrac{y^2}{36} = 1$ **17.** $\dfrac{x^2}{64} + \dfrac{y^2}{100} = 1$ **19.** $\dfrac{x^2}{25} + \dfrac{y^2}{16} = 1$

EXERCISE 8.4

1. Vertices: $(-4, 0), (4, 0)$
Foci: $(-5, 0), (5, 0)$
Conjugate axis: 6
Eccentricity: $\frac{5}{4}$

3. Vertices: $(-5, 0), (5, 0)$
Foci: $(-\sqrt{41}, 0), (\sqrt{41}, 0)$
Conjugate axis: 8
Eccentricity: $\dfrac{\sqrt{41}}{4}$

5. Vertices: $(0, -3), (0, 3)$
Foci: $(0, -5), (0, 5)$
Conjugate axis: 8
Eccentricity: $\frac{5}{3}$

7. Vertices: $(0, -3), (0, 3)$
Foci: $(0, -\sqrt{34}), (0, \sqrt{34})$
Conjugate axis: 10
Eccentricity: $\dfrac{\sqrt{34}}{3}$

9. Vertices: $(-\sqrt{2}, 0), (\sqrt{2}, 0)$
Foci: $(-\sqrt{5}, 0), (\sqrt{5}, 0)$
Conjugate axis: $2\sqrt{3}$
Eccentricity: $\dfrac{\sqrt{5}}{\sqrt{2}}$

11. $\dfrac{x^2}{16} - \dfrac{y^2}{9} = 1$

13. $\dfrac{y^2}{9} - \dfrac{x^2}{16} = 1$

15. $x^2 - \dfrac{y^2}{3} = 1$

17. $\dfrac{y^2}{25} - \dfrac{16x^2}{225} = 1$

19. $\dfrac{x^2}{9} - \dfrac{y^2}{16} = 1$

EXERCISE 8.5

1.

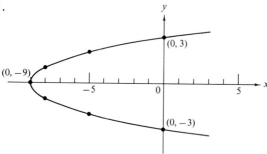

3.

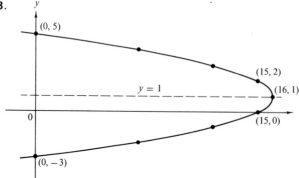

5.

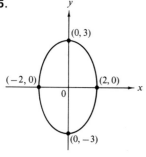

7.

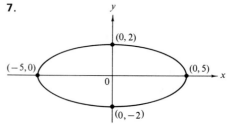

9.

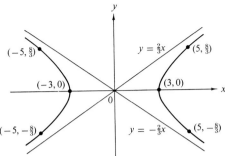

11.

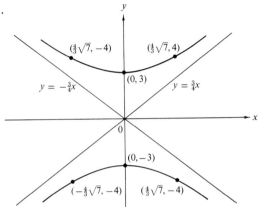

REVIEW EXERCISES, Chapter 8

1. $x^2 + y^2 - 2x - 4y - 4 = 0$ **2.** $(-2, 3); 7$ **3.** $y^2 = -16x$
4. $y^2 - 8y - 16x + 64 = 0$ **5.** $x^2 - 4x - 14y + 74 = 0$
6. Vertex: $(1, 4)$ Directrix: $y = -6$ Axis: $x = 1$
7. $a = 3, b = 2$
   Foci: $(-\sqrt{5}, 0), (\sqrt{5}, 0)$
   Vertices: $(0, -3), (0, 3)$
8. $a = \sqrt{7}, b = \sqrt{5}$
   Foci: $(-\sqrt{2}, 0), (\sqrt{2}, 0)$
   Vertices: $(-\sqrt{7}, 0), (\sqrt{7}, 0)$

9. $\dfrac{x^2}{9} + \dfrac{y^2}{49} = 1$

10. $\dfrac{x^2}{169} + \dfrac{y^2}{144} = 1$

11. Vertices: $(-2, 0), (2, 0)$
   Foci: $(-\sqrt{13}, 0), (\sqrt{13}, 0)$

12. Vertices: $(0, -\sqrt{7}), (0, \sqrt{7})$
   Foci: $(0, -2\sqrt{3}), (0, 2\sqrt{3})$

13. $\dfrac{y^2}{16} - \dfrac{x^2}{9} = 1$ **14.** $\dfrac{x^2}{144} - \dfrac{y^2}{25} = 1$

15.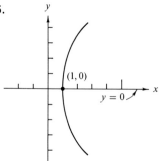

16.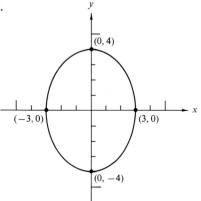

17.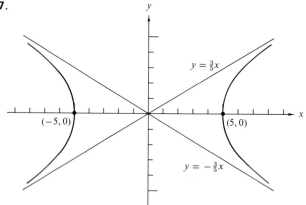

Chapter 9

EXERCISE 9.1

1.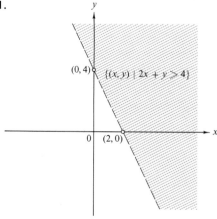

3.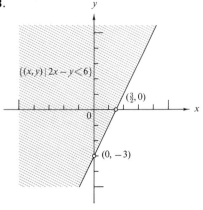

5.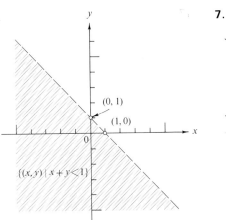

7.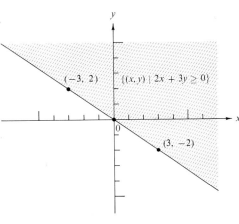

9.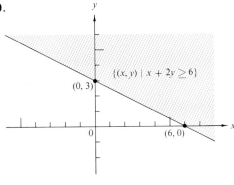

11.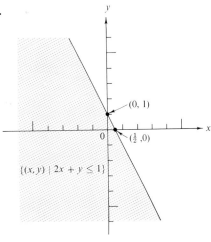

13.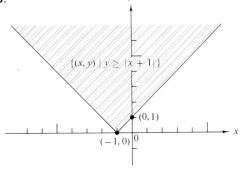

398 ANSWERS TO ODD-NUMBERED EXERCISES

15.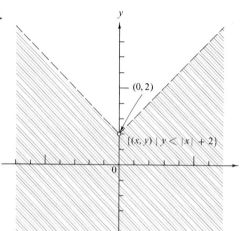

17.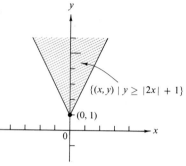

19.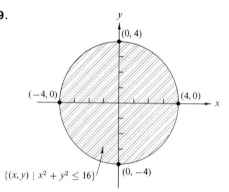

21.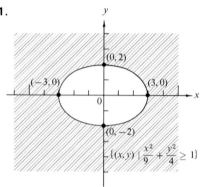

23.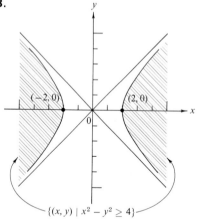

25.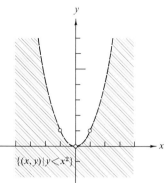

27.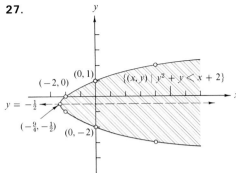

29.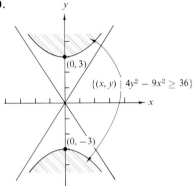

31.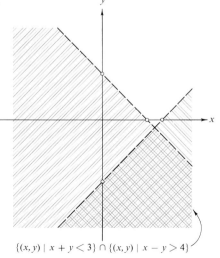

EXERCISE 9.2

1. $d = kt$ **3.** $V = \dfrac{k}{P}$ **5.** $s = kt^2$
7. $S = kwd$ **9.** $L = \dfrac{kwd^2}{x}$ **11.** $\tfrac{5}{3}$
13. $\tfrac{3}{2}$ **15.** 8, 32 **17.** 10 **19.** 15 **21.** 1600 feet **23.** 3456
25. $y = 3x - 4$

EXERCISE 9.3

1. $\{(-3,0), (3,1), (2,2), (5,3), (2,4)\}$ **3.** $y = \dfrac{x-3}{4}$ **5.** $y = \dfrac{x-2}{3}$
7. $y = \dfrac{2x+8}{3}$ **9.** $|y| = \sqrt{x},\ x \geq 0$ **11.** $|y| = \sqrt{x^2 + 4}$
13. $|y| = \sqrt{x^2 + 16}$

EXERCISE 9.4

1.

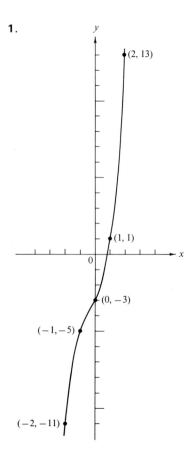

3.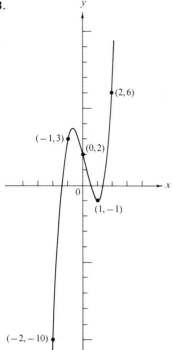

5.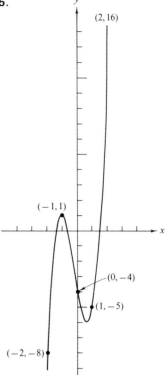

7.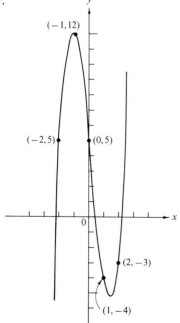

9.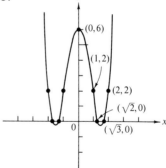

EXERCISE 9.5

1.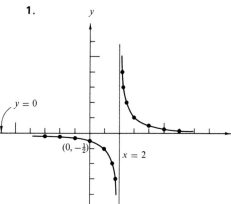

3.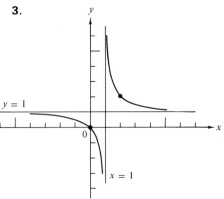

5.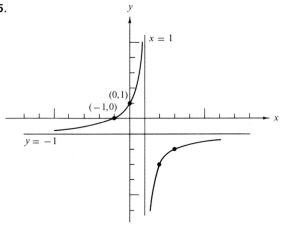

7.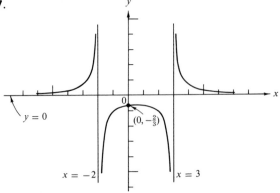

9.

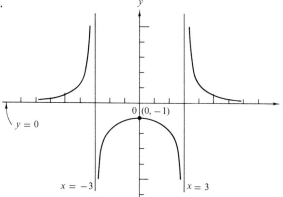

11.

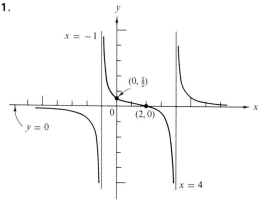

REVIEW EXERCISES, Chapter 9

1.

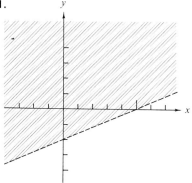

2.

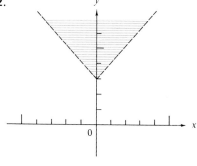

3.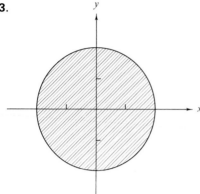

4. 16   **5.** $\frac{32}{27}$

6. 105 ounces   **7.** $\frac{3}{2}$

8. $y = \frac{x-3}{2}$ **9.** $|y| = x^2 + 9$

10. $|y| = \sqrt{x^2 + 1}$ **11.** 8

12.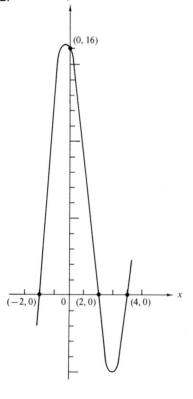

13.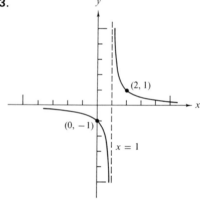

14.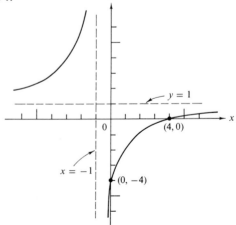

Chapter 10

EXERCISE 10.1

1. Dependent 3. Inconsistent 5. Inconsistent 7. Independent 9. Dependent
11. Inconsistent 13. $\{(4,7)\}$ 15. $\{(3,2)\}$ 17. $\{(2,6)\}$ 19. $\{(5,3)\}$
21. $\{(\frac{3}{5},\frac{14}{5})\}$ 23. $\{(7,2)\}$ 25. $\{(-2,1,3)\}$ 27. $\{(1,1,-1)\}$ 29. $\{(0,1,2)\}$

EXERCISE 10.2

1. $\{(4,7)\}$ 3. $\{(3,2)\}$ 5. $\{(2,6)\}$ 7. $\{(5,3)\}$ 9. $\{(\frac{3}{5},\frac{14}{5})\}$ 11. $\{(7,2)\}$
13. $\{(3,3)\}$ 15. $\{(7,5)\}$ 17. $\{(\frac{2}{3},\frac{3}{2})\}$ 19. $\{(2,3)\}$ 21. $\{(2,3)\}$
23. $\{(80,50)\}$ 25. $\{(1,1,0)\}$ 27. $\{(-2,0,2)\}$ 29. $\{(2,2,1)\}$ 31. $\{(2,-3,1)\}$

EXERCISE 10.3

1. $\{(-3,-4),(4,3)\}$ 3. $\{(-7,-1),(7,1)\}$ 5. $\{(2,3),(10,-13)\}$
7. $\{(-3,5),(5,-3)\}$ 9. $\{(\frac{13}{5},\frac{22}{5}),(3,4)\}$ 11. $\{(1,2),(\frac{29}{4},-\frac{17}{4})\}$
13. $\{(1,2),(2,1)\}$ 15. $\{(1,-4),(8,-\frac{1}{2})\}$ 17. $\{(-\frac{3}{5},\frac{5}{3}),(1,-1)\}$ 19. $\{(\frac{1}{3},\frac{1}{2}),(\frac{1}{2},\frac{1}{3})\}$

EXERCISE 10.4

1. $\{(-5,0),(5,0)\}$ 3. $\{(-3,0),(3,0)\}$
5. $\{(-4,-1),(-4,1),(4,-1),(4,1)\}$ 7. $\{(-3,-5),(-3,5),(3,-5),(3,5)\}$
9. $\{(4\sqrt{2},0),(-4\sqrt{2},0)\}$ 11. $\{(-2,-1),(-2,1),(2,-1),(2,1)\}$
13. $\{(-\sqrt{11},-\sqrt{11}),(\sqrt{11},\sqrt{11}),(-2,-4),(2,4)\}$
15. $\left\{\left(-\frac{3\sqrt{2}}{2},-\frac{\sqrt{2}}{2}\right),\left(\frac{3\sqrt{2}}{2},\frac{\sqrt{2}}{2}\right),(-2,-1),(2,1)\right\}$
17. $\{(-2\sqrt{6},2\sqrt{6}),(2\sqrt{6},-2\sqrt{6}),(-3,-2),(3,2)\}$
19. $\left\{\left(-\frac{\sqrt{14}}{2},-\frac{\sqrt{14}}{2}\right),\left(\frac{\sqrt{14}}{2},\frac{\sqrt{14}}{2}\right)\right\}$ 21. $\{(-9,-3),(9,3)\}$
23. $\left\{\left(-\frac{9\sqrt{47}}{47},-\frac{\sqrt{47}}{94}\right),\left(\frac{9\sqrt{47}}{47},\frac{\sqrt{47}}{94}\right),(-1,-1),(1,1)\right\}$ 25. $\{(-2,3),(2,-3)\}$

EXERCISE 10.5

1.
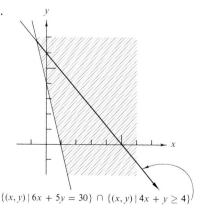
$\{(x,y)\mid 6x+5y=30\}\cap\{(x,y)\mid 4x+y\geq 4\}$

3.
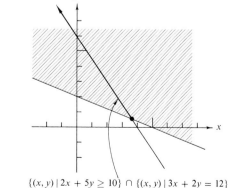
$\{(x,y)\mid 2x+5y\geq 10\}\cap\{(x,y)\mid 3x+2y=12\}$

5.

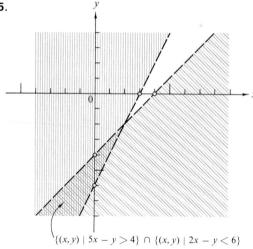

$\{(x,y) \mid 5x - y > 4\} \cap \{(x,y) \mid 2x - y < 6\}$

7.

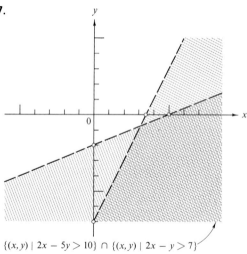

$\{(x,y) \mid 2x - 5y > 10\} \cap \{(x,y) \mid 2x - y > 7\}$

9.

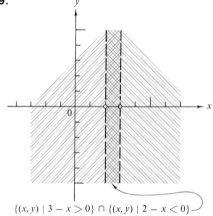

$\{(x,y) \mid 3 - x > 0\} \cap \{(x,y) \mid 2 - x < 0\}$

11.

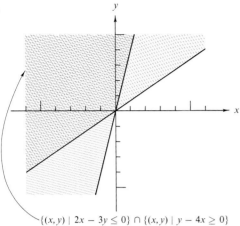

$\{(x, y) \mid 2x - 3y \leq 0\} \cap \{(x, y) \mid y - 4x \geq 0\}$

13.

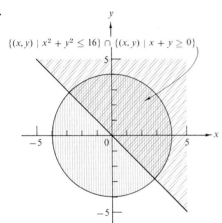

$\{(x, y) \mid x^2 + y^2 \leq 16\} \cap \{(x, y) \mid x + y \geq 0\}$

15.

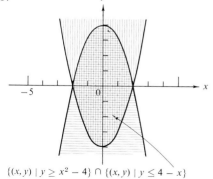

$\{(x, y) \mid y \geq x^2 - 4\} \cap \{(x, y) \mid y \leq 4 - x\}$

17.

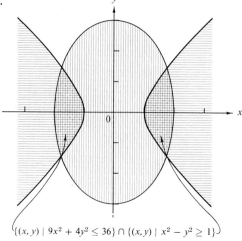

$\{(x, y) \mid 9x^2 + 4y^2 \leq 36\} \cap \{(x, y) \mid x^2 - y^2 \geq 1\}$

19.

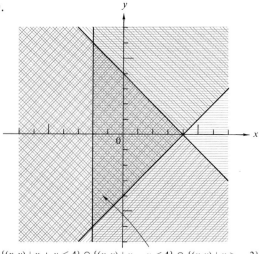

$\{(x, y) \mid x + y \leq 4\} \cap \{(x, y) \mid x - y \leq 4\} \cap \{(x, y) \mid x \geq -2\}$

21.

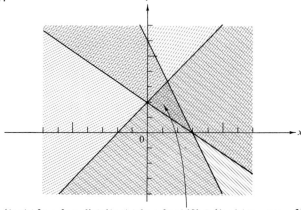

$\{(x, y) \mid 2x + 3y \geq 6\} \cap \{(x, y) \mid 6x + 3y \leq 18\} \cap \{(x, y) \mid x - y \geq -2\}$

23.

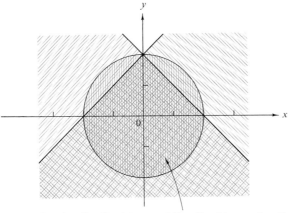

$\{(x,y) \mid x^2 + y^2 \le 4\} \cap \{(x,y) \mid x + y \le 2\} \cap \{(x,y) \mid x - y \ge -2\}$

25.

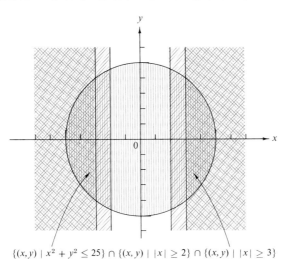

$\{(x,y) \mid x^2 + y^2 \le 25\} \cap \{(x,y) \mid |x| \ge 2\} \cap \{(x,y) \mid |x| \ge 3\}$

EXERCISE 10.6

1. (1) Let x represent one number and y the other. (2) $x + y = 95$, $x - y = 15$ (3) $\{(55, 40)\}$ (4) The numbers are 55 and 40.
3. (1) Let x represent one number and y the other. (2) $3x + 5y = 94$; $3y - 2x = 7$ (3) $\{(13, 11)\}$ (4) The numbers are 13 and 11.
5. (1) Let x represent the length and y the width. (2) $x - 2y = 36$; $2x + 2y = 504$ (3) $\{(180, 72)\}$ (4) The length is 180 inches and the width is 72 inches.
7. (1) Let t represent the tens digit and u the units digit.
(2) $t + u = 12$; $10u + t - (10t + u) = 18$ (3) $\{(5, 7)\}$ (4) The number is 57.
9. (1) Let t represent the tens digit and u the units digit.
(2) $t + u = 14$; $10t + u + 3t = 10u + t$ (3) $\{(6, 8)\}$ (4) The number is 68.
11. (1) Let t represent the tens digit and u the units digit.
(2) $3t = u + 1$; $2(10t + u) = 10u + t - 2$ (3) $\{(2, 5)\}$ (4) The number is 25.
13. (1) Let x represent the number of ounces of the 18% solution and y the number of ounces of the 45% solution.

(2) $x + y = 12$; $0.18x + 0.45y = 0.36(12)$ (3) $\{(4, 8)\}$
(4) 4 ounces of the 18% solution and 8 ounces of the 45% solution are needed.
15. (1) Let x represent the number of pounds of the 10% brand and y the number of pounds of the 19% brand.
(2) $x + y = 450$; $0.10x + 0.19y = 0.14(450)$ (3) $\{(250, 200)\}$
(4) 250 pounds of the 10% brand and 200 pounds of the 45% brand are needed.
17. (1) Let x represent the number of pounds of the 60¢ candy and y the number of pounds of the 50¢ candy.
(2) $x + y = 120$; $60x + 50y = 56(120)$ (3) $\{(72, 48)\}$
(4) 72 pounds of the 60¢ candy and 48 pounds of the 50¢ candy are needed.
19. (1) Let x represent the number of dimes and y the number of nickels.
(2) $x + y = 76$; $10x + 5y = 640$ (3) $\{(52, 24)\}$
(4) The collection contains 52 dimes and 24 nickels.
21. (1) Let h represent the hundreds digit, t the tens digit, and u the units digit.
(2) $h + t + u = 12$; $h = t + 2$; $t = u + 5$ (3) $\{(7, 5, 0)\}$ (4) The number is 750.
23. (1) Let x represent the length of the longest side, y the length of the next longest, and z the length of the shortest.
(2) $x + y + z = 30$; $x = y + 1$; $x + 2 = 3z$
(3) $\{(13, 12, 5)\}$ (4) The lengths of the sides are 13 inches, 12 inches, and 5 inches.
25. (1) Let x represent the number of pennies, y the number of nickels, and z the number of dimes.
(2) $x + 5y + 10z = 185$; $y + 6 = z$; $x + 3 = y$ (3) $\{(5, 8, 14)\}$
(4) There are 5 pennies, 8 nickels, and 14 dimes.
27. (1) Let x represent one integer and y the other.
(2) $x + y = 11$; $x^2 + y^2 = 73$ (3) $\{(3, 8), (8, 3)\}$ (4) The integers are 3 and 8.
29. (1) Let x represent one integer and y the other.
(2) $2x^2 + y^2 = 211$; $x^2 + 2y^2 = 179$ (3) $\{(-9, -7), (-9, 7), (9, -7), (9, 7)\}$
(4) The two positive integers are 9 and 7.
31. (1) Let x represent one of the numbers and y the other.
(2) $x + y = 24$; $\dfrac{1}{x} + \dfrac{1}{y} = \dfrac{2}{9}$ (3) $\{(6, 18), (18, 6)\}$
(4) The two numbers are 6 and 18.
33. (1) Let x represent the speed of the boat and y the speed of the current.
(2) $x = 4y$; $\dfrac{12}{x - y} = \dfrac{15}{x + y} + 1$ (3) $\{(4, 1)\}$
(4) The speed of the boat is 4 mph and the speed of the current is 1 mph.
35. (1) Let x represent the length of one leg and y the length of the other.
(2) $x + y = 28$; $\dfrac{xy}{2} = 96$ (3) $\{(12, 16), (16, 12)\}$
(4) The length of the hypotenuse is 20 inches.
37. $k = -7$

REVIEW EXERCISES, Chapter 10

1. Dependent   2. Independent   3. Inconsistent   4. $\{(1, 2)\}$ 5. $\{(\tfrac{2}{3}, \tfrac{3}{2})\}$
6. $\{(2, 2)\}$ 7. $\{(2, -2, 0)\}$ 8. $\{(3, -2)\}$ 9. $\{(1, 2)\}$ 10. $\{(\tfrac{1}{5}, \tfrac{1}{2})\}$
11. $\{(1, 0, 1)\}$ 12. $\{(3, 4), (-\tfrac{7}{5}, -\tfrac{24}{5})\}$ 13. $\{(-3, 4), (4, -3)\}$
14. $\{(-\tfrac{45}{22}, \tfrac{51}{22}), (0, 3)\}$ 15. $\{(-2, -3), (-2, 3), (2, -3), (2, 3)\}$

16. $\{(-1, -2), (-1, 2), (1, -2), (1, 2)\}$ **17.** $\{(-2, -1), (-2, 1), (2, -1), (2, 1)\}$
18. $\{(-2\sqrt{7}, \sqrt{7}), (-3, -1), (3, 1), (2\sqrt{7}, -\sqrt{7})\}$
19. $\{(-4, 1), (-2\sqrt{7}, -\sqrt{7}), (2\sqrt{7}, \sqrt{7}), (4, -1)\}$

20.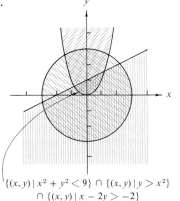

$\{(x, y) \mid x^2 + y^2 < 9\} \cap \{(x, y) \mid y > x^2\}$
$\cap \{(x, y) \mid x - 2y > -2\}$

21.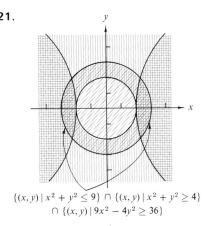

$\{(x, y) \mid x^2 + y^2 \leq 9\} \cap \{(x, y) \mid x^2 + y^2 \geq 4\}$
$\cap \{(x, y) \mid 9x^2 - 4y^2 \geq 36\}$

22. (1) Let x represent one half the length and y one half the width.
   (2) $x^2 + y^2 = 25$; $4xy = 48$
   (3) $\{(-4, -3), (-3, -4), (3, 4), (4, 3)\}$
   (4) The dimensions of the rectangle are 8 inches and 6 inches.
23. (1) Let x represent the amount of the investment and y the rate of interest.
   (2) $xy = 120$; $(x + 600)(y - 0.005) = 135$
   (3) $\{(-30{,}000, -0.004), (2400, 0.05)\}$
   (4) The amount of the investment is \$2400 and the rate of interest is 5%.

Chapter 11

EXERCISE 11.1

1. i **3.** 2   **5.** i **7.** -6 **9.** 3   **11.** i **13.** $8 + 7i$ **15.** $-9 - 9i$
17. $9 + 0i$ **19.** $0 + 0i$ **21.** $3i\sqrt{2}$ **23.** $3i\sqrt{6}$ **25.** $\dfrac{i\sqrt{10}}{5}$ **27.** $\dfrac{2i\sqrt{6}}{3}$
29. $-\sqrt{15}$ **31.** $-3\sqrt{5}$ **33.** $\{(2, -2)\}$ **35.** $\{(2, -3)\}$ **37.** $\{(2, -2)\}$
39. $\{(1, -5)\}$ **41.** $\{(4, -5)\}$ **43.** $\{(5, 2)\}$ **45.** $\{(2, 2)\}$
47. $x = -3$ or 2, $y = -3$ or -2

EXERCISE 11.2

1. $10 - i$ **3.** $-2 - 6i$ **5.** $(x + 2) + (y + 4)i$ **7.** $(2x + 3) + (y - 2)i$
9. $10 + i$ **11.** $2\sqrt{2} + 3i\sqrt{3}$ **13.** $3 + 2i$ **15.** $-1 - i$
17. $(3 - x) + (y - 2)i$ **19.** $(x - 4) + (y + 3)i$ **21.** $2 + 5i$
23. $-\sqrt{5} + 3i\sqrt{3}$

EXERCISE 11.3

1. $4 + 32i$ **3.** $37 - 10i$ **5.** $52 + 0i$ **7.** $(4x + 3y) + (xy - 12)i$
9. $(5y - 3x) + (xy + 15)i$ **11.** $(3x + 4y) + (3y - 4x)i$ **13.** $14 + i\sqrt{2}$
15. $14 + 0i$ **17.** $(5 - \sqrt{6}) + (\sqrt{10} + \sqrt{15})i$ **19.** $\dfrac{1}{26} - \dfrac{5i}{26}$ **21.** $\dfrac{-1}{10} - \dfrac{i}{5}$
23. $\dfrac{3}{14} - \dfrac{i\sqrt{5}}{14}$ **25.** $\dfrac{\sqrt{2}}{5} + \dfrac{i\sqrt{3}}{5}$ **27.** $\dfrac{\sqrt{5}}{9} - \dfrac{2i}{9}$ **29.** $\dfrac{x}{x^2 + 4} + \dfrac{2i}{x^2 + 4}$

EXERCISE 11.4

1. $\dfrac{12}{5} + \dfrac{i}{5}$ **3.** $\dfrac{13}{29} - \dfrac{11i}{29}$ **5.** $2 + 0i$ **7.** $-\dfrac{1}{19} - \dfrac{5i\sqrt{3}}{19}$ **9.** $\dfrac{8}{17} - \dfrac{2i}{17}$
11. $\dfrac{26}{25} - \dfrac{7i}{25}$ **13.** $-\dfrac{15}{37} - \dfrac{16i}{37}$ **15.** $\dfrac{7}{28} + \dfrac{7i\sqrt{3}}{28}$ **17.** $\dfrac{6}{5} + \dfrac{8i}{5}$ **19.** $\dfrac{23}{25} - \dfrac{14i}{25}$
21. $-3 + 0i$ **23.** $\dfrac{1}{3} + \dfrac{2i\sqrt{5}}{3}$

EXERCISE 11.5

1. $\{1 - 2i, 1 + 2i\}$ **3.** $\{2 - 2i, 2 + 2i\}$ **5.** $\left\{\dfrac{1 - i}{2}, \dfrac{1 + i}{2}\right\}$
7. $\{-\sqrt{2} - i, -\sqrt{2} + 1\}$ **9.** $\left\{\dfrac{-\sqrt{7} - 3i}{4}, \dfrac{-\sqrt{7} + 3i}{4}\right\}$
11. $\left\{\dfrac{3 - i\sqrt{23}}{2}, \dfrac{3 + i\sqrt{23}}{2}\right\}$ **13.** $x^2 + 1 = 0$ **15.** $x^2 - 2x + 2 = 0$
17. $x^2 - 4x + 13 = 0$ **19.** $x^2 - 4x + 9 = 0$ **21.** Sum: 2 Product: $\tfrac{5}{2}$
23. Sum: $-\tfrac{3}{4}$ Product: -2 **25.** Sum: $\tfrac{3}{7}$ Product: $-\tfrac{4}{7}$

REVIEW EXERCISES, Chapter 11

1. $3 + 2i$ **2.** $2 + 0i$ **3.** $3i\sqrt{3}$ **4.** $4i\sqrt{3}$ **5.** -4 **6.** -6
7. $x = 3, y = -8$ **8.** $x = 1, y = 2$ **9.** $7 - 4i$ **10.** $8 + i$
11. $-1 - 2i$ **12.** $2 + i$ **13.** $31 + i$ **14.** $(\sqrt{6} + \sqrt{15}) + (\sqrt{10} - 3)i$
15. $40 + 0i$ **16.** $5 + 0i$ **17.** $-\dfrac{2}{13} - \dfrac{3i}{13}$ **18.** $\dfrac{\sqrt{5}}{8} + \dfrac{i\sqrt{3}}{8}$
19. $\dfrac{7}{29} + \dfrac{26i}{29}$ **20.** $-1 - i$ **21.** $\dfrac{5}{19} - \dfrac{6i\sqrt{3}}{19}$ **22.** $\{1 - i, 1 + 1\}$
23. $\{3 - 2i, 3 + 2i\}$ **24.** $\left\{\dfrac{3 - 3i\sqrt{3}}{2}, \dfrac{3 + 3i\sqrt{3}}{2}\right\}$
25. $\left\{\dfrac{1 - i\sqrt{15}}{4}, \dfrac{1 + i\sqrt{15}}{4}\right\}$
26. Sum: -4 Product: 8   **27.** Sum: 9 Product: 13
28. Sum: -2 Product: $-\tfrac{10}{5}$ **29.** Sum: $\tfrac{5}{2}$ Product: $\tfrac{7}{2}$

Chapter 12

EXERCISE 12.1

1. $\sqrt{5}$ 3. 3   5. 5   7. $5c$ 9. 5   11. 10   13. 5

15.

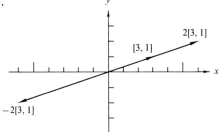

17.

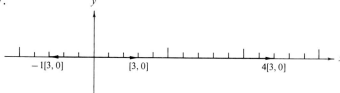

19.

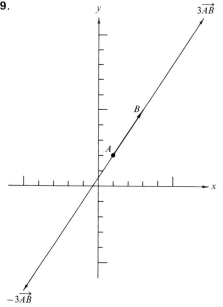

21.

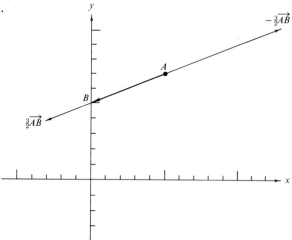

23. $\left[-\dfrac{3}{\sqrt{10}}, \dfrac{1}{\sqrt{10}}\right]$

25. $\left[-\dfrac{2}{\sqrt{13}}, \dfrac{3}{\sqrt{13}}\right]$

27. $\left[-\dfrac{\sqrt{2}}{2}, \dfrac{\sqrt{2}}{2}\right]$

29. $[-\tfrac{3}{5}, -\tfrac{4}{5}]$

31. $[1, 0]$

EXERCISE 12.2

1. $[5, 6]$ **3.** $[2, 3]$ **5.** $[-5, -6]$

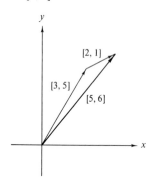 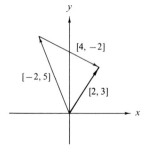 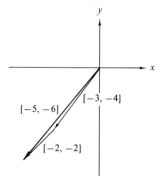

7. $5\mathbf{i} + 6\mathbf{j}$ **9.** $-\mathbf{i} - 5\mathbf{j}$ **11.** $-2\mathbf{i} - 8\mathbf{j}$

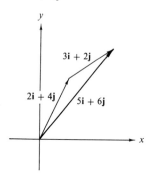 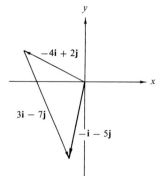 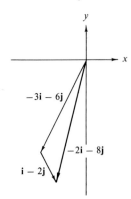

13. [2, 4]

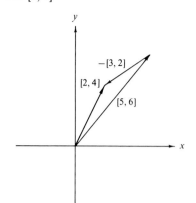

15. [6, −3]

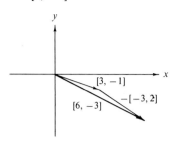

17. $-2\mathbf{i} - 4\mathbf{j}$

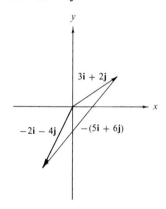

19. $-7\mathbf{i} - 3\mathbf{j}$

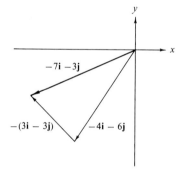

EXERCISE 12.3

1. $\theta = 45°$, $\|\mathbf{v}\| = 10$

3. $\theta = 61°$, $\|\mathbf{v}\| = 9$

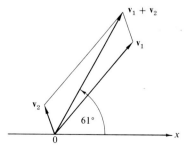

5. $\theta = 295°$, $\|\mathbf{v}\| = 5$

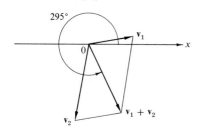

7. $\theta = 0°$, $\|\mathbf{v}\| = 9$

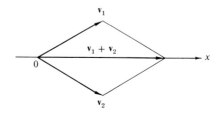

9. $\theta = 23°$, $\|\mathbf{v}\| = 13$

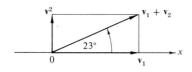

11. $\theta = 329°$, $\|\mathbf{v}\| = 6$

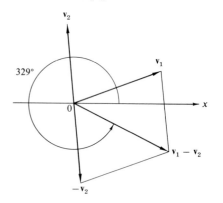

13. $\theta = 22°$, $\|\mathbf{v}\| = 6$

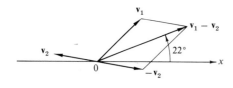

15. $\theta = 55°$, $\|\mathbf{v}\| = 9$

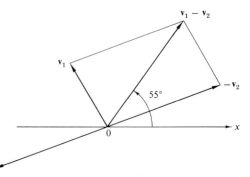

17. $\theta = 172°$, $\|\mathbf{v}\| = 5$

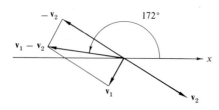

Answers to Odd-Numbered Exercises **417**

19. $\theta = 86°$, $\|\mathbf{v}\| = 13$

21. $\theta = 113°$, $\|\mathbf{v}\| = 8$

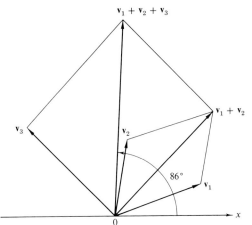

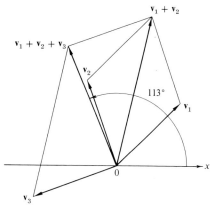

23. $\theta = 198°$, $\|\mathbf{v}\| = 11$

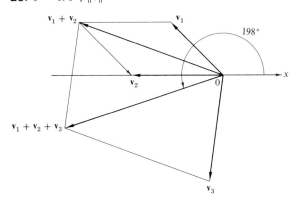

25. $\theta = 204°$, $\|\mathbf{v}\| = 8$

27. 37° South of East, 25 pounds

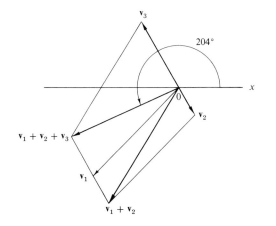

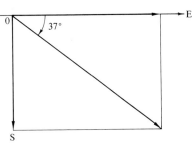

29. 26° North of West, 2 pounds

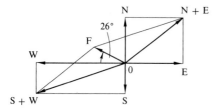

REVIEW EXERCISES, Chapter 12

1. $2\sqrt{5}$ **2.** $\sqrt{13}$ **3.** 5   **4.** $\sqrt{29}$ **5.** $[\frac{4}{5}, -\frac{3}{5}]$ **6.** $\left[\frac{1}{\sqrt{2}}, \frac{1}{\sqrt{2}}\right]$

7. $-\frac{5}{13}\mathbf{i} + \frac{12}{13}\mathbf{j}$ **8.** $\frac{2}{\sqrt{53}}\mathbf{i} - \frac{7}{\sqrt{53}}\mathbf{j}$ **9.** $[7, -4]$ **10.** $7\mathbf{i} - 2\mathbf{j}$

11. $[3, 19]$ **12.** $8\mathbf{i} + 14\mathbf{j}$

13. $\theta = 53°$, $\|\mathbf{v}\| = 5$ **14.** $\theta = 67°$, $\|\mathbf{v}\| = 10$

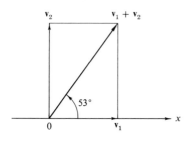

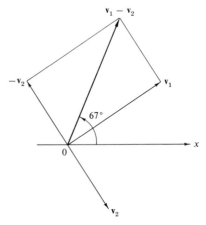

15. $\theta = 184°$, $\|\mathbf{v}\| = 8.5$

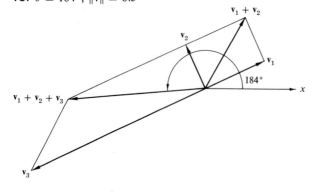

Chapter 13

EXERCISE 13.1

1. $\{4\}$ **3.** $\{1\}$ **5.** $\{(3,6)\}$ **7.** $\{(4,4)\}$ **9.** $\{(-1,-3,2)\}$
11. $\{(1,4,2)\}$ **13.** $\{(4,2)\}$ **15.** $\{(2,1)\}$

EXERCISE 13.2

1. $\begin{bmatrix} 12 & 13 \\ 8 & 12 \end{bmatrix}$ **3.** $\begin{bmatrix} -1 & 3 \\ -5 & 1 \end{bmatrix}$ **5.** Not conformable **7.** $\begin{bmatrix} 7 & -2 \\ 3 & 1 \\ -1 & 2 \end{bmatrix}$ **9.** $\begin{bmatrix} 9 & -2 & 2 \\ 1 & -1 & 3 \\ 5 & -1 & 7 \end{bmatrix}$

11. $\begin{bmatrix} 2 & 0 & -2 \\ -1 & 6 & 1 \\ 0 & 1 & -5 \end{bmatrix}$ **13.** $\begin{bmatrix} 6 & 3 & 3 & 4 \\ 3 & -1 & 1 & 1 \\ 2 & 3 & 0 & 0 \\ 3 & 1 & 3 & 2 \end{bmatrix}$ **15.** $\begin{bmatrix} 7 & 5 \\ 7 & 1 \end{bmatrix}$ **17.** $\begin{bmatrix} -1 & 1 \\ 9 & 4 \end{bmatrix}$

19. $\begin{bmatrix} 5 & 1 & 3 & 8 \\ 5 & 1 & 7 & 6 \\ 5 & 9 & 3 & 6 \end{bmatrix}$ **21.** $\begin{bmatrix} 2 & -9 & 3 \\ 2 & 1 & -9 \\ 4 & -1 & 7 \end{bmatrix} + \begin{bmatrix} 1 & 5 & -5 \\ 4 & 4 & 8 \\ -1 & 3 & -8 \end{bmatrix}$

EXERCISE 13.3

1. $[-1]$ **3.** $\begin{bmatrix} 1 & 10 \\ 23 & 6 \end{bmatrix}$ **5.** $\begin{bmatrix} 5 & 8 \\ 4 & 7 \end{bmatrix}$ **7.** $\begin{bmatrix} 1 & -4 & -8 \\ 4 & 1 & -6 \\ 4 & 0 & 1 \end{bmatrix}$ **9.** $\begin{bmatrix} 1 & -8 & 12 \\ -2 & 1 & 16 \\ 0 & -4 & 1 \end{bmatrix}$

11. $\begin{bmatrix} 1 & 8 \\ 1 & 7 \end{bmatrix}$ **13.** No **17.** $AB = \begin{bmatrix} 1 & 0 & 0 \\ 0 & 1 & 0 \\ 0 & 0 & 1 \end{bmatrix}$ $BA = \begin{bmatrix} 1 & 0 & 0 \\ 0 & 1 & 0 \\ 0 & 0 & 1 \end{bmatrix}$ Yes, multiplicative inverse.

EXERCISE 13.4

1. 1 **3.** 0 **5.** -2 **7.** -2 **9.** -26 **11.** -22 **15.** $4x + 5y - 7 = 0$

EXERCISE 13.5

1. $\{(3,2)\}$ **3.** $\{(-1,3)\}$ **5.** $\{(3,\frac{1}{2})\}$ **7.** $\{(2,1,3)\}$
9. $\{(-2,2,-1)\}$ **11.** $\{(\frac{1}{2},\frac{1}{3},0)\}$

REVIEW EXERCISES, Chapter 13

1. $x = 5, y = 3, z = 0$ **2.** $x = -\frac{5}{2}, y = -\frac{3}{2}$ **3.** $\begin{bmatrix} 0 & 3 \\ -2 & 0 \\ 3 & 1 \end{bmatrix}$ **4.** Not conformable

5. $\begin{bmatrix} 3 & -1 & -4 \\ 0 & 0 & -4 \\ -1 & -1 & 1 \end{bmatrix}$ **6.** $\begin{bmatrix} 2 & -5 \\ 2 & 4 \end{bmatrix}$ **7.** $\begin{bmatrix} 3 & 2 & -3 \\ 0 & 0 & 0 \\ 3 & -1 & 3 \end{bmatrix}$ **8.** $\begin{bmatrix} x+2y+3z \\ 2x & -z \\ x-3y-2z \end{bmatrix}$

9. 17 **10.** -4 **11.** 1 **12.** 0 **13.** $\{(\tfrac{7}{6}, \tfrac{7}{12})\}$ **14.** $\{(1, -3, 2)\}$

Chapter 14

EXERCISE 14.1

1. 2 **3.** 4 **5.** $\tfrac{1}{2}$ **7.** -3 **9.** 5 **11.** 4 **13.** 49 **15.** 32 **17.** -2
19. 4 **21.** $\log_b 3 + \log_b x$ **23.** $\log_b 2 - \log_b x$ **25.** $\log_b x + \log_b y - \log_b z$
27. $3 \log_b x$ **29.** $\log_b x + \tfrac{1}{2} \log_b y$ **31.** $2 \log_b x + 3 \log_b y$ **33.** $\log_b \dfrac{x}{y}$
35. $\log_b \dfrac{xy}{z}$ **37.** $\log_b x^2 y^5$ **39.** $\log_b \dfrac{x^3 y^2}{z^2}$ **41.** $\log_b \sqrt[3]{\dfrac{x^2 y}{z^4}}$
43. $\log_b (x+1)$

EXERCISE 14.2

1. 5.93×10^2 **3.** 2.51×10^0 **5.** 1.25×10^{-1} **7.** 1.30×10^{-3}
9. 6.25×10^4 **11.** 217 **13.** 0.000735 **15.** 2,220,000 **17.** 4.75
19. 2 **21.** 0 **23.** 1 **25.** 3 **27.** -2 **29.** -4 **31.** 2.4393
33. 1.6284 **35.** $\bar{1}.2601$ or $9.2601 - 10$ **37.** $\bar{2}.6831$ or $8.6831 - 10$
39. 3.5411 **41.** $\bar{2}.5529$ or $8.5529 - 10$ **43.** 1.6310
45. $\bar{2}.6286$ or $8.6286 - 10$ **47.** 20.3 **49.** 2840 **51.** 0.0339 **53.** 0.0793
55. 64,340 **57.** 394.5 **59.** 0.03737 **61.** 0.06839

EXERCISE 14.3

1. 89 **3.** 6.86 **5.** 0.0259 **7.** 7.89 **9.** 0.0102 **11.** 52,700
13. 0.0229 **15.** 2.66 **17.** 3.20 **19.** 0.502 **21.** 7805 **23.** 1.119
25. 9263 **27.** 1.382 **29.** 3.564

REVIEW EXERCISES, Chapter 14

1. $\log_3 N = 5$ **2.** $\log_6 N = 8$ **3.** $\log_5 N = -3$ **4.** $N = 2^6$ **5.** $N = 5^4$
6. $N = 3^{-5}$ **7.** 2 **8.** -2 **9.** $\tfrac{1}{4}$ **10.** $\log_b x - (\log_b z + \log_b y)$
11. $4 \log_b x + \log_b y$ **12.** $\dfrac{1}{n} \log_b x - \dfrac{1}{n} \log_b y$ **13.** $\log_b 2xy$ **14.** $\log_b \sqrt{\dfrac{x^3}{y}}$
15. 3.24×10^3 **16.** 6.30×10^7 **17.** 2.56×10^{-3} **18.** 1.3892 **19.** 2.5156
20. $\bar{1}.0969$ or $9.0969 - 10$ **21.** $\bar{2}.0124$ or $8.0124 - 10$ **22.** 6519 **23.** 355.5
24. 0.0441 **25.** 0.007615 **26.** 849 **27.** 4.048 **28.** 2.922
29. 16,580,000 **30.** 0.5946

Answers to Odd-Numbered Exercises   **421**

Chapter 15

EXERCISE 15.1

1. $\{6, 7, 8, 9\}$ 3. $\{1, \frac{1}{2}, \frac{1}{3}, \frac{1}{4}\}$ 5. $\{\frac{3}{2}, 3, \frac{11}{2}, 9\}$ 7. $\{1, \frac{2}{3}, \frac{3}{5}, \frac{4}{7}\}$
9. $\{0, \frac{2}{3}, \frac{3}{2}, \frac{12}{5}\}$ 11. $\{2, 4, 8, 16\}$ 13. $\{\frac{1}{2}, \frac{1}{4}, \frac{1}{8}, \frac{1}{16}\}$ 15. $\{-1, 1, -1, 1\}$
17. $\{1, 0, -\frac{1}{3}, \frac{1}{2}\}$ 19. $f(n) = 2n$ 21. $f(n) = n^2$
23. $f(n) = 2^n$ 25. $f(n) = 4^{n-1}$ 27. $f(n) = (-1)^n n$
29. $f(n) = \dfrac{(-1)^n(n+1)}{n}$

EXERCISE 15.2

1. $\{4, 7, 10, 13, 16, \ldots\}$ 3. $\{1, -4, -9, -14, -19, \ldots\}$
5. $\{-3, 0, 3, 6, 9, \ldots\}$ 7. $\{0, \frac{1}{2}, 1, \frac{3}{2}, 2, \ldots\}$
9. $\{3, 3 + \sqrt{2}, 3 + 2\sqrt{2}, 3 + 3\sqrt{2}, 3 + 4\sqrt{2}, \ldots\}$
11. $\{5, 6 + \sqrt{2}, 7 + 2\sqrt{2}, 8 + 3\sqrt{2}, 9 + 4\sqrt{2}, \ldots\}$ 13. 35 15. 53 17. 33
19. $2x - 8$ 21. $19a + 11b$ 23. $5, 9, 13, 17$ 25. $17, 13, 9, 5, 1$
27. 32 29. $\frac{603}{2}$ 31. 592 33. 350.4 35. 195 37. $k = 10$

EXERCISE 15.3

1. $\{3, 6, 12, 24, \ldots\}$ 3. $\{81, 27, 9, 3, \ldots\}$ 5. $\{\frac{1}{3}, \frac{1}{6}, \frac{1}{12}, \frac{1}{24}, \ldots\}$
7. $\{\sqrt{2}, 2, 2\sqrt{2}, 4, \ldots\}$ 9. $\{16, 4, 1, \frac{1}{4}, \ldots\}$ 11. $\left\{1, \dfrac{x}{2}, \dfrac{x^2}{4}, \dfrac{x^3}{8}, \ldots\right\}$
13. 32 15. $\frac{5}{64}$ 17. 256 19. 32 21. $\dfrac{y^6}{x^5}$ 23. 6, 12
25. $2, 4, 8, 16, 32, 64$ 27. 93 29. -42 31. $\frac{665}{81}$ 33. $\frac{20}{3}$ 35. $\frac{8}{3}$
37. 63 39. $k = -14$ or 3 41. $6, 4, 2, 1$ 43. 210 45. 42 feet

EXERCISE 15.4

1. $8!$ 3. $\dfrac{6!}{2!}$ 5. $(n-2)!$ 7. $\dfrac{n!}{(n-3)!}$ 9. 5040 11. 40,320 13. 60
15. 126 17. 720 19. 1260 21. 140
23. $\frac{31}{29}$ 25. $x^6 + 6x^5y + 15x^4y^2 + 20x^3y^3 + \cdots$
27. $x^4 - 8x^3 + 24x^2 - 32x + \cdots$ 29. $64x^6 - 192x^5y + 240x^4y^2 - 160x^3y^3 + \cdots$
31. $\dfrac{x^6}{729} + \dfrac{2x^5}{27} + \dfrac{5x^4}{3} + 20x^3 + \cdots$
33. $x^{16} - 16x^{14}y + 112x^{12}y^2 - 448x^{10}y^3 + \cdots$ 35. $x^4 + 4x^2 + 6 + \dfrac{4}{x^2} + \cdots$
37. $-3003x^{10}y^5$ 39. $11{,}520x^2y^8$ 41. $-560x^4y^3$ 43. $-252x^5y^5$
45. $70x^8y^{16}$ 47. $1 - x + x^2 - x^3 + \cdots$ 49. 1.01

REVIEW EXERCISES, Chapter 15

1. $\{1, \frac{4}{3}, \frac{9}{5}, \frac{16}{7}\}$ **2.** $\{-1, -\frac{1}{2}, 1, -\frac{5}{4}\}$ **3.** $f(n) = 3n - 2$
4. $f(n) = \frac{(-1)^{n+1}}{n^2 + n}$ **5.** $\{-2, 1, 4, 7, 10, \ldots\}$ **6.** 5; 21
7. 5, 12, 19, 26, 33 **8.** 95 **9.** $\{64, 32, 16, 8, 4, \ldots\}$ **10.** 32; 256
11. 1, 3, 9, 27, 81 or 1, $-3, 9, -27, 81$ **12.** 189 **13.** $\frac{15}{4}$ **14.** 720
15. 10,080 **16.** $\frac{21}{19}$ **17.** $x^{12} + 12x^{11}y + 66x^{10}y^2 + 220x^9y^3 + \cdots$
18. $x^{10} - 20x^9y + 180x^8y^2 - 960x^7y^3 + \cdots$ **19.** $70x^4y^4$ **20.** $-64{,}064x^5y^9$

Appendix 1

1. $(x + 1)(x + 2)$ **2.** $(x + 1)(x + 3)$ **3.** Not factorable **4.** Not factorable
5. $(x - 3)(2x + 1)$ **6.** $(x - 1)(5x - 1)$ **7.** $(x + 2)(2x + 5)$
8. $(x - 3)(3x - 5)$ **9.** $(4x - 3)(2x + 3)$ **10.** $(3x - 7)(2x + 1)$
11. Not factorable **12.** Not factorable **13.** $(x - 7)(x - 2)$
14. $(x - 5)(x + 11)$ **15.** $(x + 6)(x + 6)$ **16.** $(x - 2)(x - 2)$
17. $(3x + 1)(3x + 1)$ **18.** $(x - 4)(x - 1)$ **19.** Not factorable
20. Not factorable **21.** Not factorable **22.** $(2x - 3)(5x - 4)$
23. $(x - 1)(2x + 7)$ **24.** $(x + 1)(7x + 6)$ **25.** $(x - 4)(2x - 5)$
26. $(x - 3)(6x - 5)$ **27.** Not factorable **28.** $(3x - 2)(4x + 9)$
29. $(2x - 5)(4x + 9)$ **30.** $(4x - 3)(5x + 6)$ **31.** $(x + 3y)(x + 5y)$
32. $(x + 2y)(x + 3y)$ **33.** $(x - y)(x - 7y)$ **34.** $(x - y)(x - 9y)$
35. Not factorable **36.** Not factorable **37.** $(x - y)(2x + 3y)$
38. $(2x + y)(3x + 2y)$ **39.** $(x - y)(2x + 7y)$ **40.** Not factorable
41. $(5x - 7y)(4x + 3y)$ **42.** $(6x - 7y)(3x - 4y)$

Index

Abscissa, 173
Absolute inequality, 46
Absolute value, 22
Addition, 25
  algebraic expressions, 71
  axioms of, 11
  complex numbers, 283
  fractions, 145
  matrices, 318
  polynomials, 95
  radical expressions, 88
  vectors, 302
Algebraic expression, 70
Algorithm, division, 98
Antilogarithm, 338
Arithmetic means, 348
Arithmetic progression, 347
Arithmetic sequence, 347
Array, 173
Asymptote, 218, 240
Axioms of addition, 11
Axioms of equality, 10
Axioms of multiplication, 11
Axis of a parabola, 198
Axis of symmetry, 188

Base of a logarithm, 332
Base of a power, 63
Basic fraction, 25
Basic numeral, 25
Binary operation, 4
Binomial, 92

Cartesian coordinates, 173
Cartesian product, 7, 169
Central rectangle, 217
Characteristic of a logarithm, 336
Circle, 196ff
Closed set, 10
Coefficient, 92
Column matrix, 315
Combined variation, 231
Commutative operation, 7, 10
Complement, 6
Completing the square, 103
Complex conjugate, 288
Complex fraction, 161
Complex number, 280
Conclusion, 13
Conditional equation, 45
Conditional inequality, 46
Conformable matrices
  for addition, 317
  for multiplication, 321
Conjugate, 83
Conjugate complex numbers, 288
Constant function, 178
Constant polynomial, 92
Constant of variation, 229
Coordinate plane, 173
Coordinates, 19, 173
Counting number, 2
Cramer's Rule, 326

Decimal notation, 29
Deductive proof, 15

Degree of a polynomial, 93, 94
Dependent equations, 247
Determinant, 323
  evaluation of, 324
Difference, 25
  algebraic expressions, 71
  complex numbers, 283
  fractions, 149
  matrices, 318
  polynomials, 95
  radical expressions, 88
  of two squares, 106
  vectors, 303
Dimension of a matrix, 314
Direct variation, 229
Direction angle, 306
Directrix, 198
Discriminant, 120
Disjoint sets, 6
Distance formula, 181
Distance between two points, 180
Distributive law, 12
Division, 28
  complex numbers, 286
  fractions, 154
  polynomials, 97
  radical expressions, 83
  synthetic, 99
Division algorithm, 98
Domain, 169

Eccentricity, 209
Ellipse, 205ff
Empty set, 2
Equal sets, 3
Equality, axioms of, 10
Equations, 40
  conditional, 45
  quadratic, 115
Equations for
  circles, 196
  ellipses, 207
  hyperbolas, 212
  lines, 179ff
    intercept form, 182
    point-slope form, 183
    slope-intercept form, 183
    two-point form, 182
  parabolas, 200, 202
  sequences, 346
Equations quadratic in form, 126
Equivalent fractions, 138
Equivalent open sentences, 40
Equivalent sets, 2

Equivalent systems of equations, 249
Equivalent vectors, 297
Evaluation of determinants, 324
Expansion of a binomial, 356
Exponents, 63
  negative integer, 64
  rational number, 67
  zero, 64
Extraneous solution, 125

Factorial notation, 357
Factoring
  by grouping, 110
  other methods of, 109ff
  quadratic trinomials, 106ff
    over the set of integers, 363
Finite set, 2
First-degree open sentence, 45
Focus of a parabola, 198
Formula
  distance, 181
  quadratic, 118, 290
Function, 171
  inverse, 235
  logarithmic, 331
  one-to-one, 235
  polynomial, 237
  rational, 239
  sequence, 345

Geometric means, 352
Geometric progression, 351
Geometric sequence, 351
Geometric vector, 296
Graph, 173
  number, 19
  polynomial function, 237
  rational function, 239
  sign, 129
Greater than, 19

Half-plane, 224
Horizontal component, 299
Hyperbola, 210ff
Hypothesis, 13

Identity, 45
Image set, 169
Imaginary number, 280
Inconsistent equations, 247
Independent equations, 247

Inequality, 40
  absolute, 46
  conditional, 46
  linear, 223
  quadratic, 115, 128, 225
  solution of, 224
Infinite decimal, 30
Infinite geometric sequence, 353
Infinite set, 2
Initial point, 295
Integers, 9
Intercept method of graphing, 177
Intercepts, 177, 185
Intersection, 6
Inverse function, 235
Inverse relation, 234
Inverse variation, 229
Irrational numbers, 9

Joint variation, 231

Leading coefficient, 93
Least common denominator (LCD), 26, 42, 147
Least common multiple (LCM), 26, 146
Less than, 19
Like terms, 70
Linear combination, 253
Linear function, 176
Linear inequality, 223
Linear interpolation, 338
Linear relation, 178
List of elementary theorems, 14
Logarithm, 331
Logarithmic function, 331

Mantissa, 326
Matrix, 314
Maximum function value, 186
Members of an open sentence, 40
Minimum function value, 186
Mirror image, 187
Monomial, 92
Multiplication, 28
  axioms of, 11
  complex numbers, 284
  fractions, 153
  matrices, 320
  polynomials, 95
  radical expressions, 82
Multiplicative inverse, 285

Natural numbers, 2, 9
Negative of a matrix, 318
Negative number, 19
Negative of a vector, 298
Norm of a vector, 297
Null set, 2
Number
  complex, 280
  imaginary, 280
  irrational, 9
  negative, 19
  positive, 19
  rational, 9
Number line, 19

One-to-one function, 235
Open sentence, 39
Order, 19
Order of a determinant, 324
Order of a matrix, 314
Ordered n-tuple, 249
Ordered pair, 7
Ordered set, 2
Ordered triple, 249
Ordinate, 173
Origin, 19

Parabola, 185, 198ff
Parallelogram law, 305
Perfect square trinomial, 102
Periodic decimal, 30
Permissible replacement, 39
Point lattice, 173
Polynomial, 92
  constant, 92
  degree of, 92, 93
  monic, 93
  standard form of, 92
  zero, 92
  zero of, 93
Polynomial function, 237
Positive number, 19
Power, 63
Principal diagonal, 315
Principal square root, 32
Product, 28
  complex numbers, 284
  fractions, 153
  matrices, 320
  polynomials, 95
  radical expressions, 82
Proper subset, 3

426 INDEX

Quadrant, 173
Quadratic equation, 115
Quadratic formula, 118, 290
Quadratic function, 185
Quadratic inequality, 115, 225
Quotients, 28
  complex numbers, 286
  fractions, 154
  polynomials, 97
  radical expressions, 83

Radical expression, 77
Radical notation, 32
Radicand, 32
Range, 169
Rational number, 9
Rationalizing factor, 79
Real polynomial, 92
Rectangular coordinates, 173
Relation, 168
Relative complement, 7
Relatively prime polynomials, 140
Repeating decimal, 30
Replacement, permissible, 39, 45
Replacement set, 2, 39
Resultant, 307
Roots of real numbers, 67
Row matrix, 315

Scalar multiple, 298
Scalar quantity, 295
Scientific notation, 335
Sequence, 345
  arithmetic, 347
  function, 345
  geometric, 351
Set, 1
Set builder notation, 2, 5
Sign graph, 129
Similar terms, 70
Simplest radical form, 77
Slope, 181
Solution, 40
Solution of an inequality, 224
Solution set, 40
Square matrix, 315, 323
Square root, 32
Standard form of
  a polynomial, 92
  a quadratic equation, 115
  a quadratic trinomial, 106

Statement, 39
Subscript, 92
Subset, 3
Substitution axiom, 10
Subtraction
  algebraic expressions, 71
  complex numbers, 283
  fractions, 149
  matrices, 318
  polynomials, 95
  radical expressions, 88
  vectors, 303
Sum, 25
  algebraic expressions, 71
  complex numbers, 283
  fractions, 145
  matrices, 318
  polynomials, 95
  radical expressions, 88
  vectors, 302
Symmetry, 187
Synthetic division, 99

Terminal point, 295
Terminating decimal, 30
Terms, 70
Theorem, 11, 13
Theorems, elementary, list of, 14
Trinomial, 92
  perfect square, 102
  quadratic, 106
Turning point, 237

Union, 5
Unit vector, 298
Universal set, 3
Universe of discourse, 3

Variable, 2, 39
Variation, 229ff
Vector, 295ff
Venn diagram, 5
Vertex of a parabola, 186
Vertical component, 299

Whole number, 9

Zero polynomial, 92
Zero of a polynomial, 93, 237

A 1
B 2
C 3
D 4
E 5
F 6
G 7
H 8
I 9
J 0

TABLE OF FOUR-PLACE LOGARITHMS TO THE BASE 10

N	0	1	2	3	4	5	6	7	8	9
1.0	.0000	.0043	.0086	.0128	.0170	.0212	.0253	.0294	.0334	.0374
1.1	.0414	.0453	.0492	.0531	.0569	.0607	.0645	.0682	.0719	.0756
1.2	.0792	.0828	.0864	.0899	.0934	.0969	.1004	.1038	.1072	.1106
1.3	.1139	.1173	.1206	.1239	.1271	.1303	.1335	.1367	.1399	.1430
1.4	.1461	.1492	.1523	.1553	.1584	.1614	.1644	.1673	.1703	.1732
1.5	.1761	.1790	.1818	.1847	.1875	.1903	.1931	.1959	.1987	.2014
1.6	.2041	.2068	.2095	.2122	.2148	.2175	.2201	.2227	.2253	.2279
1.7	.2305	.2330	.2355	.2381	.2406	.2430	.2455	.2480	.2504	.2529
1.8	.2553	.2577	.2601	.2625	.2648	.2672	.2695	.2718	.2742	.2765
1.9	.2788	.2810	.2833	.2856	.2878	.2900	.2923	.2945	.2967	.2989
2.0	.3010	.3032	.3054	.3075	.3096	.3118	.3139	.3160	.3181	.3202
2.1	.3222	.3243	.3263	.3284	.3304	.3324	.3345	.3365	.3385	.3404
2.2	.3424	.3444	.3464	.3483	.3503	.3522	.3541	.3560	.3579	.3598
2.3	.3617	.3636	.3655	.3674	.3692	.3711	.3729	.3748	.3766	.3784
2.4	.3802	.3820	.3838	.3856	.3874	.3892	.3909	.3927	.3945	.3962
2.5	.3979	.3997	.4014	.4031	.4048	.4065	.4082	.4099	.4116	.4133
2.6	.4150	.4166	.4183	.4200	.4216	.4233	.4249	.4265	.4281	.4298
2.7	.4314	.4330	.4344	.4362	.4378	.4393	.4409	.4425	.4440	.4456
2.8	.4472	.4487	.4503	.4518	.4533	.4548	.4564	.4579	.4594	.4609
2.9	.4624	.4639	.4654	.4669	.4684	.4698	.4713	.4728	.4742	.4757
3.0	.4771	.4786	.4800	.4814	.4829	.4843	.4857	.4871	.4886	.4900
3.1	.4914	.4928	.4942	.4955	.4969	.4983	.4997	.5011	.5024	.5038
3.2	.5052	.5065	.5079	.5092	.5106	.5119	.5132	.5146	.5159	.5172
3.3	.5185	.5198	.5211	.5224	.5238	.5250	.5263	.5276	.5289	.5302
3.4	.5315	.5328	.5340	.5353	.5366	.5378	.5391	.5403	.5416	.5428
3.5	.5441	.5453	.5465	.5478	.5490	.5502	.5515	.5527	.5539	.5551
3.6	.5563	.5575	.5587	.5599	.5611	.5623	.5635	.5647	.5659	.5670
3.7	.5682	.5694	.5705	.5717	.5729	.5740	.5752	.5763	.5775	.5786
3.8	.5798	.5809	.5821	.5832	.5843	.5855	.5866	.5877	.5888	.5900
3.9	.5911	.5922	.5933	.5944	.5955	.5966	.5977	.5988	.5999	.6010
4.0	.6021	.6031	.6042	.6053	.6064	.6075	.6085	.6096	.6107	.6117
4.1	.6128	.6138	.6149	.6160	.6170	.6181	.6191	.6201	.6212	.6222
4.2	.6233	.6243	.6253	.6263	.6274	.6284	.6294	.6304	.6314	.6325
4.3	.6335	.6345	.6355	.6365	.6375	.6385	.6395	.6405	.6415	.6425
4.4	.6434	.6444	.6454	.6464	.6474	.6484	.6493	.6503	.6513	.6523
4.5	.6532	.6542	.6551	.6561	.6571	.6580	.6590	.6599	.6609	.6618
4.6	.6628	.6637	.6646	.6656	.6665	.6675	.6684	.6693	.6703	.6712
4.7	.6721	.6730	.6740	.6749	.6758	.6767	.6776	.6785	.6794	.6803
4.8	.6812	.6822	.6831	.6840	.6849	.6857	.6866	.6875	.6884	.6893
4.9	.6902	.6911	.6920	.6929	.6937	.6946	.6955	.6964	.6972	.6981
5.0	.6990	.6998	.7007	.7016	.7024	.7033	.7042	.7050	.7059	.7067
5.1	.7076	.7084	.7093	.7101	.7110	.7118	.7127	.7135	.7143	.7152
5.2	.7160	.7168	.7177	.7185	.7193	.7202	.7210	.7218	.7226	.7235
5.3	.7243	.7251	.7259	.7267	.7275	.7284	.7292	.7300	.7309	.7316
5.4	.7324	.7332	.7340	.7348	.7356	.7364	.7370	.7380	.7388	.7396
N	0	1	2	3	4	5	6	7	8	9